Economic Studies in Inequality, Social Exclusion and Well-Being

Volume 9

Series Editor

Jacques Silber

For further volumes:
http://www.springer.com/series/7140

Valérie Bérenger · Florent Bresson
Editors

Poverty and Social Exclusion Around the Mediterranean Sea

Springer

Editors
Valérie Bérenger
University of Nice-Sophia Antipolis
Nice
France

Florent Bresson
University of Orleans
Orléans Cedex
France

ISBN 978-1-4614-5262-1 ISBN 978-1-4614-5263-8 (eBook)
DOI 10.1007/978-1-4614-5263-8
Springer New York Heidelberg Dordrecht London

Library of Congress Control Number: 2012950856

Printed on acid-free paper

Springer is part of Springer Science+Business Media (www.springer.com)

Foreword

> You ask me what is poverty?... Poverty is living in a smell that never leaves.... Poverty is being tired.... Poverty is dirt.... Poverty is staying up all night on cold nights to watch the fire, knowing one spark on the newspaper covering the walls means your sleeping child dies in flames.... Poverty is asking for help.... Poverty is remembering. It is remembering quitting school in junior high because "nice" children had been so cruel about my clothes and my smell.... Mostly I remember being married.... I knew my husband was leaving the day he left, but there were no goodbye between us.... Poverty is looking into a black future. Your children won't play with my boys. They will turn to other boys who steal to get what they want. I can already see them behind the bars of their prison instead of behind the bars of my poverty. Or they will turn to the freedom of alcohol or drugs, and find themselves enslaved..... My children have no extra books, no magazines, no extra pencils, or crayons, or paper and most important of all, they do not have health.... They do not suffer from hunger...but they do suffer from malnutrition.... Poverty is an acid that drips on pride until all pride is worn away. Poverty is a chisel that chips on honor until honor is worn away.... Even the poor can dream,.... a dream of when asking for help does not eat away the last bit of pride. When the office you visit is as nice as the offices of other governmental agencies, when there are enough workers to help you quickly, when workers do not quit in defeat and despair. When you have to tell your story to only one person, and that person can send you for other help and you don't have to prove your poverty over and over and over again...
>
> From *America's Other Children: Public Schools Outside Suburbs*, by George Henderson, 1971, The University of Oklahoma Press.

The tragic sentences that appear in the citation above are taken from an anonymous account about living in poverty, written by a woman living in the United States around the 1960s.

A close look at the excerpts of this woman's description of her daily life clearly shows that though poverty starts as lack of income and resources, it quite quickly affects all aspects of life, including education, health, social inclusion and even expectations for one's descendants. This quotation also stresses the irreversibility of such a situation. In short poverty is a multidimensional phenomenon, not only because it encompasses all aspects of life at a given time but also because, when it is severe, it is also a lasting experience.

Although everyone tends to acknowledge the fact that poverty is multidimensional, several economists would recommend quantifying poverty via multiple indices rather than through a synthetic multidimensional poverty index. Such a position was strongly defended by Martin Ravallion in a recent exchange of views with James Foster and Sabina Alkire, the inventors of the Multidimensional Poverty Index (MPI).[1]

The MPI is however only one among many ways of devising a summary measure derived from a multidimensional approach to poverty.[2] Since most of the existing empirical illustrations of multidimensional poverty analysis have focused their attention on developing countries, Valérie Bérenger and Florent Bresson should be lauded for having taken the initiative to edit a book on *Monetary Poverty and Social Exclusion around the Mediterranean Sea*, an area which includes mainly so-called middle income countries. There are some important theoretical contributions in this book, always complemented by an interesting empirical illustration, but also studies with a more applied emphasis. Such a well-balanced combination of theoretical and applied work is especially attractive because all the chapters deal with Middle East or North African countries, an area to which relatively few studies of poverty were devoted. The only notable exception I know of is a book published in 2006 in the CROP[3] series, edited by M. Petmesidou and C. Papatheodorou, and entitled *Poverty and Social Deprivation in the Mediterranean. Trends, Policies and Welfare Prospects in the New Millennium.* This book however covered countries on both sides of the Mediterranean Sea and included only few chapters based on a multidimensional approach to poverty measurement.

Bérenger and Bresson are hence filling a research gap and one can only hope that their pioneering initiative will give rise to a series of additional studies of poverty in the Middle Eastern and North African countries, in particular to works taking a multidimensional view of poverty.

Jacques Silber
Department of Economics
Bar-Ilan University
52900, Ramat Gan
Israel

CEPS/INSTEAD
Esch-sur-Alzette
Luxembourg
e-mail: jsilber_2000@yahoo.com

---

[1] See, *Journal of Economic Inequality*, Volume 9, Number 2, June 2011, and Number 3, September 2011.

[2] For an illustration of this variety of views, see, for example, *Quantitative Approaches to Multidimensional Poverty Measurement*, N. Kakwani and J. Silber, editors, Palgrave-Macmillan, 2008.

[3] Comparative Research Programme on Poverty.

Contents

Introduction

The recent events taking place in several southern Mediterranean countries—the so-called "Arab Spring"—make it clear that the overall prosperity of a nation does not necessarily imply that the nation's populations enjoy worthwhile lives. These events also highlight the fact that wellbeing should not simply be viewed through the narrow lens of income or expenditure as conveyed by the standard approach to poverty. Street protests echo people's need for bread, jobs, freedom and dignity. They demonstrate the aspirations of the population for better living conditions and for human rights which are held in greater respect. Poverty is not only about a lack of resources; it is also reflected in a lack of basic capabilities and in the inability to apply these basic capabilities in the pursuit of economic and political opportunities, these same opportunities that give value to a person's life. The events have illuminated hidden deprivations, inequalities and the exclusion of segments of the population from economic and civil life, all of which can be considered as multiple and complex aspects of poverty (of opportunity). They remind us that development is not a matter of economic growth, but of wealth redistribution and the inclusion of the whole population in the growth process.

However, one of the striking facts about this "Arab Spring" is that during the last few decades, poverty has not been given the same emphasis in southern Mediterranean countries as in other areas of the developing/emerging world. This neglect can be illustrated by the relative lack of studies on poverty in Middle East and North African countries. Most of these countries are middle income countries and, during the 1980s, serious macroeconomic imbalances compelled the introduction of structural adjustment programmes aimed at stabilising the economy and committing it to an outward looking direction. Despite the adoption of similar patterns that prioritized economic growth, structural economic reforms have been implemented by these countries with varying levels of intensity. It suggests possible differences in performance achieved in terms of standards of living and poverty reduction for the population, in particular, with respect to the provision of social goods, such as education and health that are generally the most severely affected by state budgetary cuts.

Since the seminal works of Sen (1981, 1985) and Townsend (1979), there has been growing evidence that it is highly desirable to move beyond the money-metric measures of poverty and boost results by using approaches that adopt a broader definition of wellbeing, a definition which includes a wide array of components ranging from nutrition to freedom to achieve certain social arrangements (Baliamoune, 2004). Though poverty is only one of the many causes of the revolutionary wave of demonstrations and protests that have been observed since 2010 in southern Mediterranean countries, we argue that it is a key feature in explaining these historic events and that it is necessary to draw a detailed picture of poverty and social exclusion for each country. Moreover, if the Union for the Mediterranean (created in 2008 and building on the earlier Euro-Mediterranean Partnership) succeeds in fostering economic cooperation within the Mediterranean area, the focus on the wellbeing of the population of its southern members will deserve scrutiny in the next few years if we are to understand the forces that shape Euro-Mediterranean relationships, in particular with regard to migration issues.

The aim of this book is therefore to bring together recent advances in the measurement of poverty as applied to southern Mediterranean countries. It introduces new tools for analyzing issues that have to be addressed when using a multidimensional approach to poverty and shows that the links between income (or expenditure) and wellbeing are not straightforward; they hinge on many determinants, such as idiosyncratic characteristics and market factors. It establishes that the efficiency of poverty-reducing policies should also be assessed on the basis of the satisfaction of non-income needs, such as health, education and participation in social life.

Through the various methodologies used by the authors to study multiple aspects of poverty, the reader will discover various ways of tackling issues faced by the researcher when adopting a multidimensional approach, such as the choice of dimension, weight, poverty line and the aggregation operators used to summarise the information conveyed by the different dimensions. The innovative approaches applied to some southern Mediterranean countries make use of comprehensive information that takes into account the interplay between the multiple deprivations which individuals experience in relation to the society they live in. They hence make it possible to highlight the context of specific areas that should be prioritized for social exclusion prevention and poverty reduction.

The book is divided into three main parts. The first part is dedicated to cross country studies. Its aim is to give a general picture of poverty in the southern Mediterranean area and to compare techniques used in developing countries with the approach used in European countries.

The Chap. 1 consists of a multidimensional poverty analysis in relation to five southern Mediterranean countries, namely Egypt, Lebanon, Morocco, Tunisia and Turkey. The analysis relies on a broader concept of poverty i.e. it pays regard to the deprivations in women's educational attainments, the possession of durable goods and the consequences of housing conditions. The study is based on an axiomatic approach to poverty and on the use of stochastic dominance tools to achieve solid results that do not depend on the choice of poverty line and weighting scheme. Valérie Bérenger and Florent Bresson then provide comparisons over time for each country and between countries that cannot be obtained when each dimension of poverty is analyzed separately, since the approaches take into account the correlations which may exist between different kinds of poverty. One of the main findings is that their approach makes it possible to nuance the performance levels reached by certain countries, as poverty dominance checks do not necessarily yield rankings close to those obtained from GDP per capita comparisons. In addition, the authors consider to what extent these findings mirror differences in patterns of development policies, as well as their progress in some countries.

The Chap. 2 provides a comparison of European countries in terms of material deprivation. At first glance, this choice may seem peculiar for a book on poverty and social exclusion in southern Mediterranean countries, but we felt that it would be fruitful to compare the methods referred to in the different chapters of the book with the one usually applied to European countries. Indeed, social indicators can play an important role, not only at a national, but also at an international level, in making it possible to compare the living conditions in different countries according to a set of commonly agreed criteria. For social indicators to play this role in a multinational context, their construction needs to follow methodological principles that ensure their relevance and comparability across countries. The paper argues that the Euro-Mediterranean countries can benefit from the EU experience in building a common framework for monitoring, understanding and also fighting poverty and social exclusion. As a concrete example, Alessio Fusco, Anne-Catherine Guio and Eric Marlier discuss methodological issues raised by the construction of indicators on material deprivation, which is defined as an enforced lack of a combination of items. These items depict some aspects of living conditions in relation to housing conditions, the possession of durable goods and the capacity to afford basic necessities. More specifically, the paper focuses on the selection of items, their dimensional structure, their aggregation into a synthetic measure and their weighting. In order to emphasize the complementarity of the two approaches when applied to a group of countries with heterogeneous standards of living—a characteristic which is shared by the Euro-Mediterranean countries—it also puts material deprivation and income-based poverty indicators into perspective. The relevant EU countries are included in the 2007 EU Statistics on Income and Living Conditions (EU-SILC) users' database.

The second part of the book brings together chapters focusing on overall poverty assessments via country case studies. It underlines the methodological requirements which have to be respected in order to perform consistent evaluations of poverty when many dimensions of poverty are simultaneously taken into account.

The Chap. 3 by Satya Chakravarty and Conchita d'Ambrosio characterizes a family of subgroup decomposable unit consistent multidimensional poverty indices. Unit consistency is a very general property that only requires poverty rankings to remain unaltered when dimensions are expressed in different measurement units. It therefore offers different ways of acknowledging inequality between poor individuals. The family described by the authors is a generalization of a family of unit consistent income poverty index suggested by Zheng (2007a). The paper illustrates the index as derived from Turkish data available for the period 2003–-2005 and shows that, when wellbeing is assessed using households' expenditure levels and housing conditions, poverty decreased during the period, regardless of the approach used to describe inequality between poor individuals. A further investigation into the evolution of poverty according to area of residence led the authors to conclude that despite a decrease in rural area poverty during the period, urban areas tended to be less poverty stricken than their rural equivalent.

The Chap. 4 also focuses on the case of Turkey. Gianni Betti, Güllü Çalik, and Murat Karakaş provide a step-by-step account of how fuzzy measures for both non-monetary deprivation and monetary poverty can be constructed using survey data, such as that available from the 2007 wave of European Union-Statistics on Income and Living Conditions (EU-SILC). Their study first illustrates how income is imperfectly correlated with non-monetary deprivations in the case of Turkey, especially in relation to basic lifestyle and environmental dimensions of deprivation. By looking at NUTS1 regions, they also stress the significant regional discrepancies in Turkey, the southern population being overwhelmingly poorer than the population of Istanbul, though a fuzzy multidimensional approach to poverty indicates smoother gaps than the traditional monetary approach.

The Chap. 5, by Valérie Bérenger, Florent Bresson, Paul Makdissi, and Myra Yazbeck, is related to Lebanon. The objective of this study is to identify what orderings of poverty are robust to changes in measurement assumptions in the case of Lebanon, especially when comparing poverty between provinces (mohafazat) and between occupational groups. Firstly, the authors perform monetary and multidimensional poverty evaluations with the most commonly used poverty indices (Foster et al. 1984; Alkire and Foster 2011) i.e. capita expenditures, housing conditions, education and access to basic services. Then they make use of stochastic dominance tests to be able to identify the extent to which the poverty orderings obtained with the usual poverty indices are robust to a change in the poverty indices/poverty domain. Their findings enrich the studies on poverty in Lebanon because the results stress that close attention should be paid to the conclusions drawn from the usual poverty profiles, as these can be dependent on several arbitrary decisions made by the researcher. They also shed light on the limits of an analysis which is exclusively based on the monetary dimension of poverty insofar as it does not necessarily corroborate the results yielded by multidimensional analysis. However, they show that poverty is unambiguously lower in Beirut than in other mohazafat. Also, their multidimensional poverty measures uncover particularities in the profiles of Nabatieh and the South. In addition, they indicate that it is not easy to extract a unique sector as exhibiting the highest level of poverty. The sole occupational group that displays a significantly different level of poverty with respect to other groups is the financial sector.

In Chap.6 Mohamed Ayadi and Naouel Chtioui use a composite welfare indicator to apply rank-dependent poverty measures in an analysis of poverty in

Tunisia. More specifically, they propose classes of ethically generalized Sen–Shorrocks–Thon (SST) poverty measures that generalize the Shorrocks (1995) poverty measure, itself a modified version of the Sen (1976) poverty measure. The poverty trends in Tunisia between 1994 and 2006 are then scrutinized using the composite welfare indicator and the generalized SST poverty measures. Taking the sampling properties of their estimates into account, the authors show that poverty decreased between 1994 and 2006, but that there were inequalities within the poorer population. They also stress that poverty was essentially rural and that it was concentrated in the North West, Centre West and South East of Tunisia.

The last chapter of this second part is also based on the use of composite welfare indicators. Abdeljaouad Ezzrari and Paolo Verme opt to use a factorial technique, namely the Multiple Correspondence Analysis (MCA), to assess poverty in Morocco between 2001 and 2007. As argued by the authors, MCA provides two major advantages in the multidimensional measurement of poverty: it generates a matrix of "weights" based on the variance-covariance matrix for all welfare dimensions selected and it provides a natural approach for constructing a composite welfare indicator that satisfies essential axioms of poverty orderings. After being carried out separately for urban and rural areas, the application indicates that the multidimensional indicators characterizing these two areas for both the poorest and the least poor households are often different. It confirms the existence of a deeply divided society along urban and rural lines. The findings show that poverty in Morocco declined according to both monetary and multidimensional indicators during the relevant period and that the decline in poverty was stronger in rural as opposed to urban areas. Moreover, these findings are robust to first order stochastic dominance tests. The authors then conclude that the sustained positive growth experienced by Morocco during the last decade has translated into improvements in living conditions which go well beyond monetary returns.

The studies presented in the third part of the book are not intended to provide overall estimations of poverty for the countries they focus on, but to shed light on some specific facets of poverty.

Chapter 8 by Daniel Gottlieb and Suleiman Abu-Bader deals with the socio-economic situation of the Arab-Bedouin population in the Negev (Israel). In comparison with the remaining Israeli Arab population, this minority is highly affected by poverty. Indeed, according to the Galilee Society's 2004 social survey, the incidence of Israeli Arab monetary poverty was found to be 52 %, with nearly two thirds in persistent poverty. Among Bedouins in villages which have gone unrecognized by the Israeli government, this figure is close to 80 %. In the same spirit, the severity of poverty, as measured by the Sen index, was about 7 times higher than that among mainstream Jewish society. The authors then emphasize the role played by education for Arab women in diminishing poverty, but stress that the lack of public infrastructure impedes the acquisition of education as well as the participation of Bedouin people in the labour force. This also has an adverse indirect effect toward the completion of schooling, thus keeping mothers' fertility relatively high and reducing the potentially diminishing effect of education on poverty. The authors finally point out the considerable mismatch between skills and employment observed among Arab academics, thereby highlighting the possibility of discrimination and segregation in their particular labour market. Consequently they argue that government intervention in infrastructure has a high expected social return and could therefore help in interrupting the vicious circle of poverty.

The next chapter by Naama Haron also deals with deprivations within the Israeli population. More specifically, the goal of this chapter is to measure the extent of social exclusion in Israel and compare the results obtained with those derived from a traditional income-based approach to poverty measurement. It is often assumed that poverty and social exclusion are related but different phenomena; while poverty is linked to all elements that hamper wellbeing, social exclusion is generally regarded as dealing with deprivations that prevent a satisfactory participation in social life. Thus, it is worth checking whether or not those individuals who are deemed poor are also those who are socially excluded. To do this, the author uses the 2002 round of the European Social Survey, a database that provides several indicators related to social-economic security and social inclusion. An index of social exclusion at the individual level is designed and compared with the corresponding income level and linked to various factors that are assumed to have an impact on income poverty and social exclusion. The author then shows how monetary poverty and social exclusion are imperfectly correlated. Moreover, risk factors such as a low level of education, poor health, young age, the type of the household, being a female and not being a Jew, seem to have a stronger impact on social exclusion than income. Finally, a cluster analysis reveals the existence of three sub-groups in Israel, sub-groups which could be labelled "rich," "middle class" and "poor." While there is a significant income gap between the "rich" and the two other clusters, the largest gap (with respect to the issue of social exclusion) is observed between the "poor" and the two other clusters.

Chapter 10 by Nader Kabbani and Zaki Mehchy addresses another component of social exclusion, namely empowerment. This is analyzed within the Syrian context. Syria experienced healthy economic growth rates between 2000 and 2010. However, as underlined by the recent uprising against Bashar al-Assad's regime, its population suffers from deficits in social and political dimensions of human development. Marginalized and disempowered groups, including rural communities, face barriers to participation and opportunities. Using original data collected in 2008 and 2009 through surveys designed with the participation of local communities, the study performed by the authors seeks to operationalize the concept of empowerment and identify its determinants and its association with money-metric poverty in twenty four villages from rural Syria. By focusing on operational elements of empowerment that include access to information, inclusion and participation, and local organizational capacity, the authors observe that gender, age, educational level, and occupation are key determinants of empowerment in rural Syria. They also call attention to a positive zero-order correlation between empowerment and monetary poverty. Consequently, they conclude that increasing empowerment does not necessarily result in improved economic livelihood. The findings suggest that policies and programmes aimed at alleviating economic, social and political dimensions of deprivation may be complimentary and should therefore be introduced in tandem in order to improve the wellbeing of the population.

The last chapter, by Shireen AlAzzawi and Mona Said, differs from the previous studies with respect to two points. First, it investigates the dynamics of poverty by analyzing horizontal mobility within the population of Egyptian workers. Secondly, it does not only consider the monetary dimension of poverty; it also considers working conditions in the light of trade liberalization policy measures that were taken during the period 1998–2006. Trade liberalization is often regarded as a powerful driver of growth, but its social consequences are generally widely debated, even at the theoretical level. The authors find a relatively low level of economic mobility in both income and non-income indicators, with the majority of those who were "poor" in 1998, whether in the monetary or non-monetary dimensions, remaining so by 2006. Moreover, trade reform in the shape of lower tariffs and increased export promotion exerted a small positive influence on the incomes of poor individuals. However, this came at the expense of a greater informalization of workers and a higher incidence of low quality jobs. Once again, this result supports the necessity of adopting a broad and complete description of poverty and social exclusion if the intention is to scrutinize the social effects of specific factors or policy measures.

To conclude, this collection of papers does not exhaust the subject of poverty and social exclusion in southern Mediterranean countries. Some populations, like the Libyan, Algerian, or Palestinian populations, should of course warrant specific studies. In the same vein, it would be worth analyzing particular dimensions of wellbeing in more depth, such as access to water or housing conditions, in conjunction with the policy measures taken by central and local authorities. This would seem to be highly relevant in the context of these countries. Our opinion is that this book should also be regarded as an invitation for researchers to pursue these investigations. The "Arab Spring" has revealed that, although the region is relatively well-off in comparison to the rest of the developing world, its population is yearning for noticeable improvements in many dimensions of wellbeing.

Acknowledgments

Most of the chapters in this book are updated version of papers presented at a workshop on multidimensional poverty and pro-poor growth in MENA countries held in June 2009 at the University of Nice-Sophia Antipolis (France). This workshop was organized within the framework of a research project entitled "Growth and Multidimensional Poverty in the South" (PROPOORSUDS funded by the French National Agency for Scientific Research (Agence Nationale de la Recherche, ANR). It provided the opportunity to bring together scholars from academic and institutional development organizations (among them the African Development bank and the World Bank). Its main objective was to illustrate the "state of the art" in multidimensional poverty measurement as well as in pro-poor growth analysis, to share knowledge and experience on poverty reduction within the specific context of MENA countries and to highlight the lessons that can be drawn from recent growth spells in terms of poverty alleviation.

We wish to express our gratitude to Jacques Silber for giving us the opportunity to benefit from his rich academic experience on the subject. We would like to thank him for his suggestion that we group the papers together for a book project and for his helpful editorial advice.

A long period of gestation followed the workshop in which each chapter was peer-reviewed and the authors were asked to revise their chapter in the light of comments made by the reviewers. We are grateful to the authors of the various chapters for their desire to contribute to the publication of this book.

Many thanks are also addressed to the reviewers for their thoughtful comments and suggestions on each chapter.

Particular thanks go to Dianne Brothwell and Rowland Hill for looking after the linguistic aspects of the book.

References

Baliamoune, M. (2004). *On the measurement of human well-being: Fuzzy set theory and Sen's capability approach (research paper).* Helsinki: Wider Conference.

Foster, J. E., Greer, J., Thorbecke, E. (1984). A class of decomposable poverty measures. *Econometrica, 42,* 761–766.

Sen, A. K. (1976). Poverty an ordinal approach to measurement. *Econometrica, 44,* 219–231.

Sen, A. K. (1981). *Poverty and famines. An essay on entitlement and deprivation.* Oxford: Clarendon Press.

Sen, A. (1985). *Commodities and capabilities.* Amsterdam: North Holland.

Shorrocks, A. F. (1995). Revisiting the Sen poverty index. *Econometrica, 63,* 1225–1230.

Townsend, P. (1979). *Poverty in the United Kingdom.* Hardsmonsworth: Penguin Books.

Zheng, B. (2007). Unit-consistent poverty indices. *Economic Theory, 31*, 113–142.

Part I
Cross Country Studies and a North–South Comparison in the Study of Poverty

Chapter 1
Axiomatic and Robust Multidimensional Poverty Measurements in Five Southern Mediterranean Countries

Valérie Bérenger and Florent Bresson

Abstract The main goal of this chapter is to undertake a multidimensional poverty analysis in relation to five southern Mediterranean countries, namely Egypt, Lebanon, Morocco, Tunisia, and Turkey. We rely on a broader concept of poverty by considering the deprivations in women's educational attainments, the possession of durable goods and the consequences of housing conditions, each of which we operationalize by making use of recent developments in multidimensional poverty measurement. This analysis is based on an axiomatic approach to poverty and on the use of stochastic dominance tools to achieve robust results that do not hinge on the choice of poverty line and weighting scheme. Our findings provide comparisons over time for each country and between countries that cannot be obtained when each dimension of poverty is analysed separately, as the approaches take into account the correlations which may exist between different kinds of poverty. Furthermore, in contrast to rankings drawn from cardinal poverty measures (for example, from the Multidimensional Poverty Index), multidimensional tests make it possible to nuance the performance levels reached by certain countries.

V. Bérenger
GREDEG, University of Nice-Sophia Antipolis, Valbonne, France
e-mail: valerie.berenger@unice.fr

F. Bresson (✉)
LEO (TOTEM), University of Orléans, Orléans, France
e-mail: florent.bresson@univ-orleans.fr

V. Bérenger and F. Bresson (eds.), *Poverty and Social Exclusion Around the Mediterranean Sea*, Economic Studies in Inequality, Social Exclusion and Well-Being 9,
DOI: 10.1007/978-1-4614-5263-8_1,

1.1 Introduction

The main goal of this chapter is to undertake a multidimensional poverty analysis in relation to five southern Mediterranean countries, namely Egypt, Lebanon, Morocco, Tunisia, and Turkey. The multidimensional nature of poverty is now well recognized in the academic community as well as in international development institutions, but studies of poverty in the above countries are still dominated by an absolute monetary approach. At best, poverty profiles are produced based on unidimensional monetary indicators from bivariate analysis or, in some cases, multivariate methods of analysis. To our knowledge, attempts to apply a multidimensional approach to poverty are rather scarce. The sole exceptions concern the work of Deutsch et al. (2005) on Israel, Ayadi et al. (2006) on Tunisia, Bibi (2004) on Egypt and Tunisia, Osman et al. (2006) on Egypt and a recent attempt made in a FEMISE research project using data from six Mediterranean countries (Bérenger et al. 2008), Bérenger, Deutsch, and Silber (2009) for Egypt, Morocco, and Turkey and Bérenger (2010) for Egypt.

The adoption of such an approach is of particular interest in the context of the "Arab Spring" that has taken place since December 2010. Revolutions in Tunisia and Egypt and protests in the other countries provide evidence that the coexistence of multiple deprivations e.g. poor living conditions, lack of political liberties, and repression of human rights, can lead to instability. In addition, these countries adopted structural adjustment programmes (1991 in Egypt, 1983–1993 in Morocco, 1986 in Tunisia, and 1999–2000 in Turkey), and Lebanon, traditionally characterized by a free market economy, was the scene of military confrontations. Despite a similar pattern in the adoption of programmes which prioritized economic growth, structural economic reforms have been implemented with varying intensity across the five countries. This suggests possible differences in the performance achieved in terms of living standards and poverty reduction for the population according to the choice of development strategy. Furthermore, very few studies have attempted to carry out comparisons of experiences of development and poverty across southern Mediterranean countries and more broadly, across MENA countries. The exception was an attempt made in a recent report published by the World Bank (2006). However, this study makes it difficult to carry out rigorous comparisons across countries as it reports national estimates of monetary poverty and growth performance alongside changes in human development indicators.

This chapter relies on an axiomatic approach to poverty measurement, using some of the indicators suggested in the literature (Chakravarty et al. 1998; Alkire and Foster 2011). Moreover, as the results of poverty comparisons based on synthetic indicators are always likely to be contingent on aggregation and weighting choices, this study makes extensive use of stochastic dominance tools to obtain robust results that are valid for a large set of poverty measures and that do not hinge on choices made regarding the poverty frontier.

The present chapter is organized as follows: Sect. 1.2 will present the contribution of the axiomatic and stochastic dominance approaches to multidimensional poverty analysis and Sect. 1.3 will provide multidimensional poverty results using data from the Demographic Health Surveys for Egypt (1995, 2000, 2005), Morocco (1992, 2004), and Turkey (1993, 1998, 2003), from NOFP (2001, 2006) for Tunisia and from Mosa LCHBS (2004–2005) for Lebanon. Given the availability of data and for comparison purposes over time and across countries, our multidimensional measures will include three dimensions, namely housing conditions, access to durable goods as a proxy for "living a comfortable life", and education. Concluding comments and policy recommendations will be given in Sect. 1.4.

1.2 Axiomatic and Stochastic Dominance Approaches to Multidimensional Poverty

During the last few decades, studies of poverty have moved from a traditional approach that relies on a single indicator of well-being, namely income or consumption, to one which increases the number of admissible attributes when measuring living conditions. The major advantage of such a wider concept of poverty is that it allows the researcher to go beyond the inclusion of only material conditions when attempting to capture the essential aspects of household living conditions.

The main areas of criticism of the traditional income and consumption data-based definition are now well known, whether they concern the limitations of income as a proxy for well-being (or its dual aspect, poverty) or the arbitrariness inherent in identifying poor individuals on the basis of a poverty line defined with reference to a whole population's mean/median income or predefined to a consumption basket.

New definitions of poverty have emerged since the seminal works of Townsend (1979) and Sen (1985), each sharing the common idea that income can only serve as an indirect indicator when assessing well-being. According to multidimensional approaches, in order to assess well-being it is necessary to take into account non-monetary indicators that provide more insight about the various facets of poverty. Simultaneously, selecting a multidimensional approach to poverty implies addressing issues that need not be faced when taking a unidimensional approach. Thus, in addition to the problem of determining a poverty line and choosing a measure of poverty, a multidimensional approach forces the researcher to select the relevant dimensions of poverty, the indicators purported to represent these dimensions and finally, the weights attributable to these various dimensions and indicators. Several approaches have been proposed in the literature in order to operationalize the multidimensionality of poverty. At the same time, a lack of consensus is revealed on the best methodology for deriving multidimensional poverty measures.

According to Thorbecke (2005), several strategies can be adopted in order to tackle this issue within the multidimensional framework. The first one involves aggregating several attributes of well-being into a single index via sophisticated techniques of aggregation and deriving a poverty measure from a threshold based on this index of well-being e.g. a unidimensional poverty approach. Several studies have followed this route using methodologies borrowed from efficiency analysis

(Lovell et al. 1994) and information theory (Maasoumi 1986, 1999), as well as inertia approaches (Klasen 2000; Sahn and Stifel 2000; Booysen et al. 2008).

All in all, these attempts are debatable and restrictive; Sen (1985, p. 33) points out that "The passion for aggregation makes good sense in many contexts but it can be futile or pointless in others. [...] When we hear of variety, we need not invariably reach for an aggregator". Another possibility is the adoption of a one-at-a-time analysis of each dimension. The advantage of this analysis lies in its simplicity, but at the same time it lacks synthesis, making it difficult to draw a clear picture of multidimensional poverty, in particular when deprivations do not change in the same direction.

Finally, between these two extremes exists another strategy that preserves the essence of the multiple facets of poverty. This strategy first defines poverty as a combination of shortfalls in each dimension of an individual's well-being and then derives a multidimensional measure. This is precisely the route adopted by the axiomatic approach to multidimensional poverty. Unlike the one-at-a-time analysis, this approach provides a comprehensive picture of poverty by revealing complexities and ambiguities arising from the interaction between various dimensions and their correlation in the sampled population. However, this kind of approach hinges on arbitrary choices regarding the poverty lines, the adequate weights for each attribute, and the functional form of the poverty index. The tools of the dominance approach developed by Duclos et al. (2006) within the multidimensional framework offer an attractive way to tackle these problems. For a class of poverty measures complying with some desirable properties, stochastic dominance techniques make it possible to perform assessments of robust poverty comparisons that are invariant to both the choice of poverty lines and the aggregation procedures. This approach is distinct from cardinalization as it only provides a qualitative appraisal of poverty comparisons and therefore may be viewed as complementary to cardinalization.

1.2.1 The Axiomatic Approach to Poverty

While the axiomatic approach has largely been developed for the unidimensional case, very few studies have attempted to axiomatically derive multidimensional indices of poverty.[1] The study by Chakravarty et al. (1998) appears to be the first publication on the axiomatic derivation of multidimensional poverty indices. New extensions and multidimensional classes of poverty indices were further proposed by Bourguignon and Chakravarty (2003), Alkire and Foster (2011) and Chakravarty and Silber (2008). Alkire and Foster (2011) is noticeable since their multidimensional counting method inspired the elaboration of the multidimensional poverty index (MPI) that was recently introduced in the latest version of the

[1] It is worth mentioning the contributions of Tsui (1995, 1999 and 2002) on axiomatic derivations of multidimensional inequality and poverty indices.

Human Development Report (UNDP 2010). Since this kind of approach provides cardinal comparisons dependent on arbitrary choices, the generalization of stochastic dominance offers another possibility in tackling the multidimensional nature of poverty.

The axiomatic approach to multidimensional poverty defines various axioms that are combined in order to derive different classes of multidimensional poverty measures. A poverty measure is deemed to be the aggregation of shortfalls for all individuals in each dimension. In a similar fashion to its unidimensional counterpart (Sen 1976), the construction of a multidimentional poverty index includes the identification and the aggregation steps that are supposed to reflect the ethical judgments of a social evaluator. In this section, we briefly present the set of axioms that a poverty measure has to satisfy in order to emphasize the main features of the classes of multidimensional poverty measures considered in our empirical application.

First we suppose that the relevant population consists of n individuals. Let $z = (z_1, \ldots, z_m)$ be the m-vector of poverty lines[2] and $x_i = (x_{i1}, \ldots, x_{im})$ the vector of achievements for the m attributes of the *i*th person. Let X be $n \times m$, the matrix of the quantities x_{ij} that denote the level of the *j*th attribute accruing to individual *i*

The identification step raises some issues that are more acute in the multidimensional case. Indeed, in some cases a person can be in such a state of deprivation that he or she is deemed to be poor while in other cases he or she should certainly not be classified as poor. However, it is not clear how the information obtained by several indicators should be used when poor individuals cannot be identified as easily as in the unidimensional case. The current practice is to define the poverty domain with respect to the number of deprivations suffered by an individual.

Let $\xi(x_{ij}, z_j)$ be a dichotomic function that returns 1 if the value x_{ij} of the attribute *j* falls short below the corresponding poverty line z_j, and 0 otherwise, that is:

$$\xi(x_{ij}, z_j) = \begin{cases} 1 & \text{if } x_{ij} \leq z_j \\ 0 & \text{if } x_{ij} > z_j \end{cases} \tag{1.1}$$

Using this simple dichotomic function, it is now possible to define a poverty domain using Alkire and Foster's (2011) approach to poverty identification.

[2] Note that there are two types of poverty line definitions which reflect different views of poverty: relative and absolute. These concepts have been widely debated in the literature as illustrated by the controversy between Sen (1981, 1983) and Townsend (1979, 1993). Indeed, Sen (1983) attempted to hierarchize these two concepts and argued that "poverty is an absolute concept in the space of capabilities but very often it will take a relative form in the space of commodities or characteristics" (p. 159, 1983) required to realize those capabilities. Despite the fact that multidimensional poverty analyses have mainly used absolute poverty lines, it would seem reasonable to include both views, depending on the nature of the attributes under consideration. Some attributes, such as income, which can be used to express the deprivation felt by an individual in relation to the living standards of a particular society can be viewed as relative while other attributes, such as health, education, and access to social services, which reflect capabilities in reference to supposedly universal human needs, are better expressed in absolute terms.

For these authors, the set of poor individuals is defined using a "double cutoff" approach. The first step consists of assessing whether or not individuals are deprived within each dimension of well-being included in the analysis. Then they are regarded as poor if the number of items for which they exhibit deprivations is greater than or equal to a given threshold c. This approach can be summarized using the following identification function ψ^{AF} that returns 1 when an individual is deemed poor relative to the set of poverty lines z and the threshold c[3]:

$$\psi^{AF}(x_i, z, c) = \begin{cases} 1 \text{ if } \sum_{j=1}^{m} \xi(x_{ij}, z_j) \geq c \\ 0 \text{ if } \sum_{j=1}^{m} \xi(x_{ij}, z_j) < c \end{cases} \tag{1.2}$$

This approach is quite flexible and embraces the two traditional identification functions used for multidimensional poverty analysis, namely the intersection and the union approach. For a given vector of poverty lines, the intersection is the most restrictive way of defining the poverty domain, as individuals are deemed poor if and only if they are deprived in each dimension, that is, when the threshold c is set so that it is equal to m. It can be regarded as a very conservative way of thinking about poverty, but the approach is interesting as it helps to place the focus on the "ultra-poor," i.e., those who cumulate deprivations in all relevant dimensions. On the contrary, the union approach states that individuals are poor if they are deprived with respect to at least one attribute. The parameter c is then set so that it is equal to 1. For a given vector z, it can therefore be shown that this approach is the most extensive way of defining the poverty domain. From our point of view, both poverty identification approaches are relevant in obtaining a complete picture of poverty via empirical analyses. For other values of the threshold c, the poverty identification approach will be termed "intermediate" in the present paper, as the corresponding poverty domain is always included in the union domain and always includes the intersection domain. In Sect. 1.3, the level of poverty will be assessed using the identification function ψ^{AF}, but it is worth mentioning that the approach can be further generalized using the Duclos et al. (2006) identification approach. [4] To our knowledge this approach has not been implemented for poverty

---

[3] The approach can be further generalized if different weights are given to each deprivation for the identification of the poverty domain (see Alkire and Foster 2011, for more details). This additional complexity of the approach is not presented here as equal weights are given to each dimension in our empirical analysis.

[4] These authors first define some real-valued and differentiable well-being function $\lambda : \Re^m \rightarrow \Re$ i.e. non-decreasing with the level of each attribute$\left(\frac{\partial \lambda(x_i)}{\partial x_{ij}} \geq 0 \; \forall j \in \{1, \ldots, m\}\right)$. Using this function, an individual is deemed poor if and only if his or her level of well-being is below the one corresponding to the set of poverty lines z. The approach can be summed up using the following identification function: $\psi^{\mathrm{DSY}}(x_i, z) = \begin{cases} 1 \text{ if } \lambda(x_i) \leq \lambda(z) \\ 0 \text{ if } \lambda(x_i) > \lambda(z) \end{cases}$

measurement but it serves as a general basis for the multidimensional stochastic dominance tools that will be presented at the end of the current section.

The aggregation step is concerned with the set of axioms that any poverty measure should satisfy. Some properties are simply an extension of the multidimensional case of axioms defined within the one-dimensional framework and constitute the core set of axioms. Other axioms are more specific since they address issues raised by the multidimensionality of poverty, such as the correlation between attributes.

In the related literature it is often assumed that a poverty index should comply at least with the following main axioms i.e. symmetry, focus, monotonicity, continuity, and the principle of population.[5] It is also often assumed that the measure should comply with two axioms for which the fulfilment is helpful in the design of socio-economic poverty measures aimed at combating poverty.

The first one is the *axiom of sub-group decomposability* (SD). It states that overall poverty is a weighted average by sub-group shares of the sub-group poverty levels. This axiom enables the identification of those groups which are the most afflicted by poverty.

The second one, the *axiom of factor decomposability* (FD) allows the poverty index to be broken down by dimensions and enables the evaluation of the contribution of each dimension to overall poverty (Chakravarty et al. 1998; Alkire and Foster 2011). This property is particularly suitable for policy targeting. However, it requires the individual poverty index to be additive across dimensions; this could prevent the fulfilment of some desirable transfer axioms.

Furthermore, following his seminal criticism of the unidimensional poverty index, Sen proposed that a reasonable poverty measure should be sensitive to inequality among poor individuals, which means that if inequality decreases then poverty should decrease. The general formulation of the *Pigou-Dalton transfer principle* (PDP) states that a progressive transfer from a poor individual to an even poorer individual should not increase poverty. This principle has been generalized in the multidimensional context following Kolm (1977) in his study of inequality in a multidimensional context. The *multidimensional transfer principle* (MTP) states that poverty should not increase if it is obtained by a redistribution of attributes among individuals according to a bistochastic transformation.[6] In other words, MTP requires that the post transfer distribution of attributes should be more

[5] Symmetry states that the only relevant information for assessing poverty at the individual level is that contained in the vector x_i. Symmetry thus implies equal treatment of the equals. With the (weak) focus axiom, it is assumed that an increment in the value of a particular attribute does not change the level of poverty if the corresponding individual does not belong to the set of poor individuals. On the contrary, (weak) monotonicity means that raising the level of some attribute for a poor individual does not increase the poverty level. A poverty index is said to be continuous if marginal variations of the value of some attribute do not yield large variations in the poverty index. Finally, the principle of population states that for a given matrix X, the poverty level should be the same when replicating each element in the matrix many times. This axiom is necessary for the comparison of populations of different sizes. For more details on the axiomatic of poverty measurement, see Zheng (1997).

[6] A bistochastic matrix is a square matrix with the sum of each column and row equal to one.

even than the initial distribution.[7] Duclos et al. (2006) also proposed a rival generalization of the Pigou-Dalton transfer principle for the multidimensional context that we will refer to as the non-ambiguous transfer principle (NATP) in the present paper. This axiom states that if we consider two poor individuals who can be unambiguously ranked in terms of poverty $\left(x_{ij} > x_{i'j}, \forall j \in \{1,\ldots,m\}\right)$, a progressive transfer between the two individuals should decrease the level of poverty. On the one hand, it differs from MTP since the recipient of the transfer should exhibit at least the same level of deprivation in each dimension as the person at the origin of the transfer. On the other hand, MTP means that progressive transfers are undertaken in each dimension and in the same proportions.

However, as Tsui (2002) points out, the MTP axiom does not take into account the essence of multidimensionality via the correlation between attributes. The axiom of non-decreasing poverty under a correlation increasing switch (NDCIS) proposed by Tsui (2002) and refined by Bourguignon and Chakravarty (2003) stress the difference between single and multidimensional poverty by taking account of the association between attributes as captured by their degree of correlation. According to NDCIS, if the dimensions correspond to similar aspects of poverty, then it is reasonable to expect that a switch in the quantity of one attribute between two individuals that increases the correlation between attributes does not decrease poverty. For two poor individuals it means that the decrease in poverty for the relatively less poor individual is not sufficient to compensate the increase in poverty for the poorer individual.[8] Owing to the existence of a trade-off between factor decomposability and NDCIS, these families can then be classified into two groups, hence making it difficult for certain classes of indices to fulfil both axioms simultaneously.[9]

1.2.2 Multidimensional Poverty Measures

For the empirical side of this chapter we use multidimensional poverty measures drawn from Alkire and Foster methodology. The counting method developed by Alkire and Foster (2011) avoids the drawbacks of the traditional approaches, i.e., the exclusion of individuals who are deprived in several important dimensions with the intersection definition, and the inclusion with the union approach of

[7] PDP and MTP both require the individual poverty function to be convex.

[8] In order for this axiom to hold, the poverty index has to be continuous, twice differentiable, and convex. Then the marginal utility of one attribute decreases when the quantity of the other increases. This corresponds to the Auspitz-Lieben-Edgeworth Pareto definition of substitutability and differs from Hicks definition.

[9] Thus, the additivity implied by factor decomposability requires that the poverty measure is insensitive to the correlation increasing switch. Conversely, poverty measures that satisfy NDCIS are sensitive to the joint relation between attributes. In this case, poverty measures are then defined by taking a nondecreasing concave function of poverty gaps for each individual in different dimensions.

people who should not be considered as multidimensionally poor. Using the identification approach described in last section, we make use of a class of "dimension-adjusted" multidimensional poverty measures based on the traditional FGT measures of poverty. This class of measures satisfies an array of desirable axioms including decomposability and dimensional monotonicity properties.

The family of dimension adjusted FGT measures is given by:

$$\theta_{\alpha}^{AF}(X, z) = \frac{1}{nm}\sum_{i=1}^{n}\psi^{AF}(x_i, z, c)\sum_{j=1}^{m}\left[1 - \frac{x_{ij}\Lambda z_j}{z_j}\right]^{\alpha_j} \tag{1.3}$$

where $(x_{ij}\Lambda z_j)$ denotes the minimum of x_{ij} and z_j. Expression (1.3) is equal to the sum of the normalized poverty gap for the multidimensional poor individuals raised to the α-power divided by the highest possible sum of normalized gap (nm).[10]

When $\alpha = 0$ for all j, the measure is the adjusted headcount ratio used for the MPI. This measure can be expressed as $\theta_0^{AF} = HA$ i.e. the product of the percentage of the multidimensional poor (H) times the average deprivation share across the poor (A). In contrast to the traditional headcount ratio, θ_0^{AF} satisfies dimension monotonicity: if a poor person becomes deprived in an additional dimension, A will increase and also θ_0^{AF}. Therefore, it takes into account both the incidence of poverty and the average extent of the multidimensional deprivation.

When $\alpha = 1$ for all j, the measure is the adjusted poverty gap; it can also be expressed as the product between the multidimensional headcount ratio (H), the average deprivation share across the poor (A), and the average depth of deprivations for poor individuals. It satisfies not only dimension monotonicity but also monotonicity. Finally, as in the case of the FGT unidimensional indices, it is possible to derive the adjusted squared poverty gap for $\alpha = 2$. It can also capture information on the incidence of poverty, the average range, and severity of deprivations among poor individuals. Along with the monotonicity properties, the measure satisfies the transfer principle as it is sensitive to the inequality of deprivations among poor individuals.

This class of indices is decomposable by population subgroup. Additionally, when the identification step has been completed, it satisfies factor decomposability. The indices are neutral to inter-dimensional interaction. Note that, due to its completely additive form, the deprivation of each individual in each dimension can be evaluated independently from the deprivation in other dimensions. Thus, the multidimensional extension of FGT indices proposed by Chakravarty et al. (1998) can be considered as a special case for this family of measures.

---

[10] This expression can be generalized by considering different weighting schemes. Let a_j be the weight attached to the jth dimension such that $\sum_{i=1}^{m} a_j = m$. In this case, the cut-off "c" can be a real number instead of being a positive integer. See Alkire and Seth (2008).

It is worth noting that the dimension decomposability property of these indices prevents the fulfilment of the axioms related to increasing correlation switches. Nevertheless, they can be seen as suitable for policy purposes since they allow for the estimation of the relative contribution of each dimension to overall poverty.

1.2.3 Ordinal Comparisons of Multidimensional Poverty

Whereas the above approach provides quantitative measures, its results depend heavily on aggregation and identification rules that entail a number of implicit assumptions. Stochastic dominance techniques can address this issue as they make it possible to establish the conditions under which poverty comparisons are valid for a wide range of poverty lines and for all the members from particular classes of poverty measures. Extending previous work on unidimensional stochastic dominance (Atkinson 1987; Foster and Shorrocks 1988), Duclos et al. (2006) proposed a method for multidimensional comparisons that is robust relative to alternative ethical assumptions. Their approach is interesting as it relies on a very small set of assumptions and therefore is likely to yield results that most people should agree on. To be more specific, the authors consider multidimensional poverty measures that take the following form:

$$\theta(X, z, \lambda) = \frac{1}{n}\sum_{i=1}^{n} \theta(x_i, z, \lambda) \tag{1.4}$$

where λ refers to the implicit well-being function used for the definition of the identification function (more details in footnote 4). In Duclos et al. (2006), the individual poverty measure θ is supposed to comply with the focus, monotonicity, continuity, symmetry, and sub-group decomposability axioms as well as the principle of population. Before comparing distributions, the first step of the procedure consists in defining a poverty domain that could be viewed as the largest subset in $\Re^m$ admissible for a definition of the poverty domain. If the stochastic dominance procedure is performed using Alkire and Foster's (2011) approach to poverty identification, the definition of the maximum poverty domain implies first the definition of a vector z^+ upon which agreement could be reached for the maximum values used for the poverty line in each dimension. Distributions are then compared using the function Ξ as defined by:

$$\Xi_\alpha(X, z) = \int_0^{z_1}\int_0^{z_2}\cdots\int_0^{z_m} \prod_{j=1}^{m} \left(z_j - x_{ij}\right)^{\alpha_j - 1} dF(x; X), \tag{1.5}$$

where $F(x; X)$ is the cumulative distributive function corresponding to the matrix X. It is worth noting the presence of the vector $\alpha = (\alpha_1, \cdots, \alpha_m)$ in the definition of the function Ξ. These parameters indicate the order of dominance for each dimension of well-being when performing the stochastic dominance checks. When all elements of α are set equal to one, the test refers to a class γ_1 of poverty measures θ that complies with the aforementioned axioms as well as the axiom NDCIS. A distribution X will then weakly dominate the distribution X' i.e. poverty can be said to be robustly lower for X than for X' for all poverty measures from the class γ_1 and all poverty domains included in the largest poverty domain corresponding to λ and z^+, if and only if:

$$\Xi_1(X,z) - \Xi_1\left(X',z\right) \leq 0 \quad \forall z \quad \text{s.t.} \quad \lambda(z) \leq \lambda(z^+), \tag{1.6}$$

and there exists at least one vector z in the maximum poverty domain such that a strict inequality is observed. From a practical point of view, the procedure involves comparing the value of the joint cumulative distribution function for every attribute between X and X' for all the points of the maximum poverty domain. Consequently, a dominance relationship is observed when the difference between the two cumulative distribution functions exhibits the same sign for all points of that set. In many cases the conditions are not met, but it is still possible to observe compliance with second-order stochastic dominance conditions. It is then necessary to turn to a less general class of poverty measures that are sensitive to transfer between poor individuals in a number of dimensions of well-being. To ease the presentation, let us assume that only two dimensions of well-being have been chosen for the analysis and that these dimensions define the class $\gamma_{2,1}$ of multidimensional poverty measures. The class is a subset of the class γ_1 and hence respects all the properties imposed for that class of poverty indices. Members of $\gamma_{2,1}$ are also sensitive to transfers since they comply with NATP for progressive transfers with respect to the first well-being attribute.[11] Moreover, the lower the value of the second attribute is for the recipient of a given transfer, the larger the poverty alleviation effect of that transfer will be. A distribution X will then weakly dominate the distribution X' i.e. poverty can be said to be robustly lower for X than for X' for all poverty measures from the class $\gamma_{2,1}$ and all poverty domains corresponding to λ and z^+, if and only if:

$$\Xi_{2,1}(X,z) - \Xi_{2,1}\left(X',z\right) \leq 0 \quad \forall z \quad \text{s.t.} \quad \lambda(z) \leq \lambda(z^+) \tag{1.7}$$

and there exists at least one vector z in the maximum poverty domain such that a strict inequality is observed. The procedure then implies the comparisons for the distributions X and X' of the aggregate values of poverty gaps observed in the first dimension and cumulated over the second dimension of well-being.

Once again it is possible to turn to higher dominance orders to obtain results that are robust from an ethical point of view if dominance is not observed using the classes $\gamma_{1,2}$ and $\gamma_{2,1}$. The price to be paid is that results are consequently valid for a

---

11 With the class $\gamma_{1,2}$, sensitivity to progressive transfers would be observed with respect to the second dimension.

smaller class of poverty indices. When all the parameters α_j are set equal to 2, the members of the corresponding class of poverty measures γ_2 then obey the principle of NATP transfers in each dimension. Moreover, the sensitivity to transfers is required to be stronger in one dimension of well-being when the level of the attributes for the other dimensions of well-being is lower. As in the previous cases, poverty can be said to be robustly lower for X than for X' for all poverty measures from the class γ_2 and all poverty domains corresponding to λ and z^+, if, and only if:

$$\Xi_2(X,z) - \Xi_2\left(X',z\right) \leq 0 \quad \forall z \quad \text{s.t.} \quad \lambda(z) \leq \lambda(z^+) \tag{1.8}$$

and there exists at least one vector z in the maximum poverty domain such that a strict inequality is observed.

Finally, at this stage we should mention that since poverty comparisons are most often performed on the basis of samples of individuals or households, statistical inference is necessary to assess the statistical robustness of the results. In line with Kakwani (1993) for the unidimensional case, as sub-group additive poverty measures can be seen as the average values of individual poverty indices, the asymptotic properties of the empirical mean can be applied to a particular class of poverty measures. Therefore, for a given poverty line, observed differences between two poverty measures asymptotically involve the normal standard distribution so that the test is based on the t-statistic. Then in order to infer the dominance of the distribution A by B, we follow the procedure devised by Davidson and Duclos (2000, 2006). This procedure is based on the rejection of the null hypothesis (H0) of the non dominance of A by B with given values of poverty lines. The statistic for the corresponding test is then the minimum value of the t-statistics estimated for each poverty line value within the poverty domain, as suggested by Kaur et al. (1994). Hence, the null hypothesis is rejected if the minimum value of the t-statistics is higher than the corresponding value at a chosen confidence level. As demonstrated by Gravel and Mukhopadhyay (2010) and Batana and Duclos (2010), the extension of this procedure to multidimensional poverty comparisons is straightforward.

1.3 Main Results for the Cases of Egypt, Lebanon, Morocco, Tunisia, and Turkey

According to the World Bank classification, all five countries are middle-income countries. During the 1980s, serious macroeconomic imbalances compelled the adoption of structural adjustment programmes (SAP) aimed at stabilizing and liberating an economy and committing it to an outward looking direction. Turkey was the first country to take up economic reforms, followed by Morocco in 1983, Tunisia in 1986, and more recently by Egypt in 1991. Despite the adoption of similar patterns which give priority to economic growth, structural economic reforms have been implemented at varying levels of intensity among the four countries. This suggests possible differences in performance achieved in terms of standards of living and poverty reduction for the population according to the type of development strategy adopted. Critics of this kind of economic reform process argued that such reforms would have a negative impact on poor individuals. However, broadly speaking, except for Tunisia and Egypt, poverty studies in these countries suffer from a lack of data and when data are collected, access to them is very limited. Moreover, a unidimensional approach has been applied within the main studies. In addition, differences in methodologies used in order to assess poverty lines led to conflicting results (see the case of Egypt, World Bank 2002) which makes any comparison difficult. Very few studies have attempted to carry out comparisons of experiences of development and poverty across southern Mediterranean countries and more broadly, across MENA countries. The exception was an attempt made in a recent report published by the World Bank (Iqbal 2006). This report frames the MENA regional experience with poverty reduction and human development before and after the period of structural economic reforms as adopted in the different countries of the region. Consequently, the report underlines the weak economic performance and the limited change in poverty despite continuous rapid improvements in human development after the period of SAPs.

Although the analysis of monetary poverty could be complemented by a survey of the changing levels of performance in several human development dimensions, no attempt has been made to look at the joint distribution of these attributes. The existing studies make it difficult to carry out rigorous comparisons, as they report national estimates alongside the change in human development indicators. They do not make it possible to introduce any nuance. To our knowledge the studies of Bibi (2004) on Egypt and Tunisia, and of Bérenger et al. (2009) on Egypt, Morocco, Tunisia, and Turkey are the only attempts at multidimensional poverty comparisons across southern Mediterranean countries.

1.3.1 Data Description

As stressed above, despite the fact that household budget surveys have been collected regularly in Egypt, Morocco, Tunisia, and Turkey, they are retained within government agencies and are not made public. However, the use of demographic and health surveys (DHS) initiated by the U.S. agency for international development (USAID) offers an alternative instrument. Although these surveys do not include data on income and expenditure, they contain significant information on the living conditions of the population in Egypt (1995, 2000, 2005), Morocco (1992, 2003), Turkey (1993, 1998, 2003). In these data bases, two main sources of information are available: a household schedule and an individual questionnaire for women of reproductive age (15–49), which can be combined in order to extract dimensions of interest. For Tunisia, the surveys used are from

NOFP (2001, 2006) and are considered as the continuation of the DHS survey conducted in 1998. For the case of Lebanon, since DHS surveys are not available, we have used data from the "2004/05 National Survey of Living Conditions and Household Budget Survey". However, this database does not make it possible to apply urban/rural breakdowns.

As our purpose is to make poverty comparisons over time (except for Lebanon) and across countries, this restricts the range of indicators to be included in a poverty measure. In the absence of monetary indicators, the use of the housing schedule/questionnaire enabled us to identify three dimensions: education, housing conditions, possession of durable goods for which data are available per year (except for Lebanon) and per country. Using the framework from Sen's capability approach we have taken any individual dimension as representing a 'functioning'. These dimensions can be viewed as proxies for those included in the Human Poverty Index.

Although using individual women's circumstances as a unit of analysis would suggest a gender approach to poverty analysis, the nature of the indicators is not sufficiently representative for this type of analysis. Indeed, only education refers to individual characteristics while housing conditions and durable goods concern the household. However, the focus on women's living conditions may be of particular interest here since it makes it possible to understand the differences in the development models followed by these countries.

The first dimension is education, measured by the number of years of schooling achieved by women respondents. Within the Human Development approach, education has a double role in terms of human well-being. Education has an intrinsic value because being educated allows people to enjoy a better life, but it also has an instrumental role in economic production, in participation, social change, and in the life of the community (Sen 1985 and 1999). In particular, numerous studies have shown that women's education can have significant multiplier effects that go beyond market activities and that spread to society at large. Considerable empirical evidence shows that female education improves human development outcomes, such as child survival, health and schooling (Strauss and Thomas 1995; Hill and King 1995; World Bank 2001; Schultz 2002; World Bank 2007), and helps to break the intergenerational transmission of poverty. Moreover, in the particular context of our countries, the focus on women's education might reveal differences in priorities given to social aspects of human development and especially to its gender dimension. Despite some significant progress during the last few decades, MENA countries generally have lower levels of women's education and labor force participation than other countries with similar income levels. Except for Turkey and Lebanon, which are secular states, gender discrimination is codified in law, family law, and civil codes. Hence, the educational attainment of women should mirror the impact of past gender social policies as well as education reforms. In order to quantify deprivation, the threshold is fixed at 8 years of schooling, which corresponds to a level of compulsory education, and will vary up to a maximum of 12 years for robustness check.

The information about housing conditions is based on two indicators. The first variable highlights the overcrowding aspect and concerns the number of rooms per capita adjusted by taking into account the household composition.[12] The second variable refers to floor construction material; this is used to obtain a coefficient that is applied to the first variable in order to take the quality of the dwelling into account. In line with Bibi (2004), the housing deprivation threshold is set at 0.3 rooms per person and varies up to 0.75. This dimension is of particular interest, as absent, limited, or low quality assets in this dimension could have indirect consequences on health and education. Moreover, it may be relevant for our countries under study; according to a survey of MENA housing sectors (Baharoglu et al. 2005), many of the region's countries are confronted with a problem of housing affordability rather than availability. This is a consequence of housing policies and land policies which in some countries are characterized by a high level of public ownership, inefficient systems of land ownership (lack of secure property rights), and some changes in rent control policies that restrict the use of housing stock. Consequently, the high demand is reflected by increases in housing prices. In many cases, scarcity exists for low income population groups and their demand is met by the informal sector. Therefore, policies for upgrading slums are becoming a major priority in some countries, especially in Morocco and Egypt. Furthermore, rent controls that lower costs for sitting tenants can lead to under-maintained housing and a shrinking private rental market; this results in the exclusion of the poorest from the formal rental sector.[13]

The last dimension refers to the availability of durable goods, such as the possession of a television, refrigerator, or car, which are all binary variables. In terms of a 'functioning', their inclusion is debatable. Nevertheless, they can be viewed as a proxy for what enables the control of resources. As argued by Brandolini et al. (1998), they can represent the "living a comfortable life" attribute. Given that the indicators are computed at household level, this dimension neglects intra-family distribution and is not entirely satisfactory when the individual is the unit of reference. The "durable goods" indicator has been constructed using the pooled distribution for each item from the most recent survey in each country. More specifically, the "durable goods" index is a weighted means, showing the availability of each item, with weights being defined on the basis of the inverse of the frequency observed for the pooled distributions.[14] The poverty cut-off is set at 0.2 according to the characteristics of the distribution and will vary up to 0.5.

---

[12] Children under 5 years are supposed to be less demanding with respect to housing surface and so to weight twice less than more than 5-year-old individuals.

[13] In addition, changes in rent control may affect the residential mobility of households. In particular, shifts from the older versions of rent control which maintained rents at fixed nominal levels to other kinds of rent control where rents are freely set but cannot go higher than a maximum rental index might be responsible for adverse effects: smaller housing area per capita and a deterioration in housing conditions.

[14] The weights are 0.13 for a television, 0.67 for a car, and 0.19 for a refrigerator.

1.3.2 Investigation of Poverty by Dimension Over Time and Across the Five Countries

First of all we begin by scrutinizing deprivation in each dimension separately using stochastic dominance tests for the whole population of the five countries and then for the sole urban and rural areas.

The analysis of the results reported in Table 1.1 for the whole population as well as for the urban and rural areas provides useful information. Improvements in education achievements can be observed at the first-order over time in all but Turkey where a statistically significant worsening is registered between 1998 and 2003. This result seems likely to be due to the worsening observed in rural areas since no dominance relationship is observed in urban areas. A statistically significant improvement in education in Tunisia is noted only in urban areas using transfer-sensitive poverty measures.

Looking at the housing conditions, the decrease in deprivation over time is only statistically significant in Morocco for the whole population and urban/rural areas with first-order dominance checks. Although not significant for the whole population in Tunisia, improvements seem to have benefited unambiguously those severely deprived urban individuals. That evolution is widening discrepancies between urban and rural areas. In contrast, while no dominance relationship is observed in Egypt and Turkey for the whole period, the results stress a non-monotonic evolution over the sub-periods when considering poverty measures that are sensitive to extreme poverty. More specifically, the situation worsens during the first sub-period in Egypt (1995–2000) and the first sub-period in Turkey (1993–1998). In this latter case it appears to be mainly due to the degradation exhibited in rural areas.

With respect to the possession of durable goods, results show an unambiguous and statistically significant decrease in poverty over time at the first order in Morocco and at the second order in Egypt and Tunisia. However, the evolution in Turkey exhibits a non-monotonic pattern over the two sub-periods, with a worsening in deprivations between 1998 and 2003 in particular for the severely deprived individuals.

The breakdown by area of residence indicates significant improvements using second-order dominance checks for the rural population in Egypt, whereas the evolution in the urban area is not conclusive. In the same way, we can observe that the worsening in Turkey seems to be confirmed in urban areas while the increase in the number of poor individuals in rural areas is not significant at the first order. Finally, in the case of Morocco we can highlight the striking improvement in access for the population to durable goods in rural areas.

These findings make it possible to underline some significant features that characterize the deprivations in each dimension over time for each country under consideration and to draw attention to certain specific aspects. In particular, we ought to mention that Egypt seemed to face a deterioration in housing conditions between 1995 and 2000 and that the improvement recorded in the possession of

Table 1.1 Unidimensional dominance tests

Dimension dom. order	Whole population						Urban population						Rural population					
	Education		Housing		Dur. goods		Education		Housing		Dur. goods		Education		Housing		Dur. goods	
	1	2	1	2	1	2	1	2	1	2	1	2	1	2	1	2	1	2
Egypt																		
2005–1995	≻*		∅	∅	∅	≻*	≻*		∅	∅	∅	≻	≻*		∅	∅	∅	≻*
	(0)		(1)	(0.852)	(1)	(0)	(0)		(0.999)	(0.618)	(1)	(0.393)	(0)		(1)	(0.981)	(1)	(0)
2000–1995	≻*		∅	≺*	∅	≻*	≻*		∅	≺*	∅	∅	≻*		∅	≺*	∅	≻*
	(0)		(1)	(0)	(1)	(0)	(0.033)		(0.961)	0	(1)	(1)	(0)		(1)	(0)	(1)	(0)
2005–2000	≻*		≻	≻*	∅	≻*	≻*		∅	≻*	∅	≻*	≻*		∅	≻*	∅	≻*
	(0)		(0.498)	(0)	(1)	(0)	(0)		(1)	(0)	(1)	(0.138)	(0)		(1)	(0)	(1)	(0)
Turkey																		
2003–1993	≻*		∅	∅	∅	∅	≻*		∅	∅	∅	≺	≻*		∅	∅	≻*	
	(0)		(1)	(1)	(1)	(1)	(0)		(0.999)	(0.998)	(0.99)	(0.458)	(0)		(0.981)	(0.891)	(0)	
1998–1993	≻*		∅	≺	≻*		≻*		∅	∅	≻*		≻*		∅	≺*	≻*	
	(0)		(1)	(0.118)	(0)		(0)		(1)	(0.865)	(0)		(0)		(0.541)	(0.019)	(0)	
2003–1998	≺*		∅	≻	≺	≺*	∅	∅	∅	∅	≺*		≺*		∅	∅	≺	≺*
	(0.048)		(1)	(0.142)	(0.189)	(0)	(0.834)	(0.825)	(1)	(1)	(0.012)		(0)		(1)	(1)	(0.185)	(0.027)
Marocco																		
2003–1992	≻*		≻*		≻*		≻*		≻*		∅	∅	≻*		≻*		≻*	
	(0)		(0)		(0)		(0)		(0)		(1)	(1)	(0)		(0)		(0)	
Tunisia																		
2006–2001	∅	∅	≻	≻	≻	≻*	≻	≻*	≻	≻*	∅	≻*	∅	∅	∅	∅	≻*	
	(0.978)	(0.946)	(0.104)	(0.104)	(0.135)	(0)	(0.228)	(0.049)	(0.07)	(0.005)	(1)	(0)	(1)	(0.987)	(1)	(1)	(0)	

Note: ≻(≺) indicates that the first distribution dominates (is dominated by) the second distribution at the corresponding order of dominance. ∅ denotes a non-conclusive test. P values are reported in parentheses and refer to the hypothesis of non-dominance of the second distribution with respect to the first one when ≺ is reported and the hypothesis of non-dominance of the first distribution with respect to the second one in the other cases. Values less than 10^{-3} or greater than $1–10^{-3}$ are respectively replaced by 0 and 1 for the sake of simplicity. The * indicates that the result is significant at the 5 % level. The results of higher order dominance checks are not reported when a significant dominance relationship at a given order is observed durable goods was mainly due to a better access by the population living in rural areas.

This result seems to corroborate certain findings from previous reports and studies (Bérenger 2010). Indeed, Egypt has had to contend with a problem of informal settlements and slums mainly in urban areas. Moreover, a new rental law was adopted in 1996 and replaced the old rental law under which the rents being paid bore no relation to market prices (Sims et al. 2008). Morocco is the only country that shows a decrease in deprivation in all dimensions over time for both urban and rural areas, with the exception of durable goods where the decrease in deprivation is only observed for rural areas. In Tunisia, barring durable goods, a decrease in poverty over time is evidenced by stochastic dominance in urban areas. Moreover, between 1998 and 2003, women in Turkey faced deterioration in their education levels (more specifically in rural areas) and an increase in deprivation in terms of the possession of durable goods; they faced no significant changes in housing conditions.[15]

Finally, since the same dimensions were used for the five countries, we have performed stochastic dominance tests across countries using the most recent dataset for each country. Results for the overall sample and by area of residence are displayed in Table 1.2. Tunisia and Lebanon have the lowest level of poor individuals in education, at least for distribution sensitive poverty measures, while Morocco records the highest level for the overall sample, regardless of the areas of residence. This result is not surprising; it is illustrative of the differences in the development model chosen by these countries.

One of the main challenges in Morocco is education. Although spending on education remained a significant percentage of GDP, the state did not have a clear strategy to eradicate illiteracy. Past public policies for formal education were uneven and focused on urban areas to the detriment of rural areas (Bergh 2005). Furthermore, education only became compulsory in the 1990s, but it is not enforced, particularly in rural areas.

In contrast, Morocco records the best performances in terms of housing conditions and this dominance relationship is supported in rural areas as well as in urban areas, though not always significantly in the latter case. However, the largest deprivation in this dimension can be found in Lebanon, followed by Egypt. At first sight, this result is surprising when we consider the characteristics of the housing market (Baharoglu et al. 2005). But, it is worth mentioning that Morocco is considered as a progressive country in its search for suitable solutions for housing its poor population. A variety of housing policies have been implemented since 1956. While growth of informal settlements began to accelerate in the 1980s, a major programme to address sub-standard housing was launched, followed in 2004 by the "2004–2010 Cities without Slums" programme aiming at achieving a slum-free target by 2010 (Martin et al. 2008). Conversely, the inferior housing

[15] Moreover, the analysis of the urban bias for each dimension shows that deprivation is mainly a rural phenomenon. The results of the tests are available upon request.

Table 1.2 Unidimensional dominance tests: cross country comparisons

Dimension	Education		Housing		Dur. goods	
Dominance order	1	2	1	2	1	2
Whole population						
Turkey vs Egypt	$\varnothing$	$\varnothing$	$\varnothing$	$\succ^*$		
	(1)	(1)	(1)	(0.017)	(0)	
Morocco vs Egypt	$\prec^*$		$\succ^*$		$\varnothing$	$\prec^*$
	(0)		(0)		(1)	(0)
Tunisia vs Egypt	$\succ^*$		$\varnothing$	$\varnothing$	$\succ^*$	
	(0)		(1)	(1)	(0)	
Lebanon vs Egypt	$\varnothing$	$\succ^*$	$\varnothing$	$\varnothing$		
	(1)	(0)	(1)	(0.85)	(0)	
Morocco vs Turkey	$\varnothing$	$\prec^*$	$\succ^*$		$\prec^*$	
	(1)	(0)	(0.007)			
Tunisia vs Turkey	$\succ^*$		$\varnothing$	$\varnothing$	$\varnothing$	$\varnothing$
	(0)		(1)	(1)	(0.0692)	(0.0692)
Lebanon vs Turkey	$\succ^*$		$\varnothing$	$\prec^*$	$\succ^*$	
	(0)		(1)	(0)	(0)	
Tunisia vs Morocco	$\succ^*$		$\prec^*$	$\prec^*$	$\succ^*$	
	(0)		(0.23)	(0)	(0)	
Lebanon vs Morocco	$\succ^*$		$\prec^*$		$\succ^*$	
	(0)		(0)		(0)	
Lebanon vs Tunisia	$\varnothing$	$\varnothing$	$\prec^*$		$\succ^*$	
	(1)	(1)	(0)		(0)	
Urban population						
Turkey vs Egypt	$\varnothing$	$\varnothing$	$\varnothing$	$\succ^*$	$\succ^*$	
	(1)	(1)	(1)	(0.004)	(0)	
Morocco vs Egypt	$\prec^*$		$\succ^*$		$\varnothing$	$\prec^*$
	(0)		(0.002)		(1)	(0)
Tunisia vs Egypt	$\varnothing$	$\succ^*$	$\succ^*$		$\succ^*$	
	(1)	(0)	(0)		(0)	
Morocco vs Turkey	$\varnothing$	$\prec^*$	$\varnothing$	$\succ$	$\prec^*$	
	(1)	(0)	(1)	(0.461)	(0)	
Tunisia vs Turkey	$\succ^*$		$\succ$	$\succ$	$\varnothing$	$\succ^*$
	(0)		(0.179)	(0.179)	(1)	(0)
Tunisia vs Morocco	$\succ^*$		$\varnothing$	$\succ$	$\succ^*$	
	(0)		(1)	(0.18)	(0.001)	
Rural population						
Turkey vs Egypt	$\varnothing$	$\varnothing$	$\varnothing$	$\prec$	$\succ^*$	
	(1)	(1)	(1)	(0.488)	(0)	
Morocco vs Egypt	$\prec^*$		$\succ^*$		$\varnothing$	$\prec^*$
	(0)		(0)		(1)	(0)
Tunisia vs Egypt	$\succ^*$		$\varnothing$	$\varnothing$	$\succ$	$\succ^*$
	(0)		(1)	(1)	(0.056)	(0.016)
Morocco vs Turkey	$\prec$	$\prec^*$	$\succ^*$		$\prec^*$	
	(0.448)	(0)	(0)		(0)	

(continued)

Table 1.2 (continued)

Dimension	Education		Housing		Dur. goods	
Dominance order	1	2	1	2	1	2
Tunisia vs Turkey	$\succ^*$		$\varnothing$	$\varnothing$	$\prec^*$	
	(0)		(1)	(1)	(0)	
Tunisia vs Morocco	$\succ^*$		$\prec^*$		$\succ^*$	
	(0)		(0)		(0)	

Note: $\succ$($\prec$) indicates that the first distribution dominates (is dominated by) the second distribution at the corresponding order of dominance. $\varnothing$ denotes a non-conclusive test. P values are reported in parentheses and refer to the hypothesis of non-dominance of the second distribution with respect to the first one when $\prec$ is reported and to the hypothesis of non-dominance of the first distribution with respect to the second one in the other cases. Values less than 10^{-3} or greater than $1–10^{-3}$ are respectively replaced by 0 and 1 for the sake of simplicity. The * indicates that the result is significant at the 5 % level. The results of second-order dominance checks are not reported when a significant first-order dominance relationship is observed

conditions in Lebanon mirror the rapid urbanization caused by massive rural migration, the impact of civil war, and the lack of clear public housing policies (UN-Habitat 2003). Moreover, although poverty gaps in housing conditions are greater in Egypt than in Turkey, the results are not supported in rural areas although being insignificant.

Finally, in terms of durable goods, Morocco once again shows the highest level of deprivation for the overall samples, regardless of the areas of residence. In contrast, the population in Lebanon is the least deprived. Although no dominance relationship can be found between Tunisia and Turkey, the breakdown by areas of residence shows that the urban population in Tunisia experiences the lowest level of poverty whereas the situation for rural populations seems to be better in Turkey relative to the three other countries. We can note that the ranking obtained in this dimension differs slightly from the one based on GNI per capita (2008 PPP US$) provided by the UNDP database. Indeed, with respect to GNI per capita we should expect the population in Turkey to be richer than in Lebanon, though a slight advantage for Lebanon is registered after 2009, just as Turkey is better ranked than Tunisia.[16]

To conclude, it is important to note that the one-at-a-time analysis provides interesting results which are not easy to interpret in some instances. It can lead to different rankings between countries and over time, since a given country can perform well in one dimension but not in another one e.g. Morocco or Turkey over time. It does not give rise to a comprehensive picture of the multiple deprivations suffered by a population since it does not take into account the interdependencies between dimensions of well-being. Thus, this analysis highlights the importance of capturing the distribution of the number of deprivations faced by a population.

[16] For further details see online data: http://hdr.undp.org.

1.3.3 *Multidimensional Poverty Measures and Multidimensional Stochastic Dominance*

Although the one-at-a-time analysis by dimension provides useful information, it does not consider the association between attribute shortfalls. Hence, multidimensional poverty measures are performed using a union, an intermediate, or an intersection approach in the identification of poor individuals. Tables 1.3, 1.4 and 1.5 display multidimensional poverty measures based on the methodology suggested by Alkire and Foster (see Sect. 2.2).

For any value of c, individuals are deemed poor when they face at least c deprivations. Hence, for $c = 1$, which corresponds to the union approach, poverty incidences (column 2 in Table 1.3) are very high. They vary from 0.58 % (Tunisia 2001) to 0.91 % (Morocco 1992). In contrast, poverty is lower when we consider the individuals that are deprived in all three dimensions ($c = 3$). The headcount ratio varies from 0.0069 % (Lebanon 2004) to 0.33 % (Morocco 1992). We can observe that all poverty measures decrease over time for each country, apart from Turkey where multidimensional poverty increases between 1998 and 2003 whatever the value of c.

More interesting are the adjusted multidimensional FGT measures that take into account the average share of deprivations among poor individuals. In particular, the decrease in the adjusted headcount ratio (columns corresponding to $\alpha = 0$) over time for each country except for Turkey is due to the simultaneous decrease in the headcount ratio and the average number of deprivations.

Decomposition by area of residence provides interesting insights. As is evident from Tables 1.4 and 1.5, multidimensional poverty is higher in rural areas than in urban areas and decreases over time in both areas except for Turkey between 1998 and 2003. In all countries the rural population appears to be poorer than the urban population. However, whatever the identification approach adopted, the difference in poverty between urban and rural areas differs across countries, hence Tunisia in 2001, Morocco, and to a lesser extent Egypt record the highest differences between rural and urban multidimensional poverty measures.

The differences are lowest in Turkey. Comparisons over time indicate that the urban–rural disparity decreases significantly in Tunisia and Egypt. However, the results for Turkey and Morocco are more ambiguous and are dependent on the identification of poor individuals. They diverge slightly for those individuals who are deprived in more than one or two dimensions, but decrease overall when we consider the most deprived. These changes are consistent with those evidenced when considering the changes in poverty measures by location over time.

These changes are significantly larger for rural areas in Tunisia and in Egypt for all values of c. On the contrary the changes in poverty for Morocco have only been higher in rural areas than in urban areas for the "intersection poor". Finally, between 1998 and 2003 Turkey showed an increase in urban poverty whereas rural poverty decreased slightly. These results can be linked to the ones obtained from the one-at-a-time analysis, as the corresponding unidimensional poverty measures

Table 1.3 Multidimensional FGT indices: Alkire and Foster (2011)

Years	c = 1				c = 2				c = 3			
	H	$\alpha = 0$	$\alpha = 1$	$\alpha = 2$	H	$\alpha = 0$	$\alpha = 1$	$\alpha = 2$	H	$\alpha = 0$	$\alpha = 1$	$\alpha = 2$
Egypt												
1995	0.82	0.53	0.29	0.25	0.57	0.45	0.25	0.21	0.21	0.21	0.11	0.092
	(0.0032)	(0.0028)	(0.0019)	(0.0019)	(0.0041)	(0.0034)	(0.0021)	(0.002)	(0.0033)	(0.0033)	(0.0019)	(0.0017)
2000	0.78	0.49	0.27	0.23	0.51	0.4	0.22	0.19	0.17	0.17	0.091	0.078
	(0.0033)	(0.0027)	(0.0019)	(0.0019)	(0.004)	(0.0032)	(0.0021)	(0.002)	(0.003)	(0.003)	(0.0018)	(0.0016)
2005	0.72	0.39	0.21	0.17	0.38	0.28	0.15	0.12	0.084	0.084	0.043	0.035
	(0.0033)	(0.0023)	(0.0015)	(0.0014)	(0.0005)	(0.0027)	(0.0016)	(0.0014)	(0.002)	(0.002)	(0.0011)	(0.00004)
Turkey												
1993	0.88	0.43	0.19	0.14	0.35	0.26	0.11	0.081	0.067	0.067	0.03	0.024
	(0.004)	(0.0002)	(0.002)	(0.0021)	(0.0059)	(0.0044)	(0.0023)	(0.002)	(0.0031)	(0.0031)	(0.0015)	(0.0013)
1998	0.78	0.38	0.15	0.099	0.31	0.22	0.09	0.06	0.048	0.048	0.02	0.015
	(0.0045)	(0.0020)	(0.0016)	(0.0015)	(0.005)	(0.0037)	(0.0017)	(0.0014)	(0.0023)	(0.0023)	(0.001)	(0.00084)
2003	0.8	0.39	0.17	0.11	0.31	0.23	0.1	0.073	0.05	0.05	0.024	0.018
	(0.0045)	(0.003)	(0.0018)	(0.0018)	(0.0052)	(0.0038)	(0.002)	(0.0017)	(0.0024)	(0.0024)	(0.0013)	(0.001)
Morocco												
1992	0.91	0.65	0.4	0.36	0.72	0.59	0.37	0.33	0.33	0.33	0.2	0.17
	(0.0020)	(0.0002)	(0.0027)	(0.0027)	(0.0047)	(0.0041)	(0.003)	(0.0020)	(0.0049)	(0.0040)	(0.0031)	(0.0028)
2003	0.81	0.48	0.33	0.3	0.5	0.37	0.26	0.24	0.12	0.12	0.071	0.062
	(0.0031)	(0.0024)	(0.002)	(0.002)	(0.0009)	(0.000)	(0.0023)	(0.0022)	(0.0025)	(0.0025)	(0.0016)	(0.0015)
Tunisia												
2001	0.58	0.28	0.086	0.046	0.22	0.16	0.052	0.03	0.054	0.054	0.018	0.011
	(0.006)	(0.0036)	(0.0015)	(0.0012)	(0.005)	(0.0039)	(0,0015)	(0.0011)	(0.0028)	(0.0028)	(0.0011)	(0.00076)
2006	0.56	0.25	0.073	0.037	0.16	0.12	0.037	0.021	0.032	0.032	0.011	0.0068
	(0.0051)	(0.0027)	(0.0011)	(0.00085)	(0.0008)	(0.0028)	(0.0011)	(0.00077)	(0.0018)	(0.0018)	(0.00071)	(0.00051)
Lebanon												
2004	0.6	0.27	0.08	0.042	0.21	0.14	0.045	0.026	0.0069	0.0069	0.0023	0.0016
	(0.004)	(0.0021)	(0.00093)	(0.00077)	(0.0003)	(0.0023)	(0.00089)	(0.00069)	(0.00068)	(0.00068)	(0.00025)	(0.00021)

Note: standard errors in parentheses

Table 1.4 Multidimensional FGT indices: Alkire and Foster (2011) in urban areas

Years	c = 1				c = 2				c = 3			
	H	$\alpha = 0$	$\alpha = 1$	$\alpha = 2$	H	$\alpha = 0$	$\alpha = 1$	$\alpha = 2$	H	$\alpha = 0$	$\alpha = 1$	$\alpha = 2$
Egypt												
1995	0.68	0.38	0.18	0.14	0.37	0.28	0.13	0.1	0.11	0.11	0.051	0.041
	(0.0059)	(0.0042)	(0.0025)	(0.0023)	(0.0061)	(0.0048)	(0.0026)	(0.0023.)	(0.0039)	(0.0039)	(0.002)	(0.0017)
2000	0.65	0.36	0.19	0.15	0.35	0.26	0.14	0.12	0.079	0.079	0.044	0.037
	(0.0056)	(0.0038)	(0.0026)	(0.0026)	(0.0056)	(0.0043)	(0.0027)	(0.0026)	(0.0032)	(0.0032)	(0.0019)	(0.0018)
2005	0.59	0.29	0.14	0.11	0.25	0.18	0.093	0.075	0.037	0.037	0.019	0.016
	(0.0055)	(0.0032)	(0.002)	(0.0019)	(0.0049)	(0.0035)	(0.002)	(0.0017)	(0.0021)	(0.0021)	(0.0012)	(0.001)
Turkey												
1993	0.82	0.37	0.16	0.11	0.26	0.19	0.078	0.055	0.033	0.033	0.014	0.011
	(0.0050)	(0.0038)	(0.0023)	(0.0023)	(0.0069)	(0.0049)	(0.0024)	(0.002)	(0.0028)	(0.0028)	(0.0013)	(0.0011)
1998	0.72	0.33	0.13	0.086	0.25	0.18	0.071	0.048	0.027	0.027	0.011	0.0082
	(0.006)	(0.0035)	(0.019)	(0.0018)	(0.0058)	(0.0041)	(0.0011)	(0.0015)	(0.0022)	(0.0022)	(0.00095)	0.00077)
2003	0.75	0.35	0.14	0.094	0.25	0.18	0.078	0.055	0.037	0.037	0.018	0.014
	(0.0056)	(0.0034)	(0.002)	(0.0019)	(0.0056)	(0.0041)	(0.0021)	(0.0017)	(0.0025)	(0.0025)	(0.0013)	(0.001)
Morocco												
1992	0.83	0.51	0.25	0.2	0.51	0.4	0.19	0.15	0.19	0.19	0.089	0.069
	(0.0056)	(0.0049)	(0.0031)	(0.003)	(0.0075)	(0.0062)	(0.0034)	(0.0031)	(0.0059)	(0.0050)	(0.003)	(0.0025)
2003	0.66	0.32	0.19	0.17	0.25	0.18	0.1	0.09	0.058	0.058	0.032	0.027
	(0.0051)	(0.0031)	(0.0022)	(0.(022)	(0.0047)	(0.0036)	(0.0023)	(0.0021)	(0.0025)	(0.0025)	(0.0015)	(0.0013)
Tunisia												
2001	0.48	0.21	0.059	0.029	0.12	0.085	0.025	0.014	0.017	0.017	0.0054	0.0033
	(0.0074)	(0.0036)	(0.0014)	(0.0011)	(0.0048)	(0.0035)	(0.0012)	(0.00088)	(0.0019)	(0.0019)	(0.0007)	(0.0005)
2006	0.47	0.19	0.05	0.022	0.085	0.06	0.017	0.0084	0.011	0.011	0.0032	0.0018
	(0.0064)	(0.0029)	(0.0011)	(0.00075)	(0.0036)	(0.0026)	(0.00086)	(0.00057)	(0.0014)	(0.0014)	(0.00044)	(0.0003)

Note: standard errors in parentheses

Table 1.5 Multidimensional FGT indices: Alkire and Foster (2011) in rural areas

Years	c = 1				c = 2				c = 3			
	H	$\alpha = 0$	$\alpha = 1$	$\alpha = 2$	H	$\alpha = 0$	$\alpha = 1$	$\alpha = 2$	H	$\alpha = 0$	$\alpha = 1$	$\alpha = 2$
Egypt												
1995	0.93	0.64	0.37	0.32	0.72	0.57	0.33	0.29	0.28	0.28	0.15	0.13
	(0.0028)	(0.0052)	(0.0025)	(0.0025)	(0.0049)	(0.0042)	(0.0029)	(0.0027)	(0.0049)	(0.0049)	(0.0029)	(0.0025)
2000	0.89	0.6	0.33	0.29	0.66	0.52	0.29	0.25	0.24	0.24	0.13	0.11
	(0.0035)	(0.0054)	(0.0026)	(0.0026)	(0.0052)	(0.0044)	(0.0029)	(0.0028)	(0.0047)	(0.0047)	(0.0028)	(0.0025)
2005	0.81	0.47	0.26	0.22	0.47	0.35	0.19	0.16	0.12	0.12	0.06	0.049
	(0.0038)	(0.000)	(0.002)	(0.0019)	(0.0048)	(0.0037)	(0.0022)	(0.002)	(0.0031)	(0.000.1)	(0.0017)	(0.0014)
Turkey												
1993	0.98	0.54	0.25	0.19	0.5	0.38	0.17	0.13	0.12	0.12	0.058	0.046
	(0.003)	(0.005)	(0.0036)	(0.0038)	(0.01)	(0.008)	(0.0043)	(0.0039)	(0.0068)	(0.0068)	(0.0034)	(0.0029)
1998	0.91	0.48	0.19	0.12	0.43	0.32	0.13	0.085	0.091	0.091	0.038	0.028
	(0.0053)	(0.0048)	(0.0029)	(0.0029)	(0.0093)	(0.007)	(0.0034)	(0.0029)	(0.0054)	(0.0054)	(0.0024)	(0.002)
2003	0.94	0.51	0.23	0.16	0.49	0.36	0.17	0.12	0.085	0.085	0.041	0.032
	(0.005)	(0.0053)	(0.0039)	(0.0039)	(0.011)	(0.0082)	(0.0046)	(0.004)	(0.0061)	(0.0061)	(0.0032)	(0,0027)
Morocco												
1992	1	0.8	0.56	0.51	0.93	0.77	0.54	0.5	0.46	0.46	0.3	0.26
	(0.00086)	(0.0051)	(0.0029)	(0.0031)	(0.0037)	(0.0009)	(0.0034)	(0.0034)	(0.0073)	(0.0073)	(0.005)	(0.0046)
2003	0.98	0.65	0.48	0.46	0.79	0.59	0.43	0.41	0.18	0.18	0.11	0.1
	(0.0016)	(0.0026)	(0.0024)	(0.0026)	(0.0047)	(0.0038)	(0.0031)	(0.0031)	(0.0044)	(0.0044)	(0.0029)	(0.0026)
Tunisia												
2001	0.79	0.45	0.15	0.083	0.43	0.33	0.11	0.067	0.14	0.14	0.048	0.029
	(0.0089)	(0.007)	(0.0035)	(0.0029)	(0.0011)	(0.0087)	(0.0037)	(0.0029)	(0.0075)	(0.0075)	(0.003)	(0.0022)
2006	0.72	0.36	0.12	0.063	0.3	0.22	0.073	0.043	0.07	0.07	0.025	0.016
	(0.0077)	(0.0051)	(0.024)	(0.0019)	(0.0079)	(0.006)	(0.0025)	(0.0018)	(0.0044)	(0.0044)	(0.0018)	(0.0016)

Note: standard errors in parentheses

Table 1.6 Multidimensional dominance tests: whole population

Comparison	Dominance order (α_{ed}, α_{hous}, α_{dur})							
	(1,1,1)	(2,1,1)	(1,2,1)	(1,1,2)	(2,2,1)	(2,1,2)	(1,2,2)	(2,2,2)
Intersection view								
Egypt								
2005–1995	∅ (1)	∅ (1)	∅ (1)	≻* (0)	≻ (0.247)			
2000–1995	∅ (1)	∅ (1)	∅ (1)	∅ (1)	∅ (1)	∅ (1)	∅ (1)	∅ (1)
2005–2000	≻*							
Turkey								
1998–1993	∅ (1)	∅ (1)	∅ (1)	≻ (0.098)	∅ (1)	≻* (0.043)	≻* (0.02)	
2003–1993	∅ (1)	∅ (1)	∅ (1)	∅ (1)	∅ (1)	∅ (1)	∅ (1)	∅ (1)
2003–1998	∅ (1)	∅ (0.995)	∅ (1)	∅ (0.923)	∅ (0.995)	∅ (0.923)	∅ (0.926)	∅ (0.925)
Morocco								
2003–1992	≻* (0)							
Tunisia								
2006–2001	∅ (1)	∅ (1)	∅ (1)	∅ (1)	∅ (1)	∅ (1)	∅ (1)	∅ (1)
Intermediate view								
Egypt								
2005–1995	∅ (1)	∅ (1)	∅ (1)	∅ (0.047)	∅ (1)			
2005–2000	∅ (1)	≻* (0)	≻* (0.019)	≻* (0)				
Turkey								
1998–1993	∅ (1)	∅ (1)	∅ (1)	∅ (1)	∅ (1)	∅ (1)	∅ (1)	≻ (0.42)
Morocco								
2003–1992	≻* (0)							
Union view								
Egypt								
2005–1995	∅ (1)	∅ (1)	∅ (1)	∅ (1)	∅ (1)	≻ (0.491)	∅ (1)	≻ (0.058)
2005–2000	∅ (1)	≻* (0)	∅ (1)	≻* (0)				

(continued)

Table 1.6 (continued)

Comparison	Dominance order (α_{ed}, α_{hous}, α_{dur})							
	(1,1,1)	(2,1,1)	(1,2,1)	(1,1,2)	(2,2,1)	(2,1,2)	(1,2,2)	(2,2,2)
Turkey								
1998–1993	∅	∅	∅	∅	∅	∅	∅	∅
	(1)	(1)	(1)	(1)	(1)	(1)	(1)	(1)
Morocco								
2003–1992	$\succ$*							
	(0)							

Note: $\succ$($\prec$) indicates that the first distribution dominates (is dominated by) the second distribution at the corresponding order of dominance. ∅ denotes a non-conclusive test. P values are reported in parentheses and refer to the hypothesis of non-dominance of the second distribution with respect to the first one when $\prec$ is reported and the hypothesis of non-dominance of the first distribution with respect to the second one in the other cases. Values less than 10^{-3} or greater than $1–10^{-3}$ are respectively replaced by 0 and 1 for the sake of simplicity. The * indicates that the result is significant at the 5 % level. The results of higher order dominance checks are not reported when a significant dominance relationship at a given order is observed

assume that in terms of well-being, each dimension operates independently from one another. Taking into account the depth of deprivations ($\alpha = 1$) and inequality ($\alpha = 2$) between the poor do not change the conclusions.

The next step is to check the sensitivity of these results to different choices of poverty lines and aggregation procedures. Here we consider the set of poverty indices γ_1^m, γ_2^m and the intermediate set for which sensitivity to extreme deprivations is imposed for only one or two dimensions (see Sect. 2.3). Therefore, the parameter α_k to be used for the function Ξ_α^m can then be equal to 1 or 2 for each dimension. Letting α_{ed}, α_{hous}, $\alpha_{dur,}$ respectively be the coefficient referring to education, housing conditions, and durable goods, we consider eight possible combinations, namely (1, 1, 1), (2, 1, 1), (1, 2, 1), (1, 1, 2), (2, 2, 1), (2, 1, 2), (1, 2, 2), (2, 2, 2).

The tri-dimensional dominance stochastic tests have been carried out on poverty domains that have been defined according to the union, the intermediate, and the intersection approaches for the whole sample. Unsurprisingly, we can observe that it is not easy to draw conclusive results about comparisons of poverty over time and across countries using an extensive (union) poverty definition (Tables 1.6, 1.7 and 1.8). In contrast, a higher number of robust orderings is provided when moving to a more restrictive view of poverty. At this stage we are reminded that if an individual is poor in all three dimensions (intersection view), he or she will also be seen as poor according to the intermediate and union views. Hence, the existence of a dominance relationship using the intersection view is a necessary (but not sufficient) condition for a dominance relationship to hold for broader definitions of the set of poor individuals. Consequently an inconclusive result using an intensive view of poverty can be interpreted directly as an inconclusive result for the other definitions based on the same poverty lines.

Table 1.7 Multidimensional dominance tests: urban areas

Comparison	Dominance order (α_{ed}, α_{hous}, α_{dur})							
	(1,1,1)	(2,1,1)	(1,2,1)	(1,1,2)	(2,2,1)	(2,1,2)	(1,2,2)	(2,2,2)
Intersection view								
Egypt								
2005–1995	$\varnothing$	$\succ$	$\varnothing$	$\varnothing$	$\succ$	$\succ$	$\varnothing$	$\succ$
	(1)	(0.375)	(1)	(1)	(0.225)	(0.113)	(1)	(0.113)
2000–1995	$\varnothing$	$\varnothing$	$\varnothing$	$\varnothing$	$\varnothing$	$\varnothing$	$\varnothing$	$\varnothing$
	(1)	(1)	(1)	(1)	(1)	(1)	(1)	(1)
2005–2000	$\succ^*$							
	(0.001)							
Turkey								
2003–1993	$\varnothing$	$\varnothing$	$\varnothing$	$\varnothing$	$\varnothing$	$\varnothing$	$\varnothing$	$\varnothing$
	(1)	(1)	(1)	(1)	(1)	(1)	(0.998)	(0.998)
1998–1993	$\varnothing$	$\varnothing$	$\varnothing$	$\varnothing$	$\varnothing$	$\varnothing$	$\varnothing$	$\varnothing$
	(1)	(1)	(1)	(1)	(1)	(1)	(1)	(1)
2003–1998	$\varnothing$	$\varnothing$	$\varnothing$	$\varnothing$	$\varnothing$	$\varnothing$	$\varnothing$	$\varnothing$
	(0.996)	(0.967)	(0.996)	(0.767)	(0.967)	(0.767)	(0.751)	(0.751)
Morocco								
2003–1992	$\varnothing$	$\succ$	$\succ^*$	$\varnothing$		$\succ$		
	(1)	(0.332)	(0.016)	(1)		(0.287)		
Tunisia								
2006–2001	$\varnothing$	$\varnothing$	$\succ$	$\succ$	$\succ$	$\succ$	$\succ$	$\succ$
	(1)	(1)	(0.425)	(0.425)	(0.373)	(0.425)	(0.425)	(0.373)
Intermediate view								
Egypt								
2005–1995	$\varnothing$	$\varnothing$	$\varnothing$	$\varnothing$	$\varnothing$	$\varnothing$	$\varnothing$	$\varnothing$
	(1)	(1)	(1)	(1)	(1)	(1)	(1)	(1)
2005–2000	$\varnothing$	$\succ^*$	$\succ$	$\varnothing$				
	(1)	(0.043)	(0.284)	(1)				
Morocco								
2003–1992	$\varnothing$	$\varnothing$	$\varnothing$	$\varnothing$	$\succ$	$\varnothing$	$\varnothing$	$\succ$
	(1)	(1)	(1)	(1)	(0.407)	(1)	(1)	(0.476)
Tunisia								
2006–2001	$\varnothing$	$\varnothing$	$\succ$	$\succ$	$\succ$	$\succ$	$\succ$	$\succ$
	(1)	(1)	(0.425)	(0.439)	(0.373)	(0.425)	(0.425)	(0.373)
Union view								
Egypt								
2005–2000	$\varnothing$	$\succ^*$	$\varnothing$	$\varnothing$			$\varnothing$	
	(1)	(0.043)	(1)	(1)			(1)	

(continued)

Table 1.7 (continued)

Comparison	Dominance order (α_{ed}, α_{hous}, α_{dur})							
	(1,1,1)	(2,1,1)	(1,2,1)	(1,1,2)	(2,2,1)	(2,1,2)	(1,2,2)	(2,2,2)
Morocco								
2003–1992	∅	∅	∅	∅	≻	∅	∅	≻
	(1)	(1)	(1)	(1)	(0.475)	(1)	(1)	(0.491)
Tunisia								
2006–2001	∅	∅	∅	∅	≻	≻	≻	≻
	(1)	(1)	(1)	(1)	(0.373)	(0.425)	(0.425)	(0.373)

Note: ≻(≺) indicates that the first distribution dominates (is dominated by) the second distribution at the corresponding order of dominance. ∅ denotes a non-conclusive test. P values are reported in parentheses and refer to the hypothesis of non-dominance of the second distribution with respect to the first one when ≺ is reported and the hypothesis of non-dominance of the first distribution with respect to the second one in the other cases. Values less than 10^{-3} or greater than $1–10^{-3}$ are respectively replaced by 0 and 1 for the sake of simplicity. The * indicates that the result is significant at the 5 % level. The results of higher order dominance checks are not reported when a significant dominance relationship at a given order is observed

In order to reduce and simplify the presentation of the results we did not report dominance tests for the broader definitions of poverty when comparisons did not yield conclusive results using the intersection view. Moreover, since individuals that cumulate deprivations in all three dimensions are only a subset of those people who are considered deprived according to the intermediate and the union definitions of poverty, we expect that dominance results may considerably differ from those observed in the univariate analyses undertaken for each dimension.

Taking the union view of poverty into consideration, the sole countries for which robust comparisons regarding the changes in poverty over time can be assessed are Morocco, and to a lesser extent Egypt (Tables 1.6, 1.7 and 1.8). Despite the fact that multidimensional poverty decreases between 1992 and 2003 in Morocco for the whole population and in rural areas, it is rather difficult to draw conclusive results concerning the change in urban poverty according to the union view (Tables 1.7 and 1.8). Urban poverty only decreases when emphasis is put on severe deprivations in education and housing conditions, although these are not statistically significant. One explanation seems to be the absence of dominance for the dimension durable goods in urban areas, as highlighted by the uni-dimensional analysis. The intersection view confirms these findings, since the decrease in poverty for extremely poor individuals is statistically significant for the whole sample and particularly in rural areas for a large set of poverty measures. On the contrary, in urban areas this conclusion is only valid for poverty measures that exhibit a strong aversion to deprivation in at least two dimensions or in one dimension (excluding the possession of durable goods).

Turning now to the case of Egypt we can observe a general improvement over the whole period 1995–2005 for poverty measures that take into account aversion to extreme deprivations. This conclusion holds when an emphasis is put on

Table 1.8 Multidimensional dominance tests: rural areas

Comparison	Dominance order (α_{ed}, α_{hous}, α_{dur})							
	(1,1,1)	(2,1,1)	(1,2,1)	(1,1,2)	(2,2,1)	(2,1,2)	(1,2,2)	(2,2,2)
Intersection view								
Egypt								
2005–1995	$\varnothing$	$\varnothing$	$\varnothing$	$\succ$*	$\succ$			
	(1)	(1)	(1)	(0)	(0.428)			
2000–1995	$\varnothing$	$\varnothing$	$\varnothing$	$\varnothing$	$\varnothing$	$\varnothing$	$\varnothing$	$\varnothing$
	(1)	(1)	(1)	(1)	(1)	(1)	(1)	(1)
2005–2000	$\succ$*							
Turkey								
2003–1993	$\varnothing$	$\varnothing$	$\varnothing$	$\varnothing$	$\varnothing$	$\varnothing$	$\varnothing$	$\succ$
	(1)	(1)	(1)	(1)	(1)	(1)	(1)	(0.421)
1998–1993	$\varnothing$	$\varnothing$	$\varnothing$	$\succ$	$\varnothing$	$\succ$*	$\succ$	
	(1)	(1)	(1)	(0.097)	(1)	(0.039)	(0.051)	
2003–1998	$\varnothing$	$\varnothing$	$\varnothing$	$\varnothing$	$\varnothing$	$\varnothing$	$\varnothing$	$\varnothing$
	(0.997)	(0.992)	(0.997)	(0.987)	(0.99)	(0.967)	(0.987)	(0.967)
Morocco								
2003–1992	$\succ$*							
	(0)							
Tunisia								
2006–2001	$\varnothing$	$\varnothing$	$\varnothing$	$\varnothing$	$\varnothing$	$\varnothing$	$\varnothing$	$\varnothing$
	(1)	(1)	(1)	(1)	(1)	(1)	(1)	(1)
Intermediate view								
Egypt								
2005–1995	$\varnothing$	$\varnothing$	$\varnothing$	$\succ$	$\varnothing$	$\succ$*	$\succ$*	
	(1)	(1)	(1)	(0.073)	(1)	(0.018)	(0.023)	
2005–2000	$\varnothing$	$\succ$*	$\succ$*	$\succ$*				
	(1)	(0)	(0)	(0)				
Turkey								
1998–1993	$\varnothing$	$\varnothing$	$\varnothing$	$\varnothing$	$\varnothing$	$\varnothing$	$\varnothing$	$\succ$*
	(1)	(1)	(1)	(1)	(1)	(1)	(1)	(0.475)
2003–1993	$\varnothing$	$\varnothing$	$\varnothing$	$\varnothing$	$\varnothing$	$\varnothing$	$\varnothing$	$\varnothing$
	(1)	(1)	(1)	(1)	(1)	(1)	(1)	(1)
Morocco								
2003–1992	$\succ$*							
	(0)							
Union view								
Egypt								
2005–1995	$\varnothing$	$\varnothing$	$\varnothing$	$\varnothing$	$\varnothing$	$\succ$	$\varnothing$	$\succ$
	(1)	(1)	(1)	(1)	(1)	(0.299)	(1)	(0.043)
2005–2000	$\varnothing$	$\succ$*	$\varnothing$	$\succ$*				
	(1)	(0)	(1)	(0)				

(continued)

Table 1.8 (continued)

Comparison	Dominance order (α_{ed}, α_{hous}, α_{dur})							
	(1,1,1)	(2,1,1)	(1,2,1)	(1,1,2)	(2,2,1)	(2,1,2)	(1,2,2)	(2,2,2)
Turkey								
1998–1993	∅	∅	∅	∅	∅	∅	∅	∅
	(1)	(1)	(1)	(1)	(1)	(1)	(1)	(1)
Morocco								
2003–1992	$\succ$*							
	(0)							

Note: $\succ$ ($\prec$) indicates that the first distribution dominates (is dominated by) the second distribution at the corresponding order of dominance. ∅ denotes a non-conclusive test. P values are reported in parentheses and refer to the hypothesis of non-dominance of the second distribution with respect to the first one when $\prec$ is reported and the hypothesis of non-dominance of the first distribution with respect to the second one in the other cases. Values less than 10^{-3} or greater than $1-10^{-3}$ are respectively replaced by 0 and 1 for the sake of simplicity. The * indicates that the result is significant at the 5 % level. The results of higher order dominance checks are not reported when a significant dominance relationship at a given order is observed

extreme deprivations with regard to (1) education for poor individuals from urban areas, though not significant, and (2) the possession of durable goods or severe deprivations in at least two dimensions in the case of the rural population. However, only mixed conclusions can be drawn for the two sub-periods. Indeed, it is not possible to assess whether or not poverty has increased or decreased between 1995 and 2000 whereas the alleviation of poverty is statistically significant during the period 2000–2005 for the whole population and in both rural and urban areas for a large set of poverty measures.

When examining the case of Turkey, we can note that even with the intersection view it is difficult to obtain robust orderings on the changes in poverty for the whole population over the period 1993–2003 and the sub-period 1998–2003 and for those living in urban areas whatever the period under consideration. Here, it should be noted that the deterioration in women's education levels/the possession of durable goods in rural areas between 1998 and 2003 that has been emphasized in the one-at-a-time analysis does not give a clear picture as to the development of multidimensional poverty. In fact, the sole comparisons concern the period 1993–1998 where the results indicate a statistically significant alleviation of multidimensional poverty primarily in rural areas once strong deprivations in at least the access to durable goods are considered. Moreover, the worsening of the standard of living, as emphasized using the Alkire and Foster's poverty measures, is not supported for all the poverty lines inside the poverty domain, but solely for a given vector of poverty lines.

Finally, very few conclusive results have been obtained for Tunisia where we can observe a decrease in the depth of multidimensional poverty only for the poorest section of the urban population, although this decrease is not statistically significant. Here again, the improvement in situation for people living in rural

Table 1.9 Multidimensional dominance tests: cross country comparisons for the whole sample

Years	Dominance order (α_{ed}, α_{hous}, α_{dur})							
	(1,1,1)	(2,1,1)	(1,2,1)	(1,1,2)	(2,2,1)	(2,1,2)	(1,2,2)	(2,2,2)
Intersection view								
Turkey vs Egypt	$\varnothing$	$\varnothing$	$\varnothing$	$\varnothing$	$\succ$	$\varnothing$	$\varnothing$	$\succ$
	(1)	(1)	(1)	(1)	(0.376)	(1)	(1)	(0.376)
Morocco vs Egypt	$\varnothing$	$\varnothing$	$\varnothing$	$\varnothing$	$\varnothing$	$\varnothing$	$\varnothing$	$\varnothing$
	(1)	(1)	(1)	(0.964)	. (1)	(0.885)	(0.964)	(0.885)
Tunisia vs Egypt	$\succ$*							
	(0.047)							
Lebanon vs Egypt	$\varnothing$	$\succ$*	$\varnothing$	$\succ$*				
	(1)	(0)	(1)	(0)				
Morocco vs Turkey	$\varnothing$	$\varnothing$	$\varnothing$	$\prec$	$\varnothing$	$\prec$*	$\prec$	
	(1)	(1)	(1)	(0.183)	(1)	(0.016)	(0.183)	
Tunisia vs Turkey	$\varnothing$	$\succ$	$\varnothing$	$\varnothing$	$\succ$	$\succ$	$\varnothing$	$\succ$*
	(1)	(0.183)	(1)	(1)	(0.176)	(0.057)	(1)	(0.036)
Lebanon vs Turkey	$\varnothing$	$\succ$	$\varnothing$	$\succ$*	$\succ$			
	(1)	(0.199)	(1)	(0)	(0.199)			
Tunisia vs Morocco	$\varnothing$	$\succ$*	$\varnothing$	$\succ$*				
	(1)	(0)	(1)	(0.033)				
Lebanon vs Morocco	$\varnothing$	$\varnothing$	$\varnothing$	$\succ$*	$\varnothing$			
	(1)	(1)	(1)	(1)	(1)			
Lebanon vs Tunisia	$\varnothing$	$\varnothing$	$\varnothing$	$\succ$	$\varnothing$	$\succ$	$\succ$	$\succ$
	(1)	(1)	(1)	(0.037)	(1)	(0.037)	(0.0323)	(0.0323)
Intermediate view								
Turkey vs Egypt	$\varnothing$	$\varnothing$	$\varnothing$	$\varnothing$	$\varnothing$	$\varnothing$	$\varnothing$	$\varnothing$
	(1)	(1)	(1)	(1)	(1)	(1)	(1)	(1)
Tunisia vs Egypt	$\succ$*							
	(0.047)							
Lebanon vs Egypt	$\varnothing$	$\varnothing$	$\varnothing$	$\varnothing$	$\varnothing$	$\succ$*	$\varnothing$	
	(1)	(1)	(1)	(1)	(1)	(0)	(1)	
Morocco vs Turkey	$\varnothing$	$\varnothing$	$\varnothing$	$\varnothing$	$\varnothing$	$\varnothing$	$\varnothing$	$\varnothing$
	(1)	(1)	(1)	(1)	(1)	(1)	(1)	(1)
Tunisia vs Turkey	$\varnothing$	$\varnothing$	$\varnothing$	$\varnothing$	$\varnothing$	$\varnothing$	$\varnothing$	$\succ$
	(1)	(1)	(1)	(1)	(1)	(1)	(1)	(0.386)
Lebanon vs Turkey	$\varnothing$	$\varnothing$	$\varnothing$	$\varnothing$	$\varnothing$	$\succ$*	$\varnothing$	
	(1)	(1)	(1)	(1)	(1)	(0.042)	(1)	
Tunisia vs Morocco	$\varnothing$	$\succ$	$\varnothing$	$\varnothing$	$\succ$	$\succ$*	$\varnothing$	
	(1)	(0.467)	(1)	(1)	(0.347)	(0.001)	(1)	
Lebanon vs Morocco	$\varnothing$	$\varnothing$	$\varnothing$	$\varnothing$	$\varnothing$	$\varnothing$	$\varnothing$	$\varnothing$
	(1)	(1)	(1)	(1)	(1)	(1)	(1)	(1)
Lebanon vs Tunisia	$\varnothing$	$\varnothing$	$\varnothing$	$\varnothing$	$\varnothing$	$\varnothing$	$\varnothing$	$\varnothing$
	(1)	(1)	(1)	(1)	(1)	(1)	(1)	(1)
Union view								
Turkey vs Egypt	$\varnothing$	$\varnothing$	$\varnothing$	$\varnothing$	$\varnothing$	$\varnothing$	$\varnothing$	$\succ$
	(1)	(1)	(1)	(1)	(1)	(1)	(1)	(0.395)

(continued)

Table 1.9 (continued)

Years	Dominance order (α_{ed}, α_{hous}, α_{dur})							
	(1,1,1)	(2,1,1)	(1,2,1)	(1,1,2)	(2,2,1)	(2,1,2)	(1,2,2)	(2,2,2)
Tunisia vs Egypt	$\succ$*							
	(0.047)							
Lebanon vs Egypt	∅	∅	∅	∅	∅	$\succ$	∅	$\succ$
	(1)	(1)	(1)	(1)	(1)	(0.08)	(1)	(0.119)
Tunisia vs Turkey	∅	∅	∅	∅	∅	∅	∅	$\succ$
	(1)	(1)	(1)	(1)	(1)	(1)	(1)	(0.386)
Lebanon vs Turkey	∅	∅	∅	∅	∅	∅	∅	∅
	(1)	(1)	(1)	(1)	(1)	(1)	(1)	(1)
Tunisia vs Morocco	∅	∅	∅	∅	∅	$\succ$	∅	$\succ$
	(1)	(1)	(1)	(1)	(1)	(0.236)	(1)	(0.314)

Note: $\succ$ ($\prec$) indicates that the rural distribution dominates (is dominated by) the urban distribution at the corresponding order of dominance. ∅ denotes a non-conclusive test. P values are report in parentheses and refer to the hypothesis of non-dominance of the urban distribution with respect to the rural one when $\prec$ is reported and to the hypothesis of non-dominance of the rural distribution with respect to the urban one in the other cases. Values less than 10^{-3} or greater than $1\text{–}10^{-3}$ are respectively replaced by 0 and 1 for the sake of simplicity. The * indicates that the result is significant at the 5 % level. The results of second-order dominance checks are not reported when a significant first-order dominance relationship is observed

areas as recorded using cardinal measures is not supported. At the same time it is important to remember that the univariate analyses by each dimension did not yield conclusive results for education and housing conditions in rural areas.

Turning now to the comparisons between the countries, again very few results are conclusive, especially when we use an extensive definition of poverty (Tables 1.9, 1.10 and 1.11).

Surprisingly, it is not possible to assess robust orderings for multidimensional poverty in rural areas between countries when we adopt a broad definition of poverty. However, if we limit our interest to those individuals who cumulate poverty in all dimensions and to more restrictive measures of poverty, Table 1.11 shows us that the rural population in Tunisia is not as poor as in Morocco or Egypt when an emphasis is put on severe deprivations, particularly with regard to education; likewise rural poverty in Turkey is less than in Morocco when we take into account severe deprivations in the possession of durable goods. However, it is not possible to rank rural poverty between the other remaining pairs of countries. For the whole sample, the striking result is Lebanon's dominance in terms of whole population over the other four countries, although this dominance is not statistically significant and it is restricted to measures that are sensitive to strong deprivations in education or durable goods (Table 1.9). Nevertheless, when we consider broader definitions of poverty, it is no longer possible to obtain an unambiguous ranking between Lebanon and Morocco. It can also be observed that urban populations achieve better living conditions in Tunisia than in Egypt,

Table 1.10 Multidimensional dominance tests: cross country comparisons in urban areas

Years	Dominance order (α_{ed}, α_{hous}, α_{dur})							
	(1,1,1)	(2,1,1)	(1,2,1)	(1,1,2)	(2,2,1)	(2,1,2)	(1,2,2)	(2,2,2)
Intersection view								
Turkey vs Egypt	$\varnothing$	$\varnothing$	$\varnothing$	$\varnothing$	$\varnothing$	$\varnothing$	$\varnothing$	$\varnothing$
	(1)	(1)	(1)	(1)	(1)	(1)	(1)	(1)
Morocco vs Egypt	$\varnothing$	$\varnothing$	$\varnothing$	$\varnothing$	$\varnothing$	$\varnothing$	$\varnothing$	$\varnothing$
	(1)	(1)	(1)	(0.974)	(1)	(0.792)	(0.974)	(0.792)
Tunisia vs Egypt	$\succ$							
	(0.001)							
Morocco vs Turkey	$\varnothing$	$\varnothing$	$\varnothing$	$\varnothing$	$\varnothing$	$\varnothing$	$\varnothing$	$\varnothing$
	(1)	(1)	(1)	(0.9.68)	(1)	(0.804)	(0.968)	(0.804)
Tunisia vs Turkey	$\succ$							
	(0.009)							
Tunisia vs Morocco	$\succ$							
	(0.028)							
Intermediate view								
Tunisia vs Egypt	$\varnothing$	$\succ$*	$\succ$*	$\succ$*				
	(1)	(0)	(0.001)	(0.001)				
Tunisia vs Turkey	$\succ$	$\succ$*	$\succ$	$\succ$*				
	(0.074)	(0.007)	(0.074)	(0.024)				
Tunisia vs Morocco	$\succ$	$\succ$*	$\succ$	$\succ$				
	(0.264)	(0.019)	(0.195)	(0.264)				
Union view								
Tunisia vs Egypt	$\varnothing$	$\succ$*	$\varnothing$	$\succ$			$\succ$*	
	(1)	(0)	(1)	(0.087)			(0.002)	
Tunisia vs Turkey	$\varnothing$	$\succ$*	$\succ$	$\succ$*				
	(1)	(0.007)	(0.074)	(0.024)				
Tunisia vs Morocco	$\succ$	$\succ$	$\succ$	$\succ$			$\succ$	
	(0.349)	(0.019)	(0.195)	(0.264)			(0.264)	

Note: $\succ$ ($\prec$) indicates that the rural distribution dominates (is dominated by) the urban distribution at the corresponding order of dominance. $\varnothing$ denotes a non-conclusive test. P values are reported in parentheses and refer to the hypothesis of non-dominance of the urban distribution with respect to the rural one when $\prec$ is reported and to the hypothesis of non-dominance of the rural distribution with respect to the urban one in the other cases. Values less than 10^{-3} or greater than $1–10^{-3}$ are respectively replaced by 0 and 1 for the sake of simplicity. The·* indicates that the result is significant at the 5 % level. The results of second-order dominance checks are not reported when a significant first-order dominance relationship is observed

Turkey, and Morocco for a large set of poverty measures (Table 1.10). In contrast, no robust comparisons can be made between the other countries. At this stage it is worth noting how fragile the results based on the single poverty measures are as well as the rankings of these countries based on their MPI.[17]

[17] The rankings of countries for which data are available using the MPI is the following: Tunisia, Egypt, Turkey, Morocco (HRD 2010).

Table 1.11 Multidimensional dominance tests: cross country comparisons in rural areas

Years	Dominance order (α_{ed}, α_{hous}, α_{dur})							
	(1,1,1)	(2,1,1)	(1,2,1)	(1,1,2)	(2,2,1)	(2,1,2)	(1,2,2)	(2,2,2)
Intersection view								
Turkey vs Egypt	∅	∅	∅	∅	∅	∅	∅	∅
	(1)	(1)	(1)	(1)	(1)	(1)	(1)	(1)
Morocco vs Egypt	∅	∅	∅	∅	∅	∅	∅	∅
	(1)	(1)	(1)	(0.907)	(1)	(0.838)	(0.907)	(0.838)
Tunisia vs Egypt	∅	$\succ$	∅	∅	$\succ$	$\succ$	∅	$\succ$
	(1)	(0.184)	(1)	(1)	(0.113)	(0.184)	(1)	(0,113)
Morocco vs Turkey	∅	∅	∅	$\prec$	∅	$\prec$	$\prec$	$\prec$
	(1)	(1)	(1)	(0.473)	(1)	(0.195)	(0.168)	(0.068)
Tunisia vs Turkey	∅	∅	∅	∅	∅	∅	∅	∅
	(1)	(1)	(1)	(1)	(1)	(1)	(1)	(1)
Tunisia vs Morocco	∅	∅	∅	∅	∅	$\succ$	∅	$\succ$
	(1)	(1)	(1)	(1)	(1)	(0.064)	(1)	(0.064)
Intermediate view								
Tunisia vs Egypt	∅	∅	∅	∅	∅	∅	∅	∅
	(1)	(1)	(1)	(1)	(1)	(1)	(1)	(1)
Morocco vs Turkey	∅	∅	∅	∅	∅	∅	∅	∅
	(1)	(1)	(1)	(1)	(1)	(1)	(1)	(1)
Tunisia vs Morocco	∅	∅	∅	∅	∅	∅	∅	∅
	(1)	(1)	(1)	(1)	(1)	(1)	(1)	(1)

Note: $\succ$ ($\prec$) indicates that the rural distribution dominates (is dominated by) the urban distribution at the corresponding order of dominance. ∅ denotes a non-conclusive test. P values are report in parentheses and refer to the hypothesis of non-dominance of the urban distribution with respect to the rural one when $\prec$ is reported and to the hypothesis of non dominance of the rural distribution with respect to the urban one in the other cases. Values less than 10^{-3} or greater than $1-10^{-3}$ are respectively replaced by 0 and 1 for the sake of simplicity

1.4 Conclusion

In this paper we have investigated multidimensional poverty in five southern Mediterranean countries. We have not only relied on a broader conception of poverty by considering deprivations in women's education levels, the possession of durable goods and housing conditions, but we have also made use of recent developments regarding multidimensional poverty measurement. Our analysis was based on an axiomatic approach to poverty. We also used stochastic dominance tools to obtain solid results that are valid for a large set of poverty measures, where the measures do not hinge on choices made for the poverty lines, and that are, in the multidimensional case, insensitive to a dimension weighting scheme.

Our findings add to the literature relating to poverty within the southern Mediterranean countries. They provide comparisons over time for each country and between countries. These comparisons cannot be captured when each dimension of poverty is analyzed separately, since the approaches take into account the correlations that may exist between different kinds of deprivation. In addition, in contrast to rankings drawn from cardinal poverty measures (for example, from the MPI), multidimensional tests make it possible to nuance the performances achieved by certain countries. More interesting is to consider to what extent these findings mirror differences in patterns of development policies as well as their evolution in some countries.

First, we should remember that Morocco performs worse than the other countries, but at the same time the situation for its population improved between 1992 and 2003, especially in rural areas. Despite the fact that interest in poverty was considered as residual during the period of SAP, Morocco underwent an integrated programme of human development and political liberalization in the second half of 1990s,[18] accelerated by the adoption of the National Human Development Initiative (NHDI) in 2005. The urban/rural divide evidenced by our results can be explained by the nature of past provision policies that focused on urban areas. However, since the 1990s there have been efforts to remedy the situation through new national programmes.[19] The ambiguous results concerning the changes in poverty disparities between urban and rural areas suggest that development programmes seem to have been aimed at the poorest of the rural poor and less so at the urban poor when an intensive view of poverty identification is considered. At the same time, this result needs to be viewed with caution. Indeed, though stochastic dominance tests cannot be used to see whether rural–urban gaps have narrowed or widened, they do demonstrate a decrease in poverty over time in rural areas but are not statistically significant for urban areas. To a lesser extent, these findings are consistent with some of the trends highlighted in recent studies (Boughzala et al. 2005; Van de Walle 2005; Harrigan et al. 2009). These analyses emphasize the lack of pro-poor targeting in programmes over time, particularly in urban areas, and indicate, among other factors, the political imperative to preserve the benefits for urban elites by spending as little as possible on the poorest population (Van de Walle 2005).

Second, the assessment of trends in multidimensional poverty in Egypt is ambiguous for the whole period 1995–2005. The decrease in poverty is only assessed for the sub-period 2000–2005 and mainly for the rural population. These findings are consistent with the results obtained from the multidimensional measures. Indeed, the highest variation of poverty occurs in rural areas between 2000 and 2005 when we adopt an intersection definition of poverty. This variation is reinforced by a catching up process with urban areas. To a lesser extent these findings support those from the study by Bérenger (2010). The absence of concluding results for the sub-period 1995–2000 can be linked to the very minor changes in poverty measures regardless of the area of residence. Moreover, the

[18] The Economic and Social Plan of 1996–2000 was the first document to declare the fight against poverty as a national priority.

[19] Among them, BAJ1, launched in 1996, focuses on basic health, basic education, and infrastructure investment in the 14 poorest provinces.

unidimensional analysis pinpointed a deterioration in housing conditions for this sub-period that coincides with the implementation of a new rental law in 1996. A brief study of the evolving economic and social context could provide some explanations. Indeed, in order to limit the social costs of the economic reform and structural adjustment (ERSA) programme adopted in 1991, Egypt has kept some forms of social justice through ad hoc interventions. The social protection system is complex; it includes subsidies for multiple private and public goods and other social transfers in cash (Osman et al. 2006) but is considered as highly inefficient. Given the concern for the widening gap between urban and rural areas, the government has established several programmes over the period. The Shorouk programme, implemented in 1994, was expected to address all aspects of development. The Social Development Fund, originating in the early 1990s, has involved the population in small and micro credit, public works projects, and human resource development. In addition to these funds, the Emergency Fund was created in 1998 to meet urgent needs in rural areas. However, despite impressive progress in rural areas, poverty remains a rural phenomenon and includes a regional dimension (World Bank 2006; Bérenger 2010).

Third, our results spotlight the erratic evolution of multidimensional poverty in Turkey. Indeed, while the analyses do not provide non-conclusive results for the whole period 1993–2003, poverty evolved in opposite directions for the two sub-periods, 1993–1998 and 1998–2003. Moreover, the quantification of multidimensional poverty reveals ambiguous trends relative to the urban–rural gaps over time that are widening slightly in most cases. The one-at-a-time analysis shows a worsening situation for the population in terms of deprivation (durable goods and educational achievements by women) between 1998 and 2003, particularly in rural areas, although these trends do not go on to yield robust multidimensional poverty comparisons during this period. The findings are indicative of Turkey's rather volatile economic development process as well as its fragmented approach to combating poverty and social exclusion. Indeed, from the beginning of the 1980s Turkey has adopted a market led growth process of development and has experienced several crises: the foreign exchange crisis of 1994 and the financial crisis of 2001. However, prior to 1980 the fight against poverty was not considered as a priority. Consequently, it was not addressed directly but was supposed to disappear with economic growth. Several programmes were implemented after 1980 which provided nourishment, health care, and shelters for people suffering from poverty (Durusoy et al. 2005). A full programme was only launched in 2001. In addition, several programmes of social assistance have been put into practice as well as an acknowledgment of the problems affecting its education system; the 1997 Education Basic Program extended compulsory education from 5 years to 8 years. However, past spending patterns of public and private expenditure in education reveal that the government has restricted its spending on primary and secondary education while augmenting spending on tertiary education (Duman 2008). Spending levels have become more and more favorable to the wealthier sections of the population. This country also faces wide gaps in income and education between urban and rural areas (World Bank 2005 and 2008). This is an alarming situation, as education seems to hold an important role in understanding discrepancies in development within the country (Duman 2008). As underlined by our estimations of multidimensional poverty, the financial crisis appears to have contributed to the worsening of the situation for individuals living in urban areas, as it has led to an increase in unemployment coupled with a rapid migration from rural areas to urban areas.

The case of Tunisia provides a clear illustration as to how the choice of development strategies has an impact when these strategies are aimed at improving the well-being of the population as well as economic growth. Indeed, it is important to remember that despite multidimensional dominance tests being unable to help in the formation of a robust assessment of patterns in poverty between 2001 and 2006 for the whole population/the population in rural areas, probably due to the very short duration of the period of analysis, our measures make it possible to pinpoint the reduction in rural–urban gaps during the same period (becomes very low in 2006 when we focus on the most deprived). But it is important to note that the population living in urban areas in Tunisia experienced a higher level of well-being in terms of education, possession of durable goods, and housing conditions relative to the other three countries. These findings capture a synthetic picture that could not have been put together from the dominance tests carried out for each dimension separately since only education emerged as the best performer. Tunisia has experienced significant economic reforms, but in keeping with its economic choices it has put social constraint at the center of its priorities in order to counteract the undesired consequences of structural adjustment (Muller 2007). Education has been one of the highest priorities in Tunisia's development experience (Ben Romdhane 2006). The role of women in the development strategy has been prioritized since the 1950s through a reform of family law bolstered by subsequent amendments in 1993 and the foundation of the family planning association in 1968.

Finally, in comparison with the other countries, Lebanon is a particular case because according to multidimensional poverty measures and dominance checks it would appear to be the best performer (alongside Tunisia) and yet at the same time, below standard housing conditions seem to be a matter of great concern.

Acknowledgments This paper has been prepared within the framework of the "PROPOOR-SUDS" project funded by the ANR "French National Agency of Scientific Research". The authors would like to thank the Ministry of Social Affairs of Lebanon for providing the access to the dataset from Lebanon Consumption Household Budget Survey: 2004–2005.

References

Alkire, S., & Foster, J. (2011). Counting and multidimensional poverty measurement. *Journal of Public Economics, 95*(7–8), 476–487.

Alkire, S. & Seth, S. (2008). *Multidimensional poverty and BPL measures in India: A comparison of methods. OPHI Working Paper Series, Oxford poverty and human Development initiative.* Oxford: University of Oxford.

Atkinson, A. (1987). On the measurement of poverty. *Econometrica, 55*(4), 749–764.

Ayadi, M., El Lahga, A. & Chtioui, N. (2006). Analyse de la pauvreté et des inégalités en Tunisie entre 1988 et 2001: une approche non monétaire. PMMA, Network Session Paper

Baharoglu D., Peltier, N. & Buckley, R. (2005). *The macroeconomic and sectoral performance of housing supply in selected MENA countries: A comparative analysis.* Washington: World Bank, Working paper n° 31174.

Batana, Y. M. & Duclos, J.-Y. (2010). *Comparing multidimensional poverty with qualitative indicators of well-being.* Working Paper 10–04, CIRPÉE

Ben Romdhane, M. (2006). Social policy and development in Tunisia since independence: A political perspective. In M. Karshenas & V. M. Moghadam (Eds.), *Social policy in the Middle East, UNRISD.* New York: Palgrave Macmillan.

Bérenger, V. & Berthomieu, C. (2008). Final report of FEMISE project FEM 31-06R: Analyse de l'impact de la croissance sur la pauvreté et identification des stratégies de croissance bénéfique aux pauvres ("pro-poor growth strategies"). Etude des cas pour six pays partenaires méditerranéens: Egypte, Israël, Liban, Maroc, Tunisie et Turquie. Available at www.femise.org/PDF/ci2006/FEM31-06R.pdf.

Bérenger V., Deutsch, J. & Silber, J. (2009). *Multidimensional poverty measurement and the order of acquisition of durable goods and access to services. The Case of Egypt, Morocco and Turkey.* Presented at the MEEA Conference, Nice-Monaco in March 2009.

Bérenger, V. (2010). Multidimensional fuzzy poverty and pro-poor growth measures in non-monetary dimensions in Egypt between 1995 and 2005. *Middle East Development Journal, 2*(1), 15–38.

Bergh, S. (2005). Explaining slow economic growth and poor social development indicators: The case of Morocco. Oxford Economy Council of Good Governance. Economy Analysis n°7, October.

Bibi, S. (2004). *Comparing multidimensional poverty between Egypt and Tunisia.* Cahiers de Recherche du CIRPEE, n° 0416.

Boughzala, M., Bibi S. & Abelkhabek T. (2005). Studies on Morocco and Tunisia. In poverty reduction strategies in North Africa: Country cases for Egypt, Morocco and Tunisia, ERF, prepared for the United Nation's Commission for Africa, June

Booysen, F., van der Berg, S., Burger, R., von Maltitz, M., & du Rand, G. (2008). Using an asset index to assess trends in poverty in seven sub-saharan African countries. *World Development, 36*(6), 1113–1130.

Bourguignon, F., & Chakravarty, S. (2003). The measurement of multidimensional poverty. *Journal of Economic Inequality, 1*(1), 25–49.

Brandolini, A. & D'Alessio, G. (1998). *Measuring well-being in the functioning space, mimeo.* Rome: Banca d'Italia.

Chakravarty, S., Mukherjee, D. & Ranade, R. (1998). On the family of subgroup and factor decomposable measures of multidimensional poverty. In D. Slottje (Ed.), *Research on economic inequality* (Vol. 8, pp. 175–194). London: JAI Press.

Chakravarty, S. and Silber, J. (2008). Measuring multidimensional poverty: The axiomatic approach. In N. Kakwani & J. Silber (Eds.), *Quantitative approaches to multidimensional poverty measurement, Chapter 11* (pp. 192–209). New York: Palgrave Macmillan.

Davidson, R., & Duclos, J.-Y. (2000). Statistical inference for stochastic dominance and for the measurement of poverty and inequality. *Econometrica, 68*(6), 1435–1464.

Deutsch, J. & Silber, J. (2005). Measuring multidimensional poverty: An empirical comparison of various approaches. *Review of Income and Wealth, 51*(1), 145–172.

Duclos, J.-Y., Sahn, D., & Younger, S. (2006). Robust multidimensional poverty comparisons. *The Economic Journal, 116*, 943–968.

Duman, A. (2008). Does schooling matter: Education and income inequality in Turkey? *Financial Theory and Practice, 3*, 369–387.

Durusoy, S., Köse, S. & Karadeniz, O. (2005). Implementing new strategies for combating poverty in Turkey. *South-East Europe Review for Labor and Social Affairs*, (04) on www.ceeol.com, 47–76.

Foster, J. E., & Shorrocks, A. F. (1988). Poverty orderings and welfare dominance. *Social Choice and Welfare, 5*, 179–198.

Gravel, N., & Mukhopadhyay, A. (2010). Is India better off today than 15 years ago? A robust multidimensional answer. *Journal of Economic Inequality, 8*(2), 173–195.

Harrigan, J. R., & El-Said, H. (2009). Economic reform, social welfare, civic society and Islamists in Morocco. In J. R. Harrigan & H. El Said (Eds.), *Economic liberalisation, social capital and Islamic welfare provision*. New York: Palgrave Macmillan.

Hill, A., & King, M. (1995). Women's education and economic well-being. *Feminist Economics, 1*(2), 21–46.

Iqbal, F. (2006). *Sustaining gains in poverty reduction and human development in the MENA orientations in development series*. Washington, DC: The World Bank.

Kakwani, N. (1993). Statistical inference in the measurement of poverty. *Review of Economics and Statistics, 75*(4), 632–639.

Kaur, A., Prakasa-Rao, B., & Singh, H. (1994). Testing for second-order stochastic dominance of two distributions. *Economic Theory, 10*(5), 849–866.

Klasen, S. (2000). Measuring poverty and deprivation in South Africa. *Review of Income and Wealth, 46*(1), 33–58.

Kolm, S. C. (1977). Multidimensional egalitarianisms. *Quarterly Journal of Economics, 91*(1), 1–13.

Lovell, C. A. K., Richardson, S., Travers, P. & Wood, L. (1994). Resources and functionings: A new view of inequality in Australia. In W. Eichhorn (Ed.) *Models and Measurement of Welfare and Inequality*. Heidelberg: Springer-Verlag.

Maasoumi, E. (1986). The measurement and decomposition of multi-dimensional inequality. *Econometrica, 54*, 991–997.

Maasoumi, E. (1999). Multidimensioned approaches to welfare analysis. In Silber, J. (Ed.), *Handbook of income inequality measurement* (pp. 437–477). Kluwer Academic Publishers.

Martin R. J. & Mathena, A. S. (2008). *Housing Finance for the poor in Morocco: Programs, Policies and Institutions*, USAID, Microreport n°96.

Muller, C. (2007). Anti-poverty transfers without riots in Tunisia. *Working Paper, DIAL*, DT 2007–08.

Osman, M., Zakareya, E. & Mahrous, W. (2006). *Targeting the poor in Egypt: A ROC approach*. Presented at the 13th ERF Annual Conference "Oil : Its impact on the global economy", Kuwait, 16–18 December 2006.

Sahn, D., & Stifel, D. (2000). Poverty comparisons over time and across countries in Africa. *World Development, 28*, 2123–2155.

Schultz, T. (2002). Why governments should invest more to educate girls. *World Development, 30*(2), 207–225.

Sen, A. (1976). Poverty: An ordinal approach to measurement. *Econometrica, 44*(2), 219–231.

Sen, A. K. (1981). *Poverty and famines. An essay on entitlement and deprivation*. Oxford: Clarendon Press.

Sen, A. (1983). Poor Relatively Speaking. *Oxford Economic Papers, 37*, 669–676.

Sen, A. (1985). *Commodities and capabilities*. Amsterdam: North Holland.

Sen, A. (1999). *Development as freedom*. In Alfred Knopf (Ed.) New York. Oxford University Press

Sims, D., Kamal, H. & Solomon, D. (2008). *Housing study for urban Egypt*. USAID, Final Report, December

Strauss, J. & Thomas, D. (1995). Human resources: Empirical modeling of household and family decisions. In Behrman and Srinivasan (Eds.), *Handbook of Development Economics. Volume 3A. Handbooks in Economics, vol. 9*. Amsterdam: New York and Oxford: Elsevier Science, 1883–2023.

Thorbecke, E. (2005). Multi-dimensional poverty: Concepts and measurement issues. Paper prepared for: *The Many Dimensions of Poverty* International Conference, UNDP International Poverty Centre, Brasilia, August 29-31

Townsend, P. (1979). *Poverty in the United Kingdom*. Hardsmonsworth: Penguin Books.

Townsend, P. (1993). *The international analysis of poverty*. Hertfordshire: Harvester/Wheatsheaf.

Tsui, K.-Y. (1995). Multidimensional generalizations of the relative and absolute inequality indices: The Atkinson-Kolm-Sen approach. *Journal of Economic Theory, 67*, 251–265.

Tsui, K.-Y. (1999). Multidimensional inequality and multidimensional generalized entropy measures: An axiomatic derivation. *Social Choice and Welfare, 16*(1), 145–157.

Tsui, K. Y. (2002). Multidimensional poverty indices. *Social Choice and Welfare, 19*(1), 69–93.

UNDP (2010). *Human Development Report 2010.* New York: Palgrave Macmillan.

UN-Habitat (2003). *The challenge of slums: Global report on human settlement, UN Human Settlement Programme*. London and Sterling, VA: EarthScan Publications Ltd.

Van de Walle, D. (2005). Do services and transfers reach Morocco's poor? Evidence from poverty and spending maps. *World Bank Policy Research Working Paper* 3478.

World Bank. (2001). *Engendering development: Through gender equality in rights, resources, and voice*. New York: Oxford University Press.

World Bank (2002). *Arab Republic of Egypt, Poverty Reduction in Egypt*. Vol. 1, n° 24234-EGT.

World Bank (2005). *Turkey: Joint poverty assessment report*, n° 29619-TU, vol. 1.

World Bank (2006). *Arab Republic of Egypt, challenges and priorities for rural development*, Policy Note, Report n° 36432-EG, June.

World Bank (2008), *Turkey: Country economic memorandum: Sustaining high growth: Selected Issues*, vol.1, n° 39194.

World Bank (2007) *Global monitoring report: MDG confronting the challenges of gender equality and fragile states*. Washington, DC: World Bank.

Zheng, B. (1997). Aggregate poverty measures. *Journal of Economic Surveys, 11*(2), 123–162.

Chapter 2
Building a Material Deprivation Index in a Multinational Context: Lessons from the EU Experience

Alessio Fusco, Anne-Catherine Guio and Eric Marlier

Abstract Social indicators can play an important role not only at national but also international level, in making it possible to compare the living conditions in different countries according to a set of commonly agreed criteria. For them to play this role in a multinational context, their construction needs to follow methodological principles that ensure their relevance and comparability across countries. This chapter argues that the Euro-Mediterranean countries can benefit from the European Union (EU) experience in building a common framework for monitoring, understanding and also fighting poverty and social exclusion. As a concrete example, it discusses methodological issues raised by the construction of indicators on *material deprivation*, defined as an *enforced* lack of a combination of items depicting some aspects of living conditions related to housing conditions, possession of durables and capacity to afford basic requirements. More specifically, this chapter focuses on the selection of items, their dimensional structure, their aggregation in a synthetic measure and their weighting. It also puts in perspective material deprivation and income-based poverty indicators to emphasise the complementarity of the two approaches when applied to a group of countries with heterogeneous standards of living. It covers 24 EU countries.

A. Fusco (✉) · A.-C. Guio · E. Marlier
CEPS/INSTEAD, Esch-sur-Alzette, Luxembourg
e-mail: alessio.fusco@ceps.lu

V. Bérenger and F. Bresson (eds.), *Poverty and Social Exclusion Around the Mediterranean Sea*, Economic Studies in Inequality, Social Exclusion and Well-Being 9,
DOI: 10.1007/978-1-4614-5263-8_2,

2.1 Introduction

Social indicators can play an important role both at national and international levels. As discussed by Atkinson and Marlier (2010), on the one hand they allow a country to measure its current social performances on the basis of an explicit set of criteria, to assess progress made in reducing poverty and social exclusion, and to compare the effectiveness of different policy measures. On the other hand, social indicators can also be used for international comparisons as well as *contextualised benchmarking*. The latter is important because "specific policies and their impacts measured through indicators can be properly understood only in the context of the broad institutional setting in which they operate, and this calls for a system-wide analysis (Op. Cit., p. 45)".

This (potential) international role for social indicators is particularly crucial in a formal multinational context, such as the European Union (EU), which gathers together countries with heterogeneous levels of living. The construction of relevant and robust social indicators of poverty and social exclusion requires identifying the dimensions to be included in the analysis and using an appropriate methodology to ensure the comparability of the measures across countries. At EU level, the need for a wide approach to the analysis of "poverty", focusing on both monetary (*financial poverty*) and non-monetary aspects (*material deprivation*), is explicitly referred by the EU Council of Ministers which, back in 1985 and in line with the approach developed by the late Peter Townsend (1979), defined the "poor" as:

> The persons whose resources (material, cultural and social) are so limited as to exclude them from the minimum acceptable way of life in the Member State to which they belong (EU Council of Ministers 1985).[1]

The measurement of material deprivation has been regularly on the EU agenda since at least 2004 and the EU has finally reached an agreement in 2009 on two indicators of material deprivation. Originally proposed by Guio (2009), these indicators are now part of the EU set of commonly agreed social indicators. They are used by the 27 EU Member States and the European Commission to monitor national and EU progress in the fight against poverty and social exclusion in the context of the so-called *Social open method of coordination* (see Sect. 2.2).

Based on the *EU statistics on income and living conditions (EU-SILC)* instrument, the newly endorsed EU indicators on material deprivation focus on the proportion of people living in households who cannot afford at least three of the following nine items: (1) to face unexpected expenses; (2) one week annual holiday away from home; (3) to pay for arrears (mortgage or rent, utility bills or hire purchase instalments); (4) a meal with meat, chicken or fish every second day; (5) to keep home adequately warm; (6) to have a washing machine; (7) to have a colour TV; (8) to have a telephone; (9) to have a personal car. Hence, these measures

---

[1] For an examination of the different approaches to "material deprivation", see Dickes (1989) and Fusco (2007).

aggregate information focused on some key aspects of *material* living conditions; they do not aim at covering all the dimensions of poverty and social exclusion (i.e., health, employment, education, social participation, etc.). It is essential to stress that the focus of the material deprivation indicators discussed in this chapter is not on the lack of items due to choice and lifestyle preferences but on the *enforced* lack—i.e., people would like to possess (have access to) the lacked items but cannot afford them. This approach, in terms of "enforced lack" due to financial pressures, makes the suggested indices more comparable with income poverty.[2]

The different steps leading to the computation of an aggregated measure of material deprivation are: (1) the choice of the relevant dimensions/domains and the set of elementary "indicators" representing them, (2) the evaluation of deprivation on each of these items and dimensions, (3) the aggregation of the elementary indicators into an aggregated index for each dimension and (4) if considered relevant, the aggregation of the different dimensions into an overall index of deprivation (Chiappero Martinetti 2000; Nolan and Whelan 1996). It is important to distinguish two different forms of aggregation. The first aggregation combines different characteristics at the individual level (e.g. persons or households), which are then summed over individuals to form an aggregate index. This is, for instance, the approach used for calculating the EU indicators of deprivation discussed here; however, the focus is then on multiple deprivation at the individual level. Instead of first aggregating across fields for an individual and then across individuals, the second approach aggregates first across people and then across fields. This second approach is thus a combination of aggregate indicators, as with the UN human development index (HDI). To avoid possible confusion between the two forms of

[2] For more information on EU-SILC, see: http://epp.eurostat.ec.europa.eu/portal/page/portal/microdata/eu_silc. For a comprehensive presentation of the two EU material deprivation indicators and of all the other commonly agreed indicators used at EU level in the context of the Social OMC (for monitoring progress towards the agreed EU objectives in the field of social inclusion as well as pensions and healthcare and long-term care), see: http://ec.europa.eu/social/main.jsp?catId=756&langId=en. In June 2010, EU Heads of State and Government adopted a social inclusion target as part of the new "Europe 2020" Strategy on smart, sustainable and inclusive growth: to lift at least 20 million people in the EU from the risk of poverty and exclusion by 2020. One of the three indicators used to monitor progress towards this target is the EU indicator of "severe material deprivation", which is the standard EU indicator of material deprivation (MD) presented above but with a threshold set at 4 rather than 3 lacks out of 9. This EU target has increased considerably the importance of MD indicators at EU level. The 2009 wave of EU-SILC included a thematic module on MD with several items specifically focused on the situation of children. Guio, Gordon and Marlier (2012) have analysed these data and proposed an analytical framework for developing robust EU MD indicators for the whole population as well as for children. This work will feed into the revision of the core EU-SILC MD items and indicators, which will take place in 2015 as part of the mid-term review of Europe 2020. (For a thorough discussion of the social challenges linked to the new Europe 2020 Strategy, see the various contributions included in Marlier, Natali and Van Dam (2010)).

aggregation, the former is referred to as "aggregated" indicators and the latter as "composite" indicators (Marlier et al. 2007[3]).

A reason why it took about 5 years, before an agreement on common measures of material deprivation could be reached at EU level, is that the "intuitive appeal" of these measures can explain their popularity but offers little guidance on their practical implementation, whether for statistical analysis or policy design: "the central problem is how to translate intuition into measurement" (Brandolini 2008, 4). This point is a crucial one. As pointed out by Bourguignon (2006, 76), the key challenge in the field of poverty "consists of building a set of instruments, starting with a satisfactory definition of poverty that would meet part or all of the critiques of the dominant [income] paradigm, while retaining at least part of its *operationality*".

This chapter argues that the Euro-Mediterranean countries can benefit from the EU experience in building a common framework for monitoring, understanding and also fighting poverty and social exclusion. As a concrete example, it discusses some of the methodological issues raised by the construction of material deprivation indicators for use in the EU context and explains the different steps leading to the development of such indicators in order to ensure their relevance and comparability in a multinational context. It does so by combining evidence drawn from two important EU datasets: EU-SILC and also a special Eurobarometer survey on the perception of poverty and social exclusion conducted throughout the EU in 2007. In addition, the chapter puts in perspective material deprivation and income-based poverty indicators to emphasise their complementarity when applied to a group of countries with heterogeneous standards of living. Sect. 2.2 explains the context and the ways in which the EU experience can be informative for Euro-Mediterranean countries. Sect. 2.3 focuses on the selection of items. Section 2.4 analyses the dimensional structure of the selected items and Sect. 2.5 the weighting of individual items in an aggregated index of deprivation. Finally, Sect. 2.6 presents the main conclusions.

2.2 Learning from the EU Experience

The EU Social open method of coordination (OMC) provides a concrete example of how a multinational entity has developed a common framework for better monitoring, understanding and also fighting poverty and social exclusion. The Social OMC[4] is a mutual feedback process of planning, monitoring, examination,

[3] Marlier et al. (2007) discuss the two forms of aggregation in detail. They stress the various technical and political issues raised by composite indicators and conclude by stating that even though composite indicators, like the HDI, undoubtedly can play a valuable role in certain contexts, they should not be employed for monitoring policies. See also Atkinson (2003).

[4] For a thorough overview of the Social OMC and the indicators needed in this context, see *inter alia* Atkinson et al. (2002); Frazer and Marlier (2008); Marlier et al. (2007, 2012).

comparison and adjustment of national (and sub-national) social policies, all of this on the basis of common objectives agreed for the EU as a whole. The aim of this peer review exercise, which involves the European Commission and all 27 Member States, is to share experiences and practices with a view to reaching a greater convergence on key EU social protection and social inclusion objectives. Central to this framework is the definition of a list of commonly agreed indicators which follow a principle-based approach (Atkinson et al. 2002).

Within this framework, the EU has been successful in bringing together countries with different standards of living toward a set of commonly agreed indicators for monitoring social performance despite the fact that the responsibility for social policy remains primarily in the hands of individual Member States.

Obviously, the Euro-Mediterranean partnership launched in 1995 under theso-called "Barcelona process" and continued in 2008 as the "Union for Mediterranean", does not have the same foundations and prerogatives as the EU. Based on a desire to go beyond traditional bilateralism, the Union for Mediterranean is a multilateral framework focusing on three main pillars: (1) a political and security pillar aimed at establishing a common area of peace and stability, (2) an economic and financial pillar aimed at creating an area of shared prosperity and (3) a social, cultural and human pillar aimed at developing human resources and promoting the understanding between cultures and exchanges between civil societies.

The last pillar, which bears some similarity with the Treaty-based EU social cohesion objective, reflects the growing awareness in the Euro-Mediterranean region that the concept of social cohesion is relevant—in particular, in view of the emergence of new sources of vulnerability in this region. In addition, social cohesion "appears not only as an objective to be pursued by [Euro-Mediterranean countries], but as an instrument to promote the convergence between [Euro-Mediterranean countries and the EU]" (Zupi et al. 2009, 9). However, despite this recognition and in contrast with the EU experience, there is no common framework for monitoring poverty and social exclusion in the Mediterranean region. It could be argued that the millennium development goals (MDGs) and their supporting indicators could play such a role in this region. However, MDGs indicators were not tailored to the specificity of the region.

The EU experience provides a concrete example of such a common framework. Indeed, even if the EU path has been a distinctive one, reflecting the history and culture of the countries involved, and there are significant differences from the path followed in other regions of the world, the EU experience is that of multi-country cooperation, and, as such may have valuable lessons for other countries and for international organisations. As mentioned by Atkinson and Marlier (2011:296), "the fight against poverty and social exclusion is a common challenge for all the 27 EU countries (and one of the key social objectives of the Union), and there is scope for mutual learning, despite the differences in national circumstances and levels of living among the countries of the Union".

We have already underlined how the EU definition of "poverty" is relative and includes both outcome elements ("the exclusion of minimum acceptable way of life...") and input elements ("...due to a lack of resources"). The income poverty indicators are based on country-specific relative income thresholds rather than a common EU threshold. The material deprivation indicators that were added to the EU portfolio of social indicators in 2009 provided new outcome measures to reflect the high difference of levels of living between EU countries [a characteristic which is also shared by the Euro-Mediterranean countries, see Conselho Economico E Social (2006)], which have become even more important since the EU 2004 and 2007 enlargements.

Reaching an agreement on material deprivation indicators for use at EU level was not an easy task and implied the use of appropriate methodology ensuring the relevance and comparability of indicators across countries. In the rest of this chapter we focus on the different steps leading to the construction of such indicators. The EU experience and the methodological issues raised in this context might be useful for developing similar indicators for use in the Euro-Mediterranean countries. The next sections explain the different steps leading to the definition and computation of such indicators.

2.3 Selection of Items

The selection of items to be included in a deprivation measure depends on the question(s) we want this measure to address. In the context of the Social OMC, the purpose of EU deprivation indicators is to capture a situation of exclusion from a minimum acceptable way of life due to a lack of resources (EU Council of Ministers 1985).

The selection of items also depends on the availability of items in the database. EU-SILC contains a limited number of deprivation items, compared to some national datasets, and it is to be hoped that the results of the thematic module on material deprivation that was included in the 2009 wave of EU-SILC will lead to the inclusion of new deprivation items in the core survey.

Three main groups of ("objective") items of deprivation can be identified in EU-SILC[5]:

- Economic strain, i.e. the household cannot afford:
  - to face unexpected expenses
  - one week annual holiday away from home

[5] Two other domains could have been included: financial access to healthcare and local environment. Items of access to healthcare available in EU-SILC refer to (self-reported) unmet needs in medical/dental examination. These items could have been used as material deprivation indicators. However, for the monitoring of the Social OMC objectives, the EU considered it important to use them separately, and in particular to develop a specific indicator on access to healthcare based on the question on unmet need for medical care. As to the items related to the local environment, an important reason why they have been excluded from the EU aggregate is that they tend to reflect a rural/urban divide rather than actual deprivation. It is important to stress that the EU measures of deprivation do not include subjective items on the feeling of poverty (e.g. items such as "difficulty in making ends meet"...). See Whelan et al. (2008) and Bossert et al. (2009) for other analysis of deprivation using EU-SILC.

- to pay for arrears (mortgage or rent, utility bills or hire purchase instalments)
- a meal with meat, chicken or fish every second day
- to keep home adequately warm.

- Enforced lack of durables, i.e. the household cannot afford (but would like to):
  - have a washing machine
  - have a colour TV
  - have a telephone
  - have a computer
  - have a personal car.
- Housing, i.e. the household's dwelling suffers from:
  - leaking roof/damp walls/floors/foundations or rot in window frames
  - accommodation too dark
  - no bath or shower
  - no indoor flushing toilet for sole use of the household
  - lack of space, as measured by the number of rooms available for each household member in the dwelling (which can be used for measures of overcrowded households).

Many authors have proposed theoretical or empirical criteria which, in addition to the availability and quality of the data, can help identify relevant items to be included in an index of deprivation. For example, Guio (2005) underlines that, to be chosen as a "lifestyle deprivation" item at EU level, an item should (1) reflect the lack of an ordinary or minimal living pattern common to a majority or large part of the population in the EU and most of its Member States; (2) allow international comparisons (i.e. convey the same information value in the various countries, and not relate specifically to a "national" context); (3) allow comparisons over time; and (4) be responsive to changes in the standard of living of people. (See also Whelan 1993.)

In this chapter, we focus on the first criterion, which can be related to the EU definition of poverty identifying the poor as the individuals who are excluded from a minimum acceptable way of life because of a lack of resources (EU Council 1985). This rather vague definition leaves room for debate about what is a minimum acceptable way of life.

An option to answer this question is to rely on the opinion of experts to define the needs of individuals. This type of approach is exposed to a risk of "ethnocentrism", i.e. experts can be wrong concerning the universality of their judgements, and of "paternalism", i.e. experts could impose their own point of view for the wellbeing of the others (Fleurbaey et al. 1997). This certainly does not mean that approaches based on experts' opinions who base their conclusions on accumulated evidence independent of their perception should be totally disregarded or considered wrong; however, whenever possible this calls for exploring alternative ways that allow cross-checking the results.

Mack and Lansley (1985) proposed an alternative methodology for the selection of items, namely: to collect the views of the general public (rather than those of "experts" only) about which items they consider necessary to have a decent standard of living. Their approach, by taking into account the consensual judgement of individuals to identify "social needs" (*social consensus criterion*), aims at excluding as much as possible value judgements as to what constitutes an acceptable standard of living; it implicitly defines poverty with respect to a minimum standard of living defined by all the citizens rather than to a norm. According to Mack and Lansley, an item supported by at least 50 % of interviewees constitutes a "socially perceived necessity" so that their approach can be seen as a consensual definition of deprivation. Furthermore, contrary to Townsend who simply regarded the lack of a necessity as implying deprivation, they also developed the concept of "enforced lack" and proposed a more adequate format of survey questions to discriminate between preferences and constraints of people.[6]

In order to assess the relevance of the items available in EU-SILC for measuring deprivation, and also to identify which other items of deprivation could be relevant to reflect minimum standard of living in the different EU countries, an EU wide Eurobarometer survey on the perception of poverty and social exclusion was carried out in 2007 (see TNS 2007 for a description of the survey).[7] This Eurobarometer is the first EU dataset that allows carrying out a comparative analysis of the items which citizens in the different Member States consider to be necessary for people to have an "acceptable" standard of living in the country where they live. It provides a rich body of information on the 27 EU countries, collected from national samples of adults (aged 15 years and above) living in private households. For identifying socially perceived necessities throughout the EU, Eurobarometer interviewees were asked a series of questions in the following way:

> In the following questions, we would like to understand better what, in your view, is necessary for people to have what can be considered as an acceptable or decent standard of living in [your country]. For a person to have a decent standard of living in [your country], please tell me how necessary do you think it is to …

The potential answers were: "absolutely necessary, no one should have to do without", "necessary", "desirable but not necessary" and "not at all necessary". They were coded from 3 ("absolutely necessary") to 0 ("not at all necessary").

As shown in Table 2.1, Eurobarometer figures largely confirm that the items currently available in EU-SILC and used in the EU deprivation measures are socially validated. Almost all these items are considered absolutely necessary

---

[6] In EU-SILC, questions regarding durable goods rely on this Mack and Lansley's format and enable distinguishing between "lack of items" (due to choice) and "enforced lack of items" (people would like to possess/access the items but cannot afford them). Only this latter group is considered as reflecting "deprivation", in order to exclude lifestyle preferences from the concept of deprivation.

[7] The Eurobarometer was conducted on behalf of the European Commission with a view to informing the preparation of a thematic module on material deprivation that was included in the 2009 wave of EU-SILC (see Guio, Gordon and Marlier 2012).

Table 2.1 Proportion of people considering absolutely necessary (AN) and necessary (AN + N) the items, EU-27

	Absolutely necessary (AN)	Necessary or absolutely necessary (AN + N)	No. of countries with consensus (AN + N)
Economic strain			
Unexpected expense	35	78	27
one week holiday away from home	18	50	12
Arrears	62	95	27
Repay Loans	52	97	
Utility bills	70	98	
Rent/mortgage	65	91	
Meat, chicken or fish every second day	43	81	27
Keep home adequately warm	63	97	27
Durable goods			
Washing machine	54	90	27
Colour TV	26	65	19
Phone	20	53	14
Mobile phone	19	51	
Fixed phone	20	54	
Computer	12	38	5
Car	23	56	16
Housing conditions			
No leaking roof, damp walls/floors	69	97	27
Not too dark	43	87	27
Bath/shower	64	94	27
Indoor flushing toilet	69	96	27
Space	31	73	27
To invite friends/family	27	69	
To read/write, etc.	35	77	

Source Special Eurobarometer No 279, Wave 67.1, authors' calculation
Note Each country, whether small or large, receives the same importance in these EU-27 averages. The 4th column provides the number of EU countries (out of 27) where at least 50 % of interviewees have identified the item as "absolutely necessary" or "necessary"

(AN) or necessary (AN + N) to have a decent standard of living by at least 50 % of the EU population[8] and also by most (and often all) of the 27 Member States.

With only 38 % of support, the enforced lack of a *computer* is the only item that does not pass the 50 % criterion at EU level even when considering both the AN and N responses. (The item "cannot afford one week holiday away from home" just meets the criterion at EU level (with exactly 50 % of support), whereas, as shown in the fourth column of Table 2.1, it meets it only for 12 out of the 27 Member States. Yet, in 14 out of the 15 countries which fail to reach the 50 % threshold the level of support is at least 30 %; in two countries, it is 47 % and in another two 43–44 %.)[9]

Another condition put forward by Mack and Lansley for having a social consensus is that the consensus should be achieved in the various social groups,i.e. there should be homogeneity of preferences within countries. As mentioned by Pantazis et al. (2006, 90) "the validity of this consensual approach to measuring poverty rests on the assumption that there is a universal minimum accepted by society that also reflects actual living conditions. The implication of this, which is central to the approach, is that differences in views between social groups, including ranked social strata, concerning what constitutes an acceptable living standard are relatively small. Otherwise, the definition of an unacceptable standard of living just becomes the opinion of one group against the other." If there is limited agreement over the list of items considered as social necessities and there are class and age differences in the rating of necessities, this could introduce noise into the measurement of deprivation, as judgements of importance may impact upon access to the items.

In the literature, the question of the homogeneity of preferences has been studied through the analysis of the answers given by different subgroups of the population. Research conducted on the 1990 Breadline Britain Survey and the 1999 Poverty and Social Exclusion survey (PSE—Pantazis et al. 2006) concluded that there was a high degree of homogeneity around what society considers necessary to have a decent standard of living. Pantazis et al. (2006, 98) explore this question by analysing the PSE answers subsequently for a set of binary socio-demographic characteristics (men/women, younger/older, richer/poorer…). We applied the same type of bivariate analysis to assess the influence of age, gender, household type, occupational status and subjective financial difficulty on the definition of socially perceived necessities.[10] At EU-27 level, AN answers are very

[8] Each country, whether small or large, receives the same importance in the EU-27 averages; these averages are thus not computed on the basis of population weighted national results (contrary to standard practice). For calculating the EU-27 averages, national samples have been reweighted so as to achieve a sample size of 1,000 for each country.For a list of the official EU countries' abbreviations, see Annex 4, Table A4.1.

[9] Detailed tables can be obtained upon request.

[10] Age is measured with four dummy variables (16–24; 25–39; 40–54; 55+), occupational status is measured with three dummies (employed, self-employed, not working), subjective financial poverty with three dummies (how is your income compared to what is necessary to make ends meets—higher, more or less the same, lower), household type is a combination of the number of adults (15+) and the number of children (below 15). Detailed tables, for each country, can be obtained upon request (see above).

homogeneous according to gender: differences between men and women never exceed 5 %. For the other variables, the gap is higher but exceeds rarely 10 % showing quite a high homogeneity of preferences at EU level. An application of the same procedure to individual countries shows that some of the socioeconomic variables analysed have a higher impact on the perception of the social necessities. For example in Bulgaria, individuals aged 55+ tend to consider most of the items less often absolutely necessary than those aged 25–39. In Hungary, individuals having an income higher or lower than what they think is needed for making ends meet tend to consider many items more often absolutely necessary than those who perceive their income as more or less what they need to make ends meet. In Bulgaria and Hungary, answers for some items are quite different for some household types. But the number of such cases is limited and on the whole the homogeneity of preferences is quite high across the EU.

By simply comparing individuals according to one single characteristic, e.g. men versus women, this bivariate analysis does not take account of other characteristics held by women or men that could explain the results. Moreover, this method cannot be applied to variables defined in a continuous (age) or categorical (marital status) manner. To address these issues, we applied, for each item and each country, ordered probit regressions.[11] These multiple regressions allow assessing the impact of the above-mentioned variables, other things being equal. Results of the country-by-country analysis show that most of the variables have a non-significant or moderate impact on the perception of necessity. This does not necessarily mean that all the groups have the same opinion. What it shows is that on average answers from individuals are not significantly differentiated and that the determinants of the perception of necessities are not significantly related to socio-demographics characteristics within each country. If we run the analysis on the pooled EU-27 data, it is interesting to note that when we include countries dummy, the country effect is higher than the socioeconomic variables effect on the probability of perceiving an item as (absolutely) necessary. These results were also found by Accardo and de Saint Pol (2009).

Hence, among the 15 items reviewed above only the "enforced lack of a computer" fails to meet the two selection criteria—social consensus and homogeneity of preferences at EU-27 level. All other 14 items meet these criteria (though with very different levels of support) and have therefore been retained for the next step, i.e. the analysis of the dimensional structure among the selected items.[12]

---

[11] We applied an ordered probit regression to tackle the ordinal nature of each item. As already mentioned, in the previous consensus surveys individuals could only choose between two answer categories: "necessary" and "desirable but not necessary". In the Eurobarometer, additional modalities were included allowing to better approach the range of views of interviewees.

[12] In a previous analysis of the Eurobarometer data, Dickes et al. (2010) assess the (in)variance of the structure of the perception of social needs between countries on the basis of an extension of the Multi-Dimensional Scaling (MDS) method. They show that there is a high level of congruence between the 27 national patterns. An important consequence of their analysis is that it

2.4 Dimensional Structure

Once the relevant items have been selected, one option is to keep them all separate and to give a detailed presentation of deprivation shares for each item. This option makes it hard to draw a comprehensive picture of deprivation in each country, which is needed for international purposes. As mentioned by Marlier et al. (2007), "the essential interest here is not so much in individual items per se as in the underlying situation of more generalised deprivation that they can help to capture". This requires an analysis of the dimensional structure of the list of selected items that will then inform the decision on how to aggregate the items in homogeneous dimensions of lifestyle deprivation. The identification of such structures makes it easier to interpret the information available in the list of items and can better highlight different patterns of deprivation determinants in different countries.

To group the items in dimensions, some technical choices have to be made. We can group items together according to the meaning of their underlying characteristics on the basis of arbitrary criteria (for example, all housing items together) or empirically through data analysis. Factor analysis is one technique that can be used to identify a limited number of unobservable dimensions of deprivation from the analysis of the correlation between a large set of manifest items of deprivation. However, this data driven technique is sometimes criticised (see for example McKay and Collard 2003) as there is a certain degree of arbitrariness in the choice of the number of factors. Despite such limitations, factor analysis remains a useful tool for exploring the underlying structure of the data.

In an exploratory factor analysis, the structure of the latent factor model or the underlying theory is not specified a priori; rather, data are used to reveal the structure of the factors. On the other hand, in a confirmatory factor analysis (CFA), the precise structure of the factor model is assumed and tested. The expected structure is tested to determine whether or not the data meet the conditions necessary for its valid application. CFA is thus more powerful than the exploratory approach as it allows testing the adequacy of the factor structure of the indicator.

A CFA was therefore performed on the 2007 EU-SILC data.[13] A three-factor solution and a two-factor solution were tested on the selected items. The three-factor solution includes the following dimensions:

(Footnote 12 continued)
tends to support the approach which consists of measuring deprivation on the basis of a same set of (validated) items across all the Member States.

[13] The analysis was conducted using SAS, proc CALIS. The matrix of tetrachoric correlations was used as the input for the CFA as it fits better with the binary nature of the items used. Oblique rotation was applied, implying the hypothesis that the dimensions are correlated. For a use of CFA in the deprivation literature, see also Whelan et al. (2001), Eurostat (2002), Jensen et al. (2002), Carle et al. (2009), Dekkers (2008).

Table 2.2 Fit statistics of the CFA, pooled EU data—three-factor and two-factor solutions

Pooled EU data	three-factor solution	two-factor solution
Goodness of fit index (GFI)	0.987	0.977
GFI adjusted for degrees of freedom (AGFI)	0.980	0.968
Root mean square residuals (RMRS)	0.058	0.076
Parsimonious GFI	0.784	0.802

Source EU-SILC 2007 cross-sectional data-files, version Users' Data-Base (UDB) of 01.08.09, authors' calculation (see Annex 1 for an explanation of the fit indices)

- economic strain,
- enforced lack of durables, and
- housing.[14]

The two-factor solution consists of merging the dimensions "economic strain" and "enforced lack of durables" in a single dimension. Table 2.2 presents the fit statistics for the two solutions.

The fit statistics of the CFA are reasonably high and confirm that the two solutions are supported by the data. The analysis was also performed at country level and confirmed this result. Moreover, this result is consistent with the previous dimensional structure highlighted on the ECHP and on the 2004 and 2006 EU-SILC data (Guio 2009).

For two main reasons, implying inevitably some individual judgement, the EU has opted for the two-factor solution. First, the covariance between the factor "economic strain" and "durables" in the three-factor solution is very high (0.76—see Table A1.1); being deprived in one dimension is positively correlated with being deprived in the other. Second, in an EU context it can be seen as an advantage in terms of parsimony to present only two aggregations: one based on a larger set of commodities and activities whose access is linked to the financial strain encountered by the household, and the other depicting the housing conditions (housing comfort and housing facilities) (see Marlier et al. 2007).

At the EU level, the Cronbach alpha coefficient (which measures the internal consistency of the scale) is reasonably high for the combined "economic strain/durables" factor (0.67—see Table A1.2 in Annex 1). By country, the majority of countries have alpha values ranging between 0.60 and 0.70 except Spain (0.57) and Luxembourg (0.54).

The results are less satisfactory for the housing deprivation scale with a Cronbach alpha of 0.37 at the EU level with national values ranging between very low values (0.10 in the Netherlands) to satisfactory ones (0.64–0.66 in the three Baltic States). The housing dimension is quite heterogeneous and should normally be split into different aspects. Housing amenities seem to form a distinct factor;

[14] The items are those presented in Sect. 2.3. The item "lack of space" that was considered as relevant to study deprivation in Sect. 2.3 has been discarded due to a lack of homogeneity with the other items.

and so do also the housing quality items (darkness, damp walls and leaking roof…). Further research on the housing dimension will therefore be needed and in this context full use ought to be made of the thematic module on housing that was included in the 2007 wave of EU-SILC.

In line with the approach followed in the two commonly agreed indicators on material deprivation adopted by the EU in 2009, it is therefore preferable to solely retain the dimension composed of the nine ("socially validated") items related to economic strain and durables.[15] By excluding the housing dimension from our analysis and combining the economic strain and durables factors, we move to a unidimensional material deprivation framework.

Factor scores resulting from the factor analysis satisfy properties that guarantee the robustness of the aggregated score obtained, so that they could be used directly as scores of deprivation (see e.g. Krishnakumar and Nagar 2008). However, as reminded by Brandolini (2008), it is necessary to be cautious when using mathematical algorithm to answer what is (also) a normative task. Hence, as this practice is not very transparent, CFA has been used solely to cluster the items and proceed to their aggregation in indices in the way presented in the next section.

2.5 Weighting or Not Weighting?

At this stage, it is useful to summarise the information from the nine items in an aggregate index of "material deprivation". Let x_{ij} be the (non-negative) level of deprivation of individual $i = 1 \ldots n$ on item $j = 1 \ldots m$. The nine items that have been retained for measuring deprivation are dichotomous items, so that $x_{ij} = 1$ if there is an enforced lack and 0 otherwise. Within each dimension, the simplest form of the deprivation score u_i for each individual in the sample is the linear aggregation method which is a (weighted) sum over the x_{ij}:

$$u_i = \sum_{j=1}^{m} w_j x_{ij}$$

The weights w_j are non-negative and can be normalised to sum to one. The interpretation of such indices is straightforward. A bad performance on the indicator implies a higher value of the index. This index allows to obtain a complete

[15] As mentioned above, the nine items concern the incapacity to afford: to face unexpected expenses; one week annual holiday away from home; to pay for arrears (mortgage or rent, utility bills or hire purchase instalments); a meal with meat, chicken or fish every second day; to keep home adequately warm; to have a washing machine; to have a colour TV; to have a telephone; to have a personal car. As shown in Sect. 2.3 above, all these items satisfactorily meet both the "social consensus" criterion and the "homogeneity of preferences" criterion.

ordering and to summarise the information—at the cost of a (unavoidable) loss of information resulting from the aggregation.

This complete ordering is sensitive to the choices made when building the index. For example, in the context of a benchmarking framework (such as in the EU Social OMC), weights can have a significant impact on the aggregate indicator and on the ranking of countries (Nardo et al. 2005). Weights express the trade off between the items that compose the indicator, i.e. the way in which the lack of an item can be compensated by another. Weights are essentially value judgements, and several procedures can be followed for defining them. Section 2.5.1 discusses the equal weighting option and Section 2.5.2, the unequal weighting option. In each of these sections, the pros and cons of the methods are presented as well as illustrative results.[16]

2.5.1 Equal Weights

The most straightforward weighting method consists in giving the same weight to each item. At the individual level, this equal weighting approach is a simple count of the number of lacked items.[17] Hence, choosing the items could be seen as a crude weighting—giving 1 to each item retained, and 0 to those not in the list.

For Brandolini (2008), equal weighting may result from a wish to reduce the researcher's interference to a minimum, or from the lack of information on some kind of consensus view. The main advantage of this approach is to make the interpretation of the results simpler; and its main drawback is that no discrimination is made about the items and that there can be a double counting when items overlap.[18] In fact, the relevance of equal weighting depends on the "absolute" character of the items and on the purpose of the indicator.

First, the issue of weighting depends on whether the focus of the aggregate indicator is solely on essential needs or on a larger set of items. If all the items are considered essential in all 27 EU countries, it can then be argued that access to each of these items has the same normative value—and this, in each country. For such items, equal weights can be more appropriate. If we follow this logic, then the validation of the set of items as "socially perceived necessities" by the answers to

[16] For Stata users—mdepriv—is a useful user written command which allows computing synthetic scores of multiple deprivation similar to that presented in this chapter (see Pi Alperin and Van Kerm 2009). Several alternative weighting rules are available.

[17] When normalised to sum to one, the weight of each item in an equal weighting framework is 1/m. If normalisation is a common practice that allows comparing indices composed of different numbers of items, it is possible to attribute a weight of 1 to each item so that the weights sum to m. In that case, we can talk of a counting approach (see Atkinson 2003).

[18] In case of redundancy, it has been proposed to introduce in the weighting scheme a correlation component (see Betti and Verma 2000).

Table 2.3 National deprivation rates, mean deprivation indices among people deprived and poverty risk thresholds (EU definitions), 2007

Country	People lacking at least three items (in %)	Mean number of items among the deprived (out of nine)	Poverty risk threshold (PPS)
LU	3	3.4	36,908
NL	6	3.4	22,325
SE	6	3.5	20,120
DK	7	3.8	21,367
FI	9	3.5	19,369
AT	10	3.5	22,848
IE	10	3.6	22,483
UK	10	3.5	23,868
ES	10	3.4	16,394
FR	12	3.6	19,661
BE	12	3.7	21,075
DE	12	3.6	21,846
SI	14	3.5	16,756
EE	15	3.6	8,524
IT	15	3.7	18,371
CZ	16	3.7	11,231
PT	22	3.7	11,255
EL	22	3.9	14,588
SK	30	3.7	8,678
LT	30	4	7,376
CY	31	3.6	22,970
HU	37	3.9	8,355
PL	38	3.9	7,187
LV	45	4	7,049

Source EU-SILC 2007 cross-sectional data-files, version UDB 01.08.09, authors' computation
Note Countries are ranked according to their deprivation rate. Poverty risk thresholds are annual amounts (in Purchasing Power Standards) calculated for a household consisting of two adults and two children aged below 14

the Eurobarometer is an argument to weight equally the nine items selected for our scale. Second, the EU income poverty is a *relative* measure, with a threshold fixed at 60 % of the national median equivalised income. Measures of material deprivation are expected to give a more "absolute" view of the standard of living than income poverty. It can be argued that this goal is better reached through equal weighting rather than through weights that would aim at reflecting the relative importance of individual items in the different countries.

The two commonly agreed EU indicators of material deprivation used in the Social OMC are equal weights indicators. The first one is a deprivation rate displaying the proportion of the population living in materially deprived households (see Fig. 2.1). Each person receives a score corresponding to the number of items that his/her household lacks because it cannot afford them; "deprived" individuals are those living in households lacking three or more items (out of

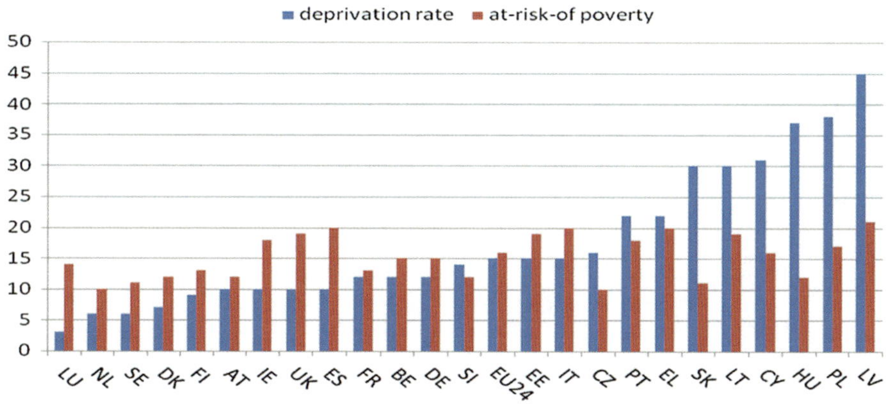

Fig. 2.1 At-risk-of-poverty rates and deprivation rates (EU definitions), 2007. *Source* EU-SILC 2007 cross-sectional data-files, version UDB 01.08.09, authors' calculation. Countries are ranked according to their deprivation rate. *Note* For each indicator, the "EU-24" provides the population weighted average for the 24 EU countries for which data are available in the UDB

nine).[19] The second indicator measures the severity of deprivation through the mean number (not normalised) of items lacked by the people deprived (see Table 2.3).

Figure 2.1 shows that national deprivation rates vary from 3 % up to as high as 45 %; the EU average (calculated for the 24 countries available in the EU-SILC users' database) is 15 %. This range is much wider than that of poverty risk rates, which vary only from 10 to 21 % (EU average: 16 %). This reflects the fact that differences in average living standards across countries and the distribution within them now come into play. The most striking examples in this respect are Hungary and Slovakia (which have high levels of deprivation but low income poverty risk rates) as well as, though to a lesser extent, Cyprus (poverty risks identical to EU average but high deprivation) and the Czech Republic (lowest poverty risk in EU, together with The Netherlands but intermediate performance on deprivation). Latvia combines both the highest poverty risk and highest proportion of deprived in the EU. Conversely, Spain has a high poverty risk, whereas it has a below average proportion deprived. In the least deprived countries deprivation rates are lower than poverty risk rates, whereas the most deprived countries face deprivation levels higher than poverty risk levels.

[19] In the case of the deprivation rate, the choice between a threshold of 2+ or 3+ enforced lacks is arbitrary and can be influenced by different considerations. First and most important, a threshold of 3 + items allows focusing on more severe deprivation and limiting the impact of potential measurement errors and misclassification. Moreover, it offers the advantage of leading to percentages which, in most EU countries, are closer to the value of the poverty rate; this makes it easier to compare the two figures.

As shown in Fig. 2.2, the mean level of deprivation is much higher for those below the poverty risk threshold than above it; this is true in all Member States even if the gap is considerably wider in some countries than in others. It is also worth noting that the mean level of deprivation for those *at risk* of poverty in some of the richest countries is lower than the corresponding figures for those *not at risk* in the poorest countries. Hence, in Spain and the UK, the mean level of deprivation for the income poor is 1.5, whereas in Hungary and Latvia the corresponding figure for those *not* at risk of poverty is 1.9. This brings out the value of complementing the indicators based on the relative EU at-risk-of-poverty measure with indicators of material deprivation, particularly in the enlarged EU where countries have very different levels of living—a characteristic which is also shared by the Euro-Mediterranean countries.

The extent of material deprivation is generally much higher in the countries with lower poverty risk thresholds, and thus with lower levels of median income (see Table 2.3). Five countries out of the six with the highest proportions deprived (Slovakia, Hungary, Lithuania, Poland and Latvia) are among the six countries with the lowest median income (their national poverty risk thresholds are below 8,678 PPS, whereas all other Member States' thresholds are above 11,000 PPS). The only exceptions are Estonia, which has a much lower proportion deprived than its median income ranking would suggest, and Cyprus which has a high level of deprivation but also a very high income threshold.

The severity of deprivation, as measured through the mean number of items lacked by people deprived, ranges from 3.4 in Luxembourg, the Netherlands and Spain to 4 in Lithuania and Latvia. These figures show that for a similar proportion of deprived, the severity of deprivation can vary between countries. For example, deprivation rates in Lithuania and in Slovakia are identical (30 %), but the severity is higher in Lithuania (4 vs. 3.7).

2.5.2 *Unequal Weights*

An alternative to equal weighting when aggregating items into a single indicator is to give different weights to the various items in order to reflect their different relative importance in the final result. Each ratio of weights can be seen as a "substitution rate" between two items, i.e. the coefficient by which a deprivation in one item can be compensated by the non-deprivation in another item (see Munda and Nardo 2005 or Decancq and Lugo 2012). Two countries with very different penetration rates might then have the same set of weights if the substitution rates between items are identical.[20]

[20] Let us take two countries A and B. In country A, the proportion of individuals possessing a car is 90 % and that of individuals having a jacuzzi is 10 %. In country B, these proportions are 45 and 5 %. The normalised weight for each item in the two countries will be the same (car: 90/100 = 45/50 = 0.9; Jacuzzi: 10/100 = 5/50 = 0.1).

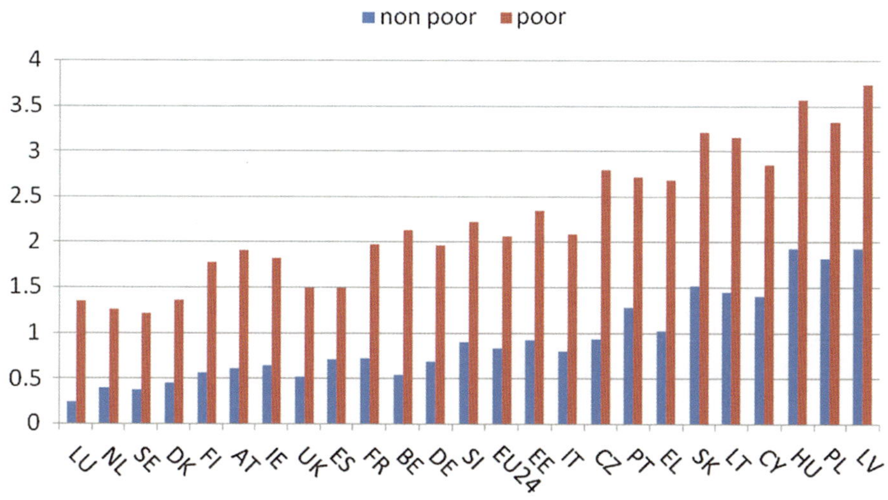

Fig. 2.2 National mean numbers of lacked items (out of nine), according to the income poverty status, 2007. *Source* EU-SILC 2007 cross-sectional data-files, version UDB 01.08.09, authors' computation. *Note* For each indicator, the "EU-24" provides the population weighted average for the 24 EU countries for which data are available in the UDB

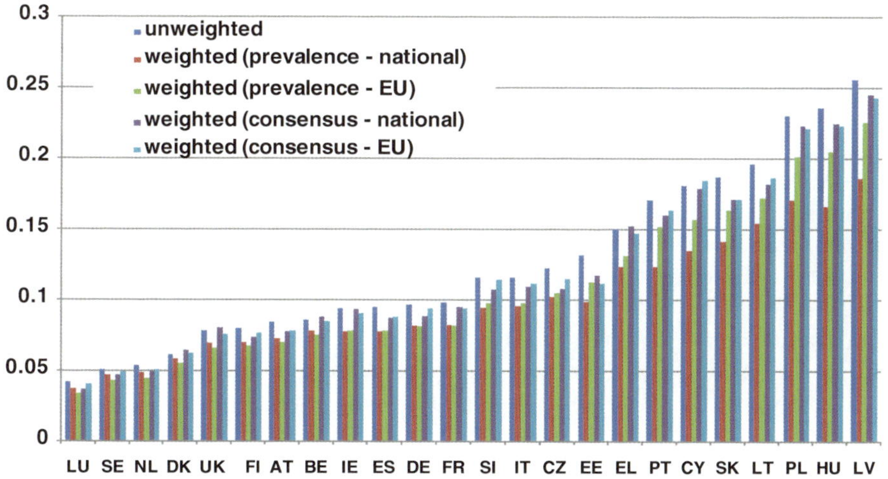

Fig. 2.3 National mean numbers of lacked items (out of nine), weighted and unweighted approaches, total population, 2007. *Source* EU-SILC 2007 cross-sectional data-files, version UDB 01.08.09, authors' calculation. *Note* Indices normalised to one; countries ranked according to "unweighted" (i.e., equal weighting) deprivation indices

When deciding on the weights to be attributed to individual items, two questions that we may want to address are: (1) Do we opt for EU or national weighting?; and (2) Should we go for "prevalence" or "consensus" weighting? (For an extensive discussion of different weighting schemes, see Guio 2009[21]).

EU or national weighting? Answering this question (which in fact also needs to be asked when measuring income poverty; see above) is related to the choice of the reference population. We can make the hypothesis that, in evaluating their material situation, respondents are influenced mostly by the perceptions they have of how they are doing compared to others in their own country, even if one may argue that, in the EU, comparisons would extend beyond national borders (Whelan et al. 2001). In the case of nationally weighted indices of deprivation, the approach would have a relative aspect as the weight of an item would reflect its relative importance in each country.

"Prevalence" or "consensus" weighting? When using aggregate indices, several options can be applied to determine the weights. Here we focus only on two of them. The first one, prevalence or frequency-based weighting, is largely used in the literature and consists in "letting the data speak". Each item is weighted according to the proportion of individuals who possess (have access to) the item.[22] This *endogenous* approach can be justified in terms of subjective perception of deprivation. The higher the proportion of people not deprived in a given item, the more likely a person unable to afford this item (but wanting it) is likely to feel deprived. Prevalence weighting could consist in using the proportion of people having the item over the whole population. For EU comparative deprivation measures, EU-SILC provides the reference data source.[23] The second option, consensus weighting, suggests that weights should be established *exogenously* on the basis of social judgements about what is necessary for a person to live a decent life (see Mack and Lansley 1985; Halleröd 1995). The latter method is more "normative" and may be seen as closer to the notion of "minimum acceptable way of life" than the prevalence approach. It may also be more stable over time as perception of needs change slowly. The Eurobarometer data presented in Sect. 2.3 constitute a unique dataset that precisely allows implementing such a consensual (and participatory) approach in the context of an international comparison. The weight of an item can then be the proportion of people considering this item (absolutely) necessary, or a transformation of this proportion.

Hence, the hypothesis underlying the prevalence approach implies that the most frequently possessed items receive a higher weight (whatever their perceived social

---

[21] See Decancq and Lugo (2012) for a survey of other options to weight the items, including regression-based methods and statistical weights. See also Haisken-DeNew and Sinning (2010) proposal to weight components by their subjective contribution to an overall measure of life satisfaction.

[22] For a similar approach, see: Desai and Shah (1988), Tsakloglou and Papadopoulos (2002), Whelan et al. (2001), Willits (2006).

[23] Guio (2009) analyses different functional forms of weights, e.g. logarithmic of exponential transformation of the proportion of non-deprived.

importance). If an essential good/need is not owned/accessed by a high proportion of the population, its weight will be low independently of its intrinsic importance. Concretely (see Annex 3, Table A3.1), a strict application of this approach would mean, for example, that in Latvia the base weight of "capacity to keep home adequately warm" would be 77.9 (100−22.1), whereas that of "having a TV" would be 98.9. In the consensus approach, it would be respectively 99 and 81 % (i.e., percentages of Eurobarometer respondents considering these items either "necessary" or "absolutely necessary"). A drawback of the prevalence method is that it can lead to a questionable and unbalanced structure of weights.[24] An additional drawback of prevalence weighting is that it is less easy to understand than consensual weighting, and therefore that it can raise serious communication difficulties. (Table A2.1 in Annex 2 displays nationally defined and EU-defined weights for both the prevalence and consensual approach; these weights have been normalised.)

The practical implications of weighting depend on the homogeneity of the items that are to be aggregated. The closer the prevalence rates of the different items in the dimension (or social judgements in the case of consensual weighting), the more equal the weights will be… and thus the closer they will be to the equal weighting approach. If the selected items are heterogeneous, the weights will differ significantly and the weighted index may then also be very different from the unweighted index.[25]

In order to show the implications of different weighting schemes, Fig. 2.3 shows the results for a same indicator of deprivation (the national mean number of lacked items (out of nine); see above for the list of items[26]) weighted according to each of the four options described above (prevalence vs. consensus; national vs. EU). In countries with the least severe deprivation, weights have little incidence on the mean indices (whatever the methodology) as the weights do not differ a lot from one item to the next and are thus closer to equal weighting. By contrast, in the most severely deprived countries there are high differences between prevalence weighted and unweighted scores, because weights give less importance to the most frequently non-possessed items. This, therefore, reduces the dispersion of the prevalence-EU weighted deprivation index. The use of consensus weighting has less impact, compared with the unweighted indices. The impact of the choice between national and EU reference appears to be more important in the prevalence weighting for the most deprived countries.

So, as could be expected, the cardinal results are sensitive to weighting. To better assess the actual impact of weighting it is useful to check whether the ordinal results

[24] Brandolini (2008) provides the example of an analysis he carried out with D'Alessio showing that in 1995 the share of Italians deprived in terms of health and education could be estimated at 19.5 and 8.6 % respectively, which would lead to education receiving a weight more than twice higher than that of health which could be seen as "a matter of disagreement".

[25] See also Boarini and Mira d'Ercole (2006) and Brandolini (2008) for a similar argument.

[26] In view of the difficulty to define a deprivation threshold on a weighted indicator, we have opted here for the presentation of the mean weighted indices rather than the deprivation rates.

in terms of comparisons of groups change. Indeed, what is important is the relative performance between countries and groups more generally. Figure 2.3 shows that the ranking of countries is quite similar under all five options.

2.6 Conclusions

This chapter has argued that the Euro-Mediterranean countries can benefit from the EU experience in building a common framework for monitoring, understanding and also fighting poverty and social exclusion as part of the EU cooperation in the social protection and social inclusion fields. While emphasising the (potential) role for indicators at both national and international levels, it has also stressed the importance of properly contextualising the analysis of indicators. Hence, whereas league tables of social indicators can easily be misused, if the interpretation of indicators is properly contextualised such tables can provide useful insight into the determinants of poverty and social exclusion and can serve as an incentive to political action. An example of a successful contextualised benchmarking exercise is provided by the 2008 in-depth EU study on child poverty and wellbeing (Social Protection Committee 2008).

As a concrete illustration of the lessons that can be drawn from the construction of EU indicators and their use in the context of the EU cooperation in the social area, the chapter has discussed various methodological issues raised by the construction of indicators on *material deprivation*, defined as an *enforced* lack of a combination of items depicting some aspects of living conditions. Highlighting these issues (and the related choices to be made) is important as measurement assumptions may considerably influence the results.

For our analysis, we have combined evidence from two important EU datasets: EU-SILC (for measuring deprivation and income poverty) and a special Eurobarometer survey on the perception of poverty and social exclusion (for assessing EU citizens' consensual judgement).

We have shown that the nine "material deprivation" items retained in the EU aggregate indicators of deprivation satisfactorily meet the two selection criteria—social consensus as well as homogeneity of preferences. Using the definitions adopted at EU level, we have also shown that the range of national material deprivation rates (from 3 to 45 %) is much larger than that of the poverty risk (10–21 %). If purely income-based indicators of poverty and inequality are essential, they are, nevertheless, not sufficient to satisfactorily reflect the fact that the 27 EU countries have very different levels of living, a characteristic which is shared by the Euro-Mediterranean countries.

As expected, the use of different structure of weights changes the cardinal value of the national results. However, for the five weighting procedures reviewed in the chapter, the impact of the weighting scheme on the ranking of countries is limited. There is no clear cut answer to the question of the choice of the weights which reflects in fact implicit value judgements and needs to take account of the purpose of the indicator. In view of these results and because of the advantages of this approach (in particular, its simplicity and transparency), an equal weighting approach seems to be well suited for the construction of an EU material deprivation index. With new data becoming available, the stability over time of the EU indicators in their current form will need to be analysed in order to assess their robustness.

In 2011, when the data collected through the special 2009 EU-SILC module on material deprivation become available, it will be important to come back to these measures with a view to refining them. In doing so, it would be very useful if EU-SILC data could again be put in perspective with (updated) social judgements collected through a new Eurobarometer survey to ensure that the selected items still accurately capture expectations of the society.

Acknowledgments We would like to thank Vincent Dautel, Daniel Defays, Joseph Deutsch, two anonymous referees, the editors and the participants to the workshop on "multidimensional poverty and pro-poor growth in the MENA countries" (CEMAFI—University of Nice Sophia Antipolis, 11–12 June 2009) for helpful comments. These persons should not, however, be held responsible in any way for the present contents.

Annex 1: Dimensional Structure: Confirmatory Factor Analysis and Cronbach Alpha

Goodness of fit index (GFI) represents the amount of variances and covariances in the sample covariance matrix that are predicted by the model. Theoretically, its maximal value is 1. However, as GFI is affected by the sample size and the number of indicators, its upper bound can be lower than one, even in the case of perfect fit. A standard rule of thumb is that the GFI for good fitting model should be greater than 0.9.

Adjusted goodness of fit index (AGFI) is the GFI adjusted for degrees of freedom. A value superior to 0.8 is more often used as a cutoff value to consider that the model is well fitted.

Root mean square residual (RMSR) is the square root of the average of the square of the residuals between the sample and modelled covariance matrix. The lower the fit between the model and the data, the larger the RMSR.

Parsimonious goodness of fit index (PGFI) is a variant of the GFI that takes the parsimony of the model into account.

Table A1.1 Covariances between factors, pooled data, three-factor model

	Economic strain	Durables	Housing
Economic strain	1	0.76	0.59
Durables	0.76	1	0.70
Housing	0.59	0.70	1

Source EU-SILC 2007 cross-sectional data-files, version UDB 01.08.09, authors' calculation

Table A1.2 National Cronbach alpha for "economic strain and durables" and for "housing"

Country	Economic strain and durables	Housing
BE	0.69	0.26
CZ	0.66	0.33
DK	0.65	0.16
DE	0.64	0.17
EE	0.62	0.64
IE	0.64	0.36
EL	0.70	0.43
ES	0.57	0.18
FR	0.64	0.29
IT	0.67	0.32
CY	0.64	0.30
LV	0.70	0.66
LT	0.69	0.65
LU	0.54	0.25
HU	0.65	0.58
NL	0.60	0.10
AT	0.63	0.33
PL	0.68	0.58
PT	0.63	0.47
SI	0.61	0.24
SK	0.65	0.53
FI	0.63	0.22
SE	0.65	0.19
UK	0.67	0.18
EU (24)	0.67	0.37

Source EU-SILC 2007 cross-sectional data-files, version UDB 01.08.09, authors' calculation

Annex 2: Weighting

Table A2.1 Prevalence and consensus weighting, country and EU level

	Unexpected expenses	Holiday	Arrears	Meat	Home warm	Washing machine	Colour TV	Phone	Car
Unweighted	0.111	0.111	0.111	0.111	0.111	0.111	0.111	0.111	0.111
Weights based on prevalence rate (people who have the item)									
BE	0.096	0.093	0.114	0.117	0.104	0.120	0.121	0.121	0.113
CZ	0.078	0.084	0.120	0.110	0.119	0.126	0.126	0.125	0.112
DK	0.097	0.107	0.113	0.115	0.106	0.117	0.118	0.118	0.108
DE	0.079	0.093	0.116	0.111	0.117	0.123	0.122	0.123	0.117
EE	0.100	0.056	0.121	0.121	0.123	0.124	0.127	0.127	0.102
IE	0.075	0.096	0.112	0.120	0.118	0.122	0.122	0.122	0.112

(continued)

Table A2.1 (continued)

	Unexpected expenses	Holiday	Arrears	Meat	Home warm	Washing machine	Colour TV	Phone	Car
EL	0.092	0.069	0.096	0.122	0.113	0.128	0.130	0.130	0.119
ES	0.087	0.079	0.114	0.120	0.114	0.123	0.123	0.122	0.118
FR	0.082	0.086	0.112	0.116	0.117	0.122	0.123	0.122	0.119
IT	0.086	0.077	0.110	0.118	0.113	0.125	0.125	0.124	0.122
CY	0.079	0.064	0.104	0.125	0.089	0.135	0.135	0.135	0.133
LV	0.056	0.052	0.134	0.105	0.116	0.139	0.148	0.145	0.105
LT	0.080	0.056	0.125	0.115	0.107	0.130	0.137	0.134	0.117
LU	0.091	0.104	0.113	0.114	0.115	0.116	0.116	0.116	0.114
HU	0.054	0.051	0.118	0.109	0.130	0.141	0.145	0.141	0.112
NL	0.093	0.101	0.112	0.116	0.115	0.117	0.117	0.117	0.111
AT	0.087	0.090	0.117	0.111	0.118	0.121	0.121	0.121	0.114
PL	0.066	0.051	0.118	0.110	0.112	0.143	0.143	0.141	0.116
PT	0.108	0.052	0.125	0.129	0.078	0.130	0.133	0.127	0.119
SI	0.073	0.089	0.108	0.113	0.120	0.125	0.125	0.125	0.121
SK	0.077	0.063	0.126	0.093	0.130	0.136	0.136	0.135	0.104
FI	0.085	0.099	0.110	0.117	0.119	0.119	0.120	0.121	0.111
SE	0.097	0.102	0.110	0.113	0.115	0.117	0.116	0.117	0.113
UK	0.088	0.095	0.110	0.116	0.115	0.120	0.120	0.120	0.115
EU (24)	0.083	0.083	0.114	0.115	0.114	0.124	0.125	0.124	0.117
Weights based on consensus rate [people considering the item as (absolutely) necessary]									
BE	0.129	0.062	0.159	0.144	0.164	0.135	0.068	0.062	0.078
CZ	0.115	0.062	0.163	0.101	0.161	0.163	0.103	0.058	0.075
DK	0.121	0.057	0.176	0.148	0.184	0.112	0.080	0.072	0.051
DE	0.120	0.046	0.163	0.111	0.173	0.157	0.088	0.078	0.064
EE	0.118	0.088	0.135	0.121	0.136	0.128	0.109	0.084	0.082
IE	0.123	0.074	0.144	0.138	0.145	0.130	0.080	0.077	0.089
EL	0.117	0.102	0.123	0.104	0.126	0.121	0.108	0.092	0.107
ES	0.126	0.065	0.145	0.141	0.144	0.144	0.077	0.073	0.086
FR	0.110	0.081	0.145	0.127	0.150	0.133	0.074	0.066	0.113
IT	0.130	0.054	0.148	0.127	0.149	0.135	0.082	0.071	0.102
CY	0.118	0.090	0.124	0.087	0.123	0.124	0.116	0.093	0.124
LV	0.121	0.084	0.137	0.122	0.139	0.125	0.114	0.086	0.073
LT	0.122	0.067	0.141	0.120	0.142	0.131	0.123	0.073	0.081
LU	0.104	0.066	0.145	0.111	0.151	0.147	0.090	0.083	0.102
HU	0.117	0.093	0.142	0.106	0.143	0.140	0.122	0.077	0.061
NL	0.121	0.065	0.170	0.149	0.188	0.164	0.048	0.057	0.038
AT	0.131	0.056	0.164	0.121	0.167	0.157	0.083	0.060	0.061
PL	0.120	0.081	0.139	0.129	0.142	0.139	0.107	0.075	0.069
PT	0.113	0.082	0.130	0.132	0.129	0.129	0.117	0.078	0.090
SI	0.101	0.079	0.140	0.103	0.145	0.144	0.095	0.084	0.109
SK	0.120	0.080	0.128	0.108	0.135	0.135	0.115	0.087	0.091
FI	0.112	0.063	0.162	0.144	0.163	0.142	0.078	0.073	0.064
SE	0.095	0.082	0.168	0.125	0.167	0.134	0.080	0.078	0.071
UK	0.130	0.070	0.165	0.132	0.170	0.146	0.065	0.067	0.054
EU (27)	0.117	0.075	0.143	0.122	0.146	0.135	0.098	0.080	0.084

Sources EU-SILC 2007 cross-sectional data-files, version UDB 01.08.09 (for prevalence weights) and Eurobarometer special No. 279, Wave 67.1 (for consensus weights). Authors' calculation

2.9 Annex 3

Table A3.1 Deprivation proportions, by items and by country (%)

Country	Unexpected expenses	Holiday	Arrears	meat	Home warm	Washing machine	Colour TV	Phone	Car
BE	20.9	23.4	6.1	3.4	14.6	1.6	0.3	0.2	6.6
CZ	38.5	33.6	5.6	13.1	6.1	0.5	0.4	1.1	11.8
DK	18.4	9.4	4.5	2.5	10.2	1.4	0.5	0.0	8.5
DE	36.0	24.0	5.7	10.1	5.1	0.4	0.5	0.3	5.1
EE	22.3	56.6	5.2	5.8	3.6	3.2	0.5	1.2	20.5
IE	39.1	21.4	8.4	2.2	3.5	0.5	0.2	0.4	9.1
EL	29.5	46.9	26.4	6.5	13.8	2.2	0.3	0.7	9.3
ES	29.0	35.9	6.8	2.1	7.2	0.2	0.1	0.3	4.1
FR	33.4	29.9	9.3	6.2	4.7	0.8	0.2	0.7	3.2
IT	32.0	39.1	12.5	6.2	10.4	0.5	0.3	1.0	2.8
CY	42.0	52.8	23.0	7.7	34.6	0.7	0.2	0.2	1.7
LV	62.6	65.2	10.6	29.6	22.1	6.7	1.1	2.7	30.0
LT	42.4	59.8	9.4	17.1	22.4	6.4	1.2	2.9	15.3
LU	21.5	10.1	2.6	1.4	0.5	0.2	0.0	0.1	1.5
HU	62.9	65.0	19.1	25.4	10.8	3.2	0.6	2.8	22.8
NL	21.1	14.2	4.2	1.4	1.6	0.1	0.0	0.0	5.5
AT	28.7	25.6	3.9	8.3	2.6	0.5	0.3	0.2	5.7
PL	54.5	64.8	18.2	23.8	22.8	1.0	0.9	2.1	19.7
PT	19.7	61.5	7.0	4.1	41.9	2.9	0.6	4.9	11.4
SI	41.6	29.6	14.2	10.3	4.2	0.3	0.7	0.4	3.5
SK	43.3	54.1	7.6	32.2	4.6	0.8	0.7	1.4	23.9
FI	29.9	18.0	9.1	2.9	1.1	1.4	1.0	0.1	8.3
SE	17.5	12.7	5.7	3.6	1.7	0.0	0.6	0.0	3.7
UK	26.7	21.4	8.5	4.0	4.6	0.3	0.1	0.2	4.9
EU24	33.5	33.7	9.5	8.3	8.8	0.8	0.4	0.8	7.0

Source EU-SILC 2007 cross-sectional data-files, version UDB 01.08.09. Authors' calculation
Reading note In Belgium, the proportion of individuals who would like to have a TV but cannot afford it (enforced lack) is 0.3 %. In a prevalence weighting approach, this item would receive a base weight of 99.7 (if weights are not normalised afterwards)

2.10 Annex 4

Table A4.1 Countries' abbreviations

"Old" member states		"New" member states	
AT	Austria	CY	Cyprus
BE	Belgium	CZ	Czech Republic
DE	Germany	EE	Estonia
DK	Denmark	HU	Hungary
EL	Greece	LT	Lithuania
ES	Spain	LV	Latvia
FI	Finland	MT	Malta
FR	France	PL	Poland
IE	Ireland	SI	Slovenia
IT	Italy	SK	Slovakia
LU	Luxembourg		
NL	The Netherlands	(BG)	(Bulgaria)
PT	Portugal	(RO)	(Romania)
SE	Sweden		
UK	United Kingdom		

EU-25: In 2004, the "old" EU-15 countries were joined by ten "new" Member States
EU-27: In 2007, Bulgaria and Romania became in turn EU Member States

References

Accardo, J., & de Saint Pol, T. (2009). Qu'est-ce qu'être pauvre aujourd'hui en Europe? L'analyse du consensus sur les privations. *Economie et Statistique, 421*, 3–27.

Atkinson, A. B. (2003). Multidimensional deprivation: Contrasting social welfare and counting approaches. *Journal of Economic Inequality, 1*, 51–65.

Atkinson, A. B., & Marlier, E. (2010). *Analysing and measuring social inclusion in a global context*, Report ST/ESA/325 produced at the request of the United Nations [Department of Economic and Social Affairs (UNDESA)]. New York: United Nations.

Atkinson, T., Cantillon, B., Marlier, E., & Nolan, B. (2002). *Social indicators: The EU and social inclusion*. Oxford: Oxford University Press.

Atkinson, A.B., & Marlier, E. (2011). Human development and indicators of poverty and social exclusion as part of the policy process. *Indian Journal of Human Development, 5*(2), 293–320

Betti, G., Verma, V. (2000). Measuring the degree of poverty in a dynamic and comparative context: The Euro-Mediterranean countries, *in* 4èmes Rencontres Euro-Méditerranéennes, *Pauvreté et Inégalités dans les Pays Riverains de la Méditerranée*, Nice, 25–27 Septembre.

Boarini, R., & Mira d'Ercole, M. (2006). Measures of material deprivation in OECD Countries. OECD Social Employment and Migration Working Papers, No. 37. Paris:OECD Publishing.

Bossert, W., Chakravarty, S. R., & D'Ambrosio, C. (2009). Multidimensional poverty and material deprivation. *ECINEQ Working Paper, 2009–129*

Bourguignon, F. (2006). From income to endowments: The difficult task of expanding the income poverty paradigm. In: D. B Grusky & R. Kanbur (Eds.), *Poverty and inequalities*. Stanford, California: Stanford University Press.

Brandolini, A. (2008). On applying synthetic indices of multidimensional well-being: Health and income inequalities in selected EU countries, Banca d'Italia, Temi di discussione 668.

Carle, A. C., Bauman, K. J., & Short, K. (2009). Assessing the measurement and structure of material hardship in the United States. *Social Indicators Research, 92*(1), 35–51.

Chiappero Martinetti, E. (2000). A multidimensional assessment of well-being based on Sen's functioning approach. *Rivista Internazionale di Scienze Sociali,* CVIII, No. 2, pp. 207–232.

Conselho Economico E Social (2006). Combating poverty in the Euro-Mediterranean countries, Working Group Report, September.

Decancq, K., & Lugo, M.-A. (2012). Weights in multidimensional indices of wellbeing: An overview, *Econometric Reviews, 32*(1), pp. 7–34.

Dekkers, G. (2008). Are you unhappy? Then you are poor! Multi-dimensional poverty in Belgium. *International Journal of Sociology and Social Policy, 28*(11/12), 502–515.

Desai, M., & Shah, A. (1988). An econometric approach to the measurement of poverty. *Oxford Economic Papers, 40*, 505–522.

Dickes, P. (1989). Pauvreté et Conditions d'Existence. Théories, modèles et mesures, *Document PSELL* No. 8, CEPS/INSTEAD, Walferdange.

Dickes, P., Fusco, A., & Marlier, E. (2010). Structure of national perceptions of social needs across EU countries, *Social Indicators Research,* Vol. 95, No. 1, January, pp. 143–167.

EU Council of Ministers. (1985). Council Decision of 19 December 1984 on Specific Community Action to Combat Poverty (85/8/EEC), OJEC, L 2, Brussels.

Eurostat. (2002). *Income, poverty and social exclusion: Second Report*, Luxembourg.

Fleurbaey, M., Herpin, N., Martinez, M., & Verger, D. (1997). Mesurer la pauvreté? *Economie et Statistiques, 308*(309/310), 113–142.

Frazer, H., & Marlier, E. (2008). Building a stronger EU social inclusion process: Analysis and recommendations of the EU network of independent national experts on social inclusion, Brussels.

Fusco, A. (2007). *La pauvreté: Un concept multidimensionnel*. Paris: L'Harmattan.

Guio, A.-C. (2005). *Material deprivation in the EU, statistics in focus, population and social conditions, living conditions and welfare, 21/2005*. Luxembourg: Eurostat.

Guio, A.-C. (2009). What can be learned from deprivation indicators in Europe? Eurostat methodologies and working paper. Luxembourg: Eurostat.

Guio, A.-C., & Gordon, D., & Marlier, E. (2012), *Measuring material deprivation in the EU: Indicators for the whole population and child-specific indicators, Eurostat methodologies and working paper*. Luxembourg: Eurostat.

Haisken-DeNew, J. P., & Sinning, M. (2010). Social deprivation of immigrants in Germany. *Review of Income and Wealth, 56*(4), 715–733.

Halleröd, B. (1995). The truly poor: Direct and indirect measurement of consensual poverty in Sweden. *Journal of European Social Policy, 5*(2), 111–129.

Jensen, J., Spittal, M., Crichton, S., Sathiyandra, S., & Krishnan, V. (2002). *Direct measurement of living standards: The New Zealand ELSI scale*. Wellington: Ministry of Social Development.

Krishnakumar, J., & Nagar, L. (2008). On exact statistical properties of multidimensional indices based on principal components, factor analysis, MIMIC and structural equation models. *Social Indicators Research, 86*, 481–496.

Mack, J., & Lansley, S. (1985). *Poor Britain*. London: Allen and Unwin.

Marlier, E., Atkinson, A. B., Cantillon, B., & Nolan, B. (2007). *The EU and social inclusion: Facing the challenges*. Bristol: Policy Press.

Marlier, E., Cantillon, B., Nolan, B., Van den Bosch K., & Van Rie T. (2012). Developing and Learning from EU Measures of Social Inclusion. In D.J. Besharov & K.A. Couch (Eds.), *Counting the poor: new thinking about european poverty measures and lessons for the United States*, International Policy Exchange Series, New York: Oxford University Press. D.J. Besharov & N. Gilbert, (Series Eds.).

Marlier, E. & Natali, D. (eds.), Van Dam, R. (2010). *Europe 2020: Towards a more Social EU?*, Brussels: P.I.E. Peter Lang.

McKay, S., & Collard, S. (2003). Developing deprivation questions for the family resources survey. Working Paper No. 13, Personal Finance Research Center, University of Bristol.

Munda, G., & Nardo, M. (2003). On the methodological foundations of composite indicators used for ranking countries", OECD/JRC Workshop on composite indicators of country performance, Ispra, Italy, May 12.

Nardo, M., Saisana, M., Saltelli, A., Tarantola, S., Hoffman, A., & Giovannini, E. (2005). Handbook on constructing composite indicators: Methodology and user guide, OECD Statistics Working Paper, STD/DOC(2005)3.

Nolan, B., & Whelan, C. T. (1996). *Resources, deprivation and poverty*. Oxford: Clarendon Press.

Pantazis, C., Townsend, P., & Gordon, D. (2006). The necessities of life. In: C. Pantazis, D. Gordon & R. Levitas (Eds.), *Poverty and social exclusion in Britain. The millennium survey* (Chapter 4, pp. 89–122). Bristol: The Policy Press.

Pi Alperin, M. N., & Van Kerm, P. (2009). mdepriv—Synthetic indicators of multiple deprivation, v1.0, CEPS/INSTEAD, Differdange, Luxembourg.

Social Protection Committee. (2008). *Child poverty and well-being in the EU*. Current status and way forward. Luxembourg: Office for Official Publications of the European Communities.

TNS (2007). Poverty and social exclusion. *Report on the Special Eurobarometer 279/Wave 67.1*.

Townsend, P. (1979). *Poverty in the United Kingdom*. Hardmonsworth: Penguin Books.

Tsakloglou, P., & Papadopoulos, F. (2002). Aggregate level and determining factors of social exclusion in twelve European countries. *Journal of European Social Policy, 12*(3), 211–225.

Whelan, B. (1993). Non-monetary indicators of poverty. In: J. Berghman & Cantillon B. (Eds.). *The European face of social security: Essays in honour of Herman Deleeck* (pp. 24–42). Avebury: Ashgate.

Whelan, C. T., Layte, R., Maître, B., & Nolan, B. (2001). Income, deprivation and economic strain. An analysis of the ECHP. *European Sociological Review, 17*(4), 357–372.

Whelan, C. T., Nolan, B., & Maître, B. (2008). Measuring material deprivation in an enlarged EU, ESRI Working Paper No. 249.

Willits, M. (2006). Measuring child poverty using material deprivation: Possible approaches, Working paper No. 28, Department of work and pensions.

Zupi M., Estruch-Puertas E., Driouchi, A. & Izzo M. (2009). *Social cohesion policies in mediterranean countries: An assessment of instruments and outcomes*, FEMISE Research Report No. FEM33-05.

Part II
Assessing Poverty

Chapter 3
A Family of Unit Consistent Multidimensional Poverty Indices

Satya R. Chakravarty and Conchita D'Ambrosio

Abstract This chapter characterizes a family of subgroup decomposable unit consistent multidimensional poverty indices. Unit consistency requires that poverty rankings should remain unaltered when dimensions are expressed in different measurement units. The characterized family is a simple generalization of a family of unit consistent income poverty index suggested by Zheng (2007b). The paper also illustrates the index numerically using Turkish data.

3.1 Introduction

Removal of poverty is still one of the primary aims of economic policy in many countries of the world. In order to judge the efficacy of a targeted poverty alleviation policy, it becomes necessary to know the dimension of poverty, that is, how much poverty is there and changes in level of poverty over time. The policy formulator also needs to identify the causal factors of poverty. All these require the quantification of the extent of poverty. More precisely, we need an indicator of poverty that becomes helpful in analyzing these issues.

In his pioneering contribution, Sen (1976) regarded the poverty measurement problem as involving two distinct but unrelated exercises: (i) the identification of

S. R. Chakravarty
Indian Statistical Institute, Kolkata, India
e-mail: satya@isical.ac.in

C. D'Ambrosio (✉)
Università di Milano-Bicocca, Milan, Italy
e-mail: conchita.dambrosio@unibocconi.it

DIW Berlin and Econpubblica, Università Bocconi, Berlin, Germany

V. Bérenger and F. Bresson (eds.), *Poverty and Social Exclusion Around the Mediterranean Sea*, Economic Studies in Inequality, Social Exclusion and Well-Being 9,
DOI: 10.1007/978-1-4614-5263-8_3,

the poor and (ii) the aggregation of the characteristics of the poor into an overall indicator that will specify the extent of poverty. In the literature, the first problem is generally solved by the income method, which requires the specification of an income poverty line, the level of income required for subsistence standard of living. A person is regarded as poor if his income falls below the poverty line. Sen (1976) criticized two crude indicators of poverty, the head-count ratio (proportion of persons in poverty) and the income gap ratio (the shortfall of the ratio between the average income of the poor and the poverty line from unity), because of their insensitivity to a transfer of income between two poor persons and the former also does not change under a reduction in the income of a poor. Sen (1976) also developed an axiomatic characterization of a more sophisticated index of poverty.[1]

However, Sen (1992) argued that the proper space for social evaluation is that of functionings, the various things, such as literacy, housing, health, provision of public goods, adequate nourishment, essential services, communing with friends, a person cares about. The capability set of a person provides information on the set of functionings that a person could achieve. This shows that well-being is intrinsically multidimensional from the capability-functioning perspective. Consequently, poverty being a manifestation of insufficient well-being, is a multidimensional phenomenon as well. In this context poverty is regarded as a problem of capability failure.

In the basic needs approach, which regards development as an improvement in the array of human needs (Streeten 1981), poverty is a consequence of lack of human needs. The social exclusion approach also regards poverty as a multidimensional issue. It refers to exclusion of individuals "from ordinary living patterns, customs and activities" (Townsend 1979, p. 31). The implicit process is generally regarded as a dynamic process whose end product is the exclusion of the concerned individuals from full participation. It a relative concept in the sense that it is defined relative to the standard of the given society (Atkinson 1998). Social exclusion incorporates the process aspect of capability failure (Sen 2002).[2]

In view of the above discussion, we assume that each person is characterized by a vector of attributes that correspond to different dimensions of human life and a direct identification method of the poor checks whether the person has "minimally acceptable levels" (Sen 1992, p. 139) or threshold levels of this set of attributes. Thus, the direct method looks at poverty from a multidimensional perspective, more precisely, in terms of shortfalls of attribute quantities from respective threshold levels. These threshold levels are determined independently of the attribute distributions.

The objective of this paper is to characterize axiomatically a family of unit consistent multidimensional poverty indices. According to unit consistency,

---

[1] Alternatives and variations of the Sen index have been suggested, among others, by Takayama (1979), Blackorby and Donaldson (1980), Kakwani (1980), Clark et al. (1981), Chakravarty (1983a, b, 1997), Thon (1983), Foster et al. (1984), Hagenaars (1987) and Shorrocks (1995).

[2] See Sen (1981, 1985), Ravallion (1996), Tsui (2002) and Bourguignon and Chakravarty (2003) for further discussion on multidimensionality of poverty.

poverty rankings remain unaffected when all the attribute quantities and threshold levels are expressed in differing measuring units. For instance, suppose that income and life expectancy are two dimensions of human life. Then unit consistency demands that the ranking of two attribute distributions when income is measured in dollars and life expectancy in years should be the same when income is measured in euros and life expectancy in months. Unit consistency is a reasonable property and demands invariance of poverty rankings when units of measurement of dimensions change. To understand this, assume, for simplicity, that there are two dimensions of well-being, income and life expectancy. Now, consider the problem of poverty rankings between two countries. Suppose income is measured in the currency of country I and life expectancy is measured in years. It turns out that country I is regarded as more poverty striken than country II. Next, suppose we decide to measure income in the currency of country II and life expectancy in months. Unit consistency then demands that the poverty ranking of the two countries remains unaltered because of these changes in the unit of measurement. Clearly, ratio scale invariance of poverty under ratio scale transformations of attribute quantities and threshold levels, implies unit consistency. But not ratio scale invariance. Thus, without unit consistency the problem of inconsistency in ranking may arise if we consider indices that are not ratio scale invariant. It is also important to note the difference between ratio scale invariance and unit consistency: while the former is a cardinal property, the latter is an ordinal requirement (see Zheng 2007a, b).

Zheng (2007b) characterized a family of income distribution-based unit consistent poverty indices. This family is a two-parameter generalization of the Clark et al. (1981) and Chakravarty (1983a) indices. Our unit consistent multidimensional index is a generalization of the Zheng index to the multivariate set up.

Section 3.2 of the paper lays down the postulates for an index of multidimensional poverty. In Sect. 3.3 we characterize the family of unit consistent indices. Section 3.4 contains an application of the index to Turkish data. Finally Sect. 3.5 concludes.

3.2 Postulates for an Index of Multidimensional Poverty

In this section we set out the properties for a multidimensional poverty index. Let $\Re^m_{++}$ be the positive orthant of the m-dimensional Euclidean space $\Re^m$. For a set of n-persons, person i possesses a vector $(x_{i1}, x_{i2}, \ldots, x_{im}) = x_i \in \Re^m_{++}$ of m attributes. The vector x_i is the *ith* row of an $n \times m$ distribution matrix $X \in M^n$, where M^n is the set of all $n \times m$ matrices whose entries are positive real numbers. The *jth* column $x_{.j}$ of $X \in M^n$ represents the distribution of attribute j $(j = 1, 2, \ldots, m)$ among the n persons. Let $M = \cup_{n \in N} M^n$, where N is the set of all positive integers. For any $n \in N$, $X \in M^n$, we write $n(X)$ (or n) for the corresponding population size. We restrict our attention to $\Re^m_{++}$ in order to avoid the problem that the welfare function $\prod_{j=1}^{m} x_{ij}^{c_j}$ of person i based on different consumption levels becomes zero if consumption of one attribute is zero. This happens irrespective of how large or small the consumption levels other attributes are.

In this multivariate set up, a threshold or cut off is defined for each attribute. These thresholds represent the minimal quantities of the $m$attributes required for maintaining a subsistence standard of living. Let $z = (z_1, \ldots, z_m) \in Z$ be the vector of given thresholds, where Z is a nonempty subset of $\Re^m_{++}$. The quantitative specification of different attributes exclude the possibility that a dimension can be of qualitative type, for instance, whether a person is ill or not.

In this structure, the *ith* person is regarded as poor or non-poor with respect to attribute j, or equivalently, attribute j is meager or non-meager for him, according as $x_{ij} < z_j$ or $x_{ij} \geq z_j$ and he is called non-poor if $x_{ij} \geq z_j$ for all j. For any $X \in M^n$, let $S_j(X)$(or S_j) be the set of persons who are poor or deprived with respect to attribute j, where $n \in N$ is arbitrary. As Bourguignon and Chakravarty (2003) argued, in order to count the number of poor in a simple way, it is convenient to define a poverty indicator variable as follows:

$$\begin{aligned} \rho(x_i; z) &= 1 \text{ if } \exists j \in \{1, 2, \ldots, m\} \ : \ x_{ij} < z_j, \\ &= 0, \text{ otherwise.} \end{aligned} \tag{3.1}$$

Then the number of poor in the multidimensional framework is given by:

$$n_p(X) = \sum_{i=1}^{n} \rho(x_i; z) \tag{3.2}$$

The identification method defined in (3.1) is known as the union method of identification. This contrasts with the intersection method which identifies a person as poor if he is deprived in all dimensions (see Bourguignon and Chakravarty 2003, 2008). Alkire and Foster (2007) considered an intermediate identification method which regards a person as poor if he is deprived in at least k dimensions, where $1 \leq k \leq m$. This method includes the union and intersection methods as special cases when $k = 1$ and $k = m$.

A multidimensional poverty index P is a non-constant real valued function defined on $M \times Z$. For any $X \in M, z \in Z$, the functional value $P(X; z)$ determines the extent of poverty associated with the distribution matrix X and the cut off vector z.

Most of the properties we consider below, following Chakravarty et al. (1998), Tsui (2002), Bourguignon and Chakravarty (2003, 2008), Chakravarty (2009) and Chakravarty and Silber (2008), are immediate generalizations of different postulates proposed for an income poverty index.[3] All properties apply for any arbitrary P and any positive integer n:

[3] For discussion on properties of an income poverty index, see Sen (1976), Foster et al. (1984), Donaldson and Weymark (1986), Cowell (1988), Seidl (1988), Chakravarty (1990, 2009), Foster and Shorrocks (1991), Bourguignon and Fields (1997) and Zheng (1997).

Focus (FOC): *For any* $(X; z) \in M \times Z$ and for any person i and attribute j such that $x_{ij} \geq z_j$, an increase in x_{ij}, given that all other attribute levels in X remain fixed, does not changes the poverty value $P(X; z)$.[4]

Normalization (NOM): *For any* $(X; z) \in M \times Z$ *if* $x_{ij} \geq z_j$ *for all i and j, then* $P(X; z) = 0$

Monotonicity (MON): *For any* $(X; z) \in M \times Z$*, any person i and attribute j such that* $x_{ij} < z_j$*, an increase in* x_{ij}*, given that other attribute levels in X remain fixed, does not increase the poverty value* $P(X; z)$.

Principle of Population (POP): *For any* $(X; z) \in M \times Z$, $P(X; z) = P\left(X^{(l)}; z\right)$, *where* $X^{(l)} = \left(X^1, X^2, \ldots X^l\right)$ *with each* $X^i = X$*, and* $l \geq 2$ *is any integer.*

Symmetry (SYM): For any $(X; z) \in M \times Z$, $P(X; z) = P(\pi X; z)$,where π is any $n \times n$ permutation matrix.[5]

Continuity (CON): $P(X; z)$ *is continuous in* $(X; z)$.

Subgroup Decomposability (SUD): *For any* $X^1, X^2, \ldots, X^l \in M$ *and* $z \in Z$, $P(X; z) = \sum_{i=1}^{l} \frac{n_i}{n} P(X^i; z)$*, where* $X = (X^1, \ldots, X^l) \in M$*,* n_i *is the population size associated with* X^i *and* $\sum_{i=1}^{l} n_i = n$.

Non-decreasingness in Subsistence Levels of Attributes (NDS): *For any* $X \in M$, $P(X; z)$ *is non-decreasing in* z_j *for all j.*

Non-poverty Growth (NPG): *For any* $(X; z) \in M \times Z$*, if Y is obtained from X by adding a rich person to the society, then* $P(Y; z) \leq P(X; z)$

Ratio Scale Invariance (SCI): *For all* $(X^1; z^1) \in M \times Z$, $P(X^1; z^1) = P(X^2; z^2)$, *where* $X^2 = X^1\Omega$, $z^2 = z^1\,\Omega$ *and* $\Omega = diag\ (\omega_1, \omega_2, \ldots \omega_m)$, $\omega_i > 0$ *for all i.*

According to **FOC** if a person is not deprived in an attribute, then giving him more of this attribute does not change the level of poverty, even if he is deprived in the other attributes. Thus, **FOC** rules out trade off between two attributes of a person who is deprived in one but non-deprived in the other. Thus, if life expectancy and a composite good are two attributes, more life expectancy above the threshold is of no use if the composite good is below its threshold. This, however, does not exclude the possibility of a trade off if a person is deprived in both the attributes. **NOM** says that if all the persons in a society are non-poor, then the value of the index is zero. **MON** states that poverty does not increase if a person becomes less deprived in a dimension. This axiom implies Dimensional Monotonicity of Alkire and Foster (2007), which says that if deprivation in a dimension is completely eliminated, then poverty should not increase. Under **POP**, poverty remains unaltered if an attribute matrix is replicated several times. Since replication enables us to transform two different sized matrices into the same size, **POP** becomes

---

[4] One may consider a stronger version of this axiom where the condition $x_{ij} \geq z_j$ applies simultaneously to all j. See Bourguignon and Chakravarty (2003).

[5] A $n \times n$ matrix with entries 0 and 1 is called a permutation matrix if each of its rows and columns sums to one.

helpful for inter-temporal and interregional poverty comparisons. **SYM** says that any characteristic other than the attribute quantities, for instance, the names of the individuals, is immaterial for poverty measurement. **CON** ensures that minor observational errors in attribute and threshold quantities will not generate an abrupt jump in the value of the poverty index. Therefore, a continuous poverty index will not be oversensitive to errors in observation on basic needs and threshold quantities. According to **SUD** for any partitioning of the population into several subgroups, say l, defined along ethnic, geographical or other lines, the overall poverty is the population share weighted average of subgroup poverty levels. The contribution of subgroup i to overall poverty is $n_i P(X^i; z)/n$ and the overall poverty will precisely reduce by this amount if poverty in subgroup i is eliminated. $(n_i P(X^i; Z)/nP(X; Z))$ 100 is the percentage contribution of subgroup i to total poverty. Each of these statistics is regarded as useful to policymakers because they become helpful for identifying subgroups of the population that are more affected by poverty (see Anand 1997; Chakravarty 1983a, 2009; Foster et al. 1984; Foster and Shorrocks 1991). By repeated application of SUD we can write the poverty index as:

$$P(X; z) = \frac{1}{n} \sum_{i=1}^{n} p\,(x_i; z) \qquad (3.3)$$

Since $p\,(x_i; z)$ depends only on person i's attributes, we call it, 'individual poverty function'. Evidently, P in (3.3) is symmetric and population replication invariant.

Between two identical communities, the one with higher threshold levels of one or more attributes should not have a lower poverty because of higher deprivation of the poor resulting from increased subsistence quantities. This is what is demanded by **NDS**. According to **NPG**, poverty should not increase if a rich person migrates to the society. **SCI** means that the poverty index should remain invariant under scale transformations of attribute quantities and threshold levels. In other words, deprivation resulting from poverty is viewed in terms of proportionate shortfalls of attribute quantities from respective cut offs.

The next postulate is concerned with the redistribution of attributes. For $X, Y \in M$, with $X \neq Y$, X is said to uniformly majorize Y if $X = BY$ for some $n \times n$ bistochastic matrix which is not a permutation matrix.[6] This means that we transform the distribution matrix Y into the matrix X by some equalizing transfers (See Kolm 1977; Tsui 2002; Savaglio 2006; Weymark 2006; Chakravarty 2009).

Transfers Principle (TRP): *For any $z \in Z$, $X, Y \in M$ if X uniformly majorizes Y then $P(X; z) \leq P(Y; z)$, given that the transfers are among the poor.*

TRP shows that if we transform the distribution matrix Y into the matrix X by some equalizing operation among the poor, then poverty under X will not be higher than that under Y.

---

[6] An $n \times n$ matrix with nonnegative entries is called a bistochastic matrix if each of its rows and columns sums to unity.

We will now consider a property which deals with the essence of multidimensional poverty measurement through correlation between attributes. By taking into account the association of attributes, as captured by the degree of correlation between them, this postulate also underlines the difference between single and multidimensional poverty measurements. To illustrate the property, consider the two-person two-attribute case, where the two persons are deprived in both the dimensions. Suppose that $x_{11} > x_{21}$ and $x_{12} < x_{22}$. Now, let us consider a switch of attribute 2 between the two persons. This switch increases the correlation between the attributes because person one who had more of attribute 1 has now more of attribute 2 as well and that is why we refer to it as a correlation increasing switch between two poor persons. Next, suppose that attributes 1 and 2 are substitutes, that is, one attribute may compensate the lack of another in the definition of individual poverty. Then increasing the correlation between the attributes should increase poverty. Indeed, the switch just defined does not modify the mean of each attribute but reduces the extent to which the lack of one attribute may be compensated by the availability of the other. An analogous argument will establish that poverty should decrease under a correlation increasing switch if the two attributes are complements. If a poverty index does not change under a correlation increasing switch, then it treats the attributes as 'independents'. In the general case, the rearrangement of attributes are made in a way such that one of the persons receives at least as much of every attribute as the other and more of at least one attribute (see Weymark 2006; Chakravarty 2009)

We state this principle formally as:

Increasing Poverty Under Correlation Increasing Switch (IPC): *Under* **SUD**, *for any* $(X; z) \in M \times Z$, *if* $Y \in M$ *is obtained from* X *by a correlation increasing switch between two poor persons, then* $P(X; z) < P(Y; z)$ *if the two attributes are substitutes.*

The corresponding property which demands decreasingness of poverty under such switch, when the attributes are complements, is denoted by DPC. [7] We say that a poverty index is sensitive to a correlation increasing switch if it satisfies either IPC or DPC.

Finally, we consider the unit consistency axiom. It allows poverty values to vary when dimensional units change, provided that poverty orderings are not affected.

Formally,

Unit Consistency (UCO): *For any* $X^1, X^2 \in M$ *and two given threshold vectors* $z^1, z^2 \in Z$, *if* $P(X^1; z^1) < P(X^2; z^2)$ *then* $P(X^1\Omega; z^1\Omega) < P(X^2\Omega; z^2\Omega)$ *for all* $\Omega = diag(\omega_1, \omega_2, \ldots \omega_m), \omega_i > 0$ *for all* i.

[7] For additional discussions on this issue, see Atkinson and Bourguignon (1982), Bourguignon and Chakravarty (1999, 2003) and Chakravarty (2009). Bourguignon and Chakravarty (1999) employed this property to examine the elasticity of substitution between proportional shortfalls of attributes from respective thresholds.

Clearly, all ratio scale invariant multidimensional poverty indices are unit consistent. However, as we will see in the next section, there exist unit consistent indices that are not ratio scale invariant.

3.3 The Characterization Theorem

For an income poverty index, Chakravarty (1983a, b) and Hagenaars (1987) interpreted poverty as the fraction of welfare losses due to the existence of poverty using the utilitarian and Gini type social welfare functions. In contrast, Chakravarty (1983b), Zheng (1993) and Chakravarty and Silber (2008) regarded it as the absolute amount of welfare loss. Here we take a similar approach.

Definition *For any arbitrary* $n \in N$, $(X\,; z) \in M^n \times Z$, a poverty index is defined as:

$$P(X; z) = W(Z_{n\times m}) - W(\hat{X}) \tag{3.4}$$

where $\hat{X}$ is the censored attribute matrix corresponding to X, that is, $\hat{x}_{ij} = \min\{x_{ij}, z_j\}$, $Z_{n\times m}$ is the $n \times m$ matrix each of whose rows is z and W is any real valued social welfare function defined on the set of all censored attribute matrices.

Thus, P is the size of the welfare loss that results from shortfall of the attribute quantities of poor persons from the respective thresholds. At this stage we do not impose any restriction on W. Note that by definition P satisfies **FOC** and **NOM**.

We can now present the following theorem.

Theorem 1 *Assume that the poverty index P is of the form* (3.4). *Then P satisfies* ***CON***, ***SUD***, ***MON*** *and* ***UCO*** *if and only if it is of the form*:

$$P(X; z) = \frac{\rho}{n\prod_{j=1}^{m} z_j^{c_j-\delta}} \sum_{i=1}^{n} \left[\prod_{j=1}^{m} z_j^{c_j} - \prod_{j=1}^{m} \hat{x}_{ij}^{c_j} \right] \tag{3.5}$$

where δ is a real number and the parameters ρ and c_j have to be chosen such that $\rho c_j > 0$ for all $1 \le j \le m$. Furthermore, ***DPC*** *(**IPC**) holds if and only if $\rho c_i c_j > 0(\rho c_i c_j < 0)$, where $i \neq j = 1, 2, \ldots, m$.*

Proof For any $X \in M^n$, $\Omega = \mathrm{diag}(\omega_1, \omega_2, \ldots \omega_m)$,let $Q(X; z) = P(X\Omega, z\Omega)$. **UCO** along with **CON** implies that if $P(X^1; z^1) = P(X^2; z^2)$ then $P(X^1\Omega; z^1\Omega) = P(X^2\Omega; z^2\Omega)$, that is, $Q(X^1; z^1) = Q(X^2; z^2)$. Furthermore, $P(X^1; z^1) < P(X^2; z^2)$ implies that $Q(X^1; z^1) < Q(X^2; z^2)$. This shows that $Q(X; z)$ is an increasing function of $P(X; z)$. Hence there exists an increasing function $g_{\omega_1,\ldots,\omega_m} : \Re^1 \to \Re^1$ such that $Q(X; z) = g_{\omega_1,\ldots,\omega_m}(P(X; z))$. Define $g : \Re^m_{++} \times \Re^1 \to \Re^1$ by $g(\omega_1, \ldots \omega_m; .) = g_{\omega_1,\ldots,\omega_m}(.)$. It then follows that:

$$P(X\Omega; z\Omega) = Q(X; z) = g(\omega_1, \ldots \omega_m; P(X; z)) \tag{3.6}$$

Since by **CON**, minor changes in ω_i's generate minor changes in $P(X\Omega; z\Omega)$, it follows that g is continuous in its first m arguments. It is also increasing in the $(m+1)^{\text{th}}$argument. [See also Propositions 1 and 3 of Zheng (2007a) and Proposition 1 of Diez et al. (2008).]

Following the structure of proof of Proposition 3 of Zheng (2007a) it can be shown that $P(X\Omega; z\Omega) = \left(\prod_{j=1}^{m} \omega_j\right)^{\delta} P(X; z)$. Define $G(X; z) = P(X; z) \Big/ \prod_{j=1}^{m} z_j^{\delta}$. Subgroup decomposability of P implies that G is subgroup decomposable.

As we have noted in (3.3), by repeated application of **SUD**, we can write any poverty index $P(X; z)$ as $\frac{1}{n}\sum_{i=1}^{n} p\,(x_i; z)$, where $X \in M^n$ and p is the individual poverty function. This in turn shows that the poverty index given by (3.4) must be of the form:

$$P\,(X;\, z) = \frac{1}{n}\sum\nolimits_{i=1}^{n}\, [h(z) - h(\hat{x}_i)] \tag{3.7}$$

where $h : \Re_{++}^{m} \to \Re^1$. By **CON**, h is continuous and **MON** demands that h is non-decreasing.

Applying now Lemma 1, Theorems 3 and 4 of Tsui (1999) [see also Diez et al. (2008)] to $G(X; z)$, when $P(X; z)$ is given by (3.4), we get the following form of the poverty index:

$$P(X; z) = G(X; z)\prod_{j=1}^{m} z_j^{\delta} = \frac{\rho}{n\prod_{j=1}^{m} z_j^{c_j-\delta}} \sum_{i=1}^{n}\left[\prod_{j=1}^{m} z_j^{c_j} - \prod_{j=1}^{m} \hat{x}_{ij}^{c_j}\right] \tag{3.8}$$

It is easy to see that for **MON** to hold we require $\rho c_j > 0$ for all $j = 1, 2, \ldots, m$. Next, we can check easily that for **DPC** (**IPC**) to hold it is necessary that $\rho c_i c_j > 0\left(\rho c_i c_j < 0\right)$, where $i \neq j = 1, 2, \ldots, m$. This establishes the necessity part of the theorem. The sufficiency is easy to check.

In the general case the restrictions on the parameters ρ and c_j for **TRP** to be fulfilled are quite complicated. However, it may be worthwhile to discuss the case for $m = 2$. The parametric restrictions that are required for fulfillment of all the postulates in this case are specified in the following corollary, which is easy to demonstrate.

Corollary 1 *Assume that the poverty index P is of the form* (3.4). *Then for* $m = 2$, *P satisfies **CON**, **SUD**, **MON** and **UCO** if and only if it is of the following form*:

$$P(X; z) = \frac{\rho}{n\prod_{j=1}^{2} z_j^{c_j-\delta}} \sum_{i=1}^{n}\left[\prod_{j=1}^{2} z_j^{c_j} - \prod_{j=1}^{2} \hat{x}_{ij}^{c_j}\right] \tag{3.9}$$

where $\rho c_j > 0$ *for* $j = 1, 2$. *Furthermore, **DPC** (**IPC**) holds if and only if* $\rho c_i c_j > 0\left(\rho c_i c_j < 0\right)$, *where* $i \neq j = 1, 2$. *Finally, for **TRP** to hold it is necessary and sufficient that* $\rho c_1(c_1 - 1) < 0$ *and* $c_1 c_2(1 - c_1 - c_2) < 0$.

When the two attributes are complements, without loss of generality we may choose $\rho = 1$. Then for any choice of $c_1, c_2 \in (0.5,\ 1)$ all the conditions stipulated in Corollary 1 are satisfied. In contrast, if the two attributes are substitutes, we can set $\rho = -1$. Then for any choice of $c_1, c_2 \in (-1, -5)$, all of our required conditions are fulfilled. To understand this, observe that when $\rho = -1$, for IPC to hold we need $c_1, c_2 > 0$. The two TRP conditions are now $c_1(c_1 - 1) > 0$ (since $\rho = -1$) and $c_1c_2(1 - c_1 - c_2) < 0$. These three inequalities are satisfied simultaneously for any choice of $c_1, c_2 \in (-1, -5)$. Note that for negative values of c_1 and c_2, the third bracketed term in (3.9) becomes non-positive.

Note that for any arbitrary number of dimensions the index satisfies **NDS** if and only if $\delta \geq 0$. **NPG** is obviously satisfied in the general case. Given c_j, P in (3.8) satisfies **SCI** if $\delta = 0$. However, unit consistency is satisfied for all real values of δ. Therefore (3.8) gives us a large class of unit consistent multidimensional poverty indices. However, the family given by (3.5) does not satisfy translation invariance for any appropriate choice of δ. and $c_j, 1 \leq j \leq m$, that is, it does not remain invariant if the attribute quantity in any dimension and the threshold limit in the corresponding dimension change by an absolute amount. Given $\rho > 0$, even in the simple case when $\delta = c_j = 1$, where $1 \leq j \leq m$, its failure to satisfy translation invariance can be checked easily. (This particular case is not appealing because it does not meet **TRP**.) The reason behind this is that in the functional representation given by (3.5) the products of threshold limits and consumption levels of different attributes appear separately.

The index given by (3.8) has close similarity with an index suggested by Diez et al. (2008) (their Eq. (9)). However, they assumed only **IPC**, which in turn means that the scale parameter ρ is negative. As we have noted in our framework ρ is positive or negative depending on whether **DPC** or **IPC** holds. Another important difference is that while our index determines the size of welfare loss under a specific structure, they do not look at their index from such a perspective.

For $m = 1$, P in (3.8) becomes the first member of the Dalton-Hagenaars class of unit consistent income poverty indices characterized by Zheng (2007b). Therefore, our index is a generalization of a single dimensional unit consistent poverty index suggested by Zheng (2007b) to the multi-attribute set up. The Zheng index itself is a two-parameter generalization of the Chakravarty (1983a) and the Clark et al. (1981) indices.

3.4 An Empirical Illustration

The purpose of this section is to illustrate the poverty index $P(X; z)$, as defined in (3.9), using the Household Income and Consumption Expenditure survey on Turkey.

We base our analysis on 3 years of available data: 2003, 2004 and 2005. The survey is collected by the State Institute of Statistics (SIS) of Turkey to provide information on household income, socio-economic status and consumption patterns of the population.

Table 3.1 Multidimensional poverty, as measured by $P(X;z)$, for different values of δ: results for the Turkey

Year	$P(X;z)$ with $\delta = 0$	$P(X;z)$ with $\delta = 0.5$	$P(X;z)$ with $\delta = 1$	$P(X;z)$ with $\delta = 1.5$	$P(X;z)$ with $\delta = 2$
2003	0.019	0.040	0.082	0.170	0.353
2004	0.016	0.034	0.070	0.146	0.303
2005	0.015	0.032	0.066	0.136	0.283

Table 3.2 Percentage contributions to multidimensional poverty by geographic area groups

Year	% Contribution of rural areas	% Contribution of urban areas	% Population of rural areas	% Population of urban areas
2003	61.303	38.697	39.24	60.76
2004	66.972	33.028	38.62	61.38
2005	57.031	42.987	38.12	61.88

Only a few of the survey's variables on individual well-being are quantitative in nature. This situation is common to many other surveys, see for example the European Community Household Panel for EU countries or the more recent EU Statistics on Income and Living Conditions, where most of the variables that could be used to measure multidimensional poverty are of qualitative type.

We decided to focus on income and number of rooms in the house. The income variable we analyze is total disposable household income equivalized using the OECD equivalence scale. In the other dimension all individuals without at least one room in the dwellings were regarded as poor. As far as income in concerned, the poverty threshold level was set equal to $4.3 in purchasing power parities per day, the threshold considered by the World Bank for medium-income countries. Purchasing power parities are from the OECD (http:/www.oecd.org/std/ppp/). We subdivide the population by the geographic areas of residence, namely, rural and urban areas. Given complementarity between the attributes, we choose $\rho = 1$. Further, in order to assign equal weight to the two dimensions we assume that $c_1 = c_2 = 0.6$. Finally, values of δ chosen to provide the estimates are: $\delta = 0,\ 0.5,\ 1,\ 1.5,\ 2$. Sample weights were used in the computation of the indices.

Results for the entire country are contained in Table 3.1. Multidimensional poverty, as measured by the index we propose in this paper increased continuously over time. We note monotonicity of the index value with respect to the parameter δ. Contributions of the rural and urban areas to total poverty, which are given as percentages of total poverty, are reported in Table 3.2. Complete elimination of poverty within an area will lower global poverty exactly by the percentage by which it contributes to total poverty. The picture is quite dismal for the rural area, for all values of δ more than 57 % contribution comes from this area. In Tables 3.3 and 3.4 area-wise poverty values are presented and, as expected, the rural area turns out to be more poverty stricken than the urban area. Thus, from the poverty reduction policy point of view the rural area deserves more attention than the urban area.

Table 3.3 Multidimensional poverty, as measured by $P(X;z)$, for different values of δ: results for the rural area

Year	$P(X;z)$ with $\delta = 0$	$P(X;z)$ with $\delta = 0.5$	$P(X;z)$ with $\delta = 1$	$P(X;z)$ with $\delta = 1.5$	$P(X;z)$ with $\delta = 2$
2003	0.012	0.234	0.050	0.104	0.216
2004	0.011	0.023	0.047	0.098	0.203
2005	0.009	0.018	0.038	0.078	0.161

Table 3.4 Multidimensional poverty, as measured by $P(X;z)$, for different values of δ: results for the urban area

Year	$P(X;z)$ with $\delta = 0$	$P(X;z)$ with $\delta = 0.5$	$P(X;z)$ with $\delta = 1$	$P(X;z)$ with $\delta = 1.5$	$P(X;z)$ with $\delta = 2$
2003	0.007	0.015	0.032	0.066	0.136
2004	0.005	0.011	0.023	0.048	0.100
2005	0.007	0.014	0.028	0.059	0.122

3.5 Conclusions

We have argued explicitly why poverty should be regarded as the failure to reach 'minimally acceptable' levels of functionings of well-being. That is, poverty should be measured in a multidimensional framework in terms of individual deprivations for different functionings from respective threshold levels. We then discussed certain desirable postulates for an indicator of poverty in such a framework.

Unit consistency is an important property of multidimensional inequality and poverty indices in the sense that use of different measurement units should not lead to inconsistent conclusions. We have made an attempt to characterize a class of unit consistent multidimensional poverty index under a specific structure. It is a generalization of one of the Zheng (2007b) single dimensional unit consistent poverty indices. An empirical illustration of the index using Turkish data was also provided.

Acknowledgments This paper was prepared for the workshop on Multidimensional Poverty and Pro-poor Growth in the MENA Countries, Nice, June 11–12, 2009. We thank participants for suggestions and the organizers, Valérie Bérenger and Florent Bresson for comments and for providing the dataset.

References

Alkire, S., & Foster, J. E. (2007). *Counting and multidimensional poverty measurement*. OPHI Working Paper.

Anand, S. (1997). Aspects of poverty in Malaysia. *Review of Income and Wealth, 23*, 1–16.

Atkinson, A. B. (1998). *Social exclusion, poverty and unemployment*. London: London School of Economics.

Atkinson, A. B., & Bourguignon, F. (1982). The comparison of multidimensioned distributions of economic status. *Review of Economic Studies, 49*, 183–201.

Blackorby, C., & Donaldson, D. (1980). Ethical indices for the measurement of poverty. *Econometrica, 58*, 1053–1060.

Bourguignon, F., & Chakravarty, S. R. (1999). A family of multidimensional poverty measures. In D. J. Slottjee (Ed.), *Advances in econometrics, income distribution and scientific methodology: Essays in honor of C. Dagum*. Berlin: Physica Verlag.

Bourguignon, F., & Chakravarty, S. R. (2003). The measurement of multidimensional poverty. *Journal of Economic Inequality, 1*, 25–49.

Bourguignon, F., & Chakravarty, S. R. (2008). Multidimensional poverty orderings theory and applications. In K. Basu & R. Kanbur (Eds.), *Welfare, development, philosophy and social science: Essays for Amartya Sen's 75th birthday (Volume 2: Development Economics and Policy)*. Oxford: Clarendon Press.

Bourguignon, F., & Fields, G. (1997). Discontinuous losses from poverty, generalized P_{∞} measures and optimal transfers to the poor. *Journal of Public Economics, 63*, 155–175.

Chakravarty, S. R. (1983a). A new index of poverty. *Mathematical Social Sciences, 6*, 307–313.

Chakravarty, S. R. (1983b). Ethically flexible measures of poverty. *Canadian Journal of Economics, 16*, 74–85.

Chakravarty, S. R. (1990). *Ethical social index numbers*. New York: Springer.

Chakravarty, S. R. (1997). On Shorrocks' reinvestigation of the Sen poverty index. *Econometrica, 65*, 1241–1242.

Chakravarty, S. R. (2009). *Inequality, polarization and poverty: Advances in distributional analysis*. New York: Springer.

Chakravarty, S. R., & Silber, J. (2008). Measuring multidimensional poverty: The axiomatic approach. In N. Kakwani & J. Silber (Eds.), *Quantitative approaches to multidimensional poverty measurement*. London: Palgrave MacMillan.

Chakravarty, S. R., Mukherjee, D., & Ranade, R. (1998). On the family of subgroup and factor decomposable measures of multidimensional poverty. *Research on Economic Inequality, 8*, 175–194.

Clark, S., Hamming, R., & Ulph, D. (1981). On indices for the measurement of poverty. *Economic Journal, 91*, 515–526.

Cowell, F. A. (1988). Poverty measures, inequality and decomposability. In D. Bos, M. Rose, & C. Seidl (Eds.), *Welfare and efficiency in public economics*. New York: Springer.

Diez, H., Lasso de la Vega, M. C., & Urrutia, A. M. (2008). Multidimensional unit- and subgroup consistent inequality and poverty measures: some characterization results. *Research on Economic Inequality, 16*, 189–211.

Donaldson, D., & Weymark, J. A. (1986). Properties of fixed-population poverty indices. *International Economic Review, 27*, 667–688.

Foster, J. E., & Shorrocks, A. F. (1991). Subgroup consistent poverty indices. *Econometrica, 59*, 687–709.

Foster, J. E., Greer, J., & Thorbecke, E. (1984). A class of decomposable poverty measures. *Econometrica, 42*, 761–766.

Hagenaars, A. (1987). A class of poverty indices. *International Economic Review, 28*, 583–607.

Kakwani, N. C. (1980). On a class of poverty measures. *Econometrica, 48*, 437–446.

Kolm, S. C. (1977). Mulditimensional egalitarianisms. *Quarterly Journal of Economics, 91*, 1–13.

Ravallion, M. (1996). Issues in measuring and modeling poverty. *Economic Journal, 106*, 1328–1343.

Savaglio, E. (2006). Three approaches to the analysis of multidimensional inequality. In F. Farina & E. Savaglio (Eds.), *Inequality and economic integration*. London: Routledge.

Seidl, C. (1988). Poverty measurement: A survey. In D. Bos, M. Rose & C. Seidl (Eds.), *Welfare and efficiency in public economics*. New York: Springer.

Sen, A. K. (1976). Poverty an ordinal approach to measurement. *Econometrica, 44*, 219–231.
Sen, A. K. (1981). *Poverty and famines*. Oxford: Clarendon Press.
Sen, A. K. (1985). *Commodities and capabilities*. Amsterdam: North Holland.
Sen, A. K. (1992). *Inequality re-examined*. Cambridge, MA: Harvard University Press.
Sen, A. K. (2002). *Rationality and freedom*. Cambridge: Harvard University Press.
Shorrocks, A. F. (1995). Revisiting the Sen poverty index. *Econometrica, 63*, 1225–1230.
Streeten, P. (1981). *First things first: Meeting basic human needs in developing countries*. New York: Oxford University Press.
Takayama, N. (1979). Poverty income inequality and their measures: Professor Sen's axiomatic approach reconsidered. *Econometrica, 47*, 749–759.
Thon, D. (1983). A poverty measure. *Indian Economic Journal, 30*, 55–70.
Townsend, P. (1979). *Poverty in the United Kingdom*. Harmondsworth: Penguin.
Tsui, K. Y. (1999). Multidimensional inequality and multidimensional generalized entropy measures: An axiomatic derivation. *Social Choice and Welfare, 16*, 145–157.
Tsui, K. Y. (2002). Multidimensional poverty indices. *Social Choice and Welfare, 19*, 69–93.
Weymark, J. A. (2006). The normative approaches to the measurement of multidimensional inequality. In F. Farina & E. Savaglio (Eds.), *Inequality and economic integration*. London: Routledge.
Zheng, B. (1993). An axiomatic characterization of the Watts poverty measure. *Economics Letters, 42*, 81–86.
Zheng, B. (1997). Aggregate poverty measures. *Journal of Economic Surveys, 11*, 123–162.
Zheng, B. (2007a). Unit-consistent decomposable inequality measures. *Economica, 74*, 97–111.
Zheng, B. (2007b). Unit-consistent poverty indices. *Economic Theory, 31*, 113–142.

Chapter 4
Multidimensional and Fuzzy Measures of Poverty and Inequality in Turkey at National and Regional Level

Gianni Betti, Güllü Çalik and Murat Karakaş

Abstract This chapter provides a step-by-step account of how fuzzy measures of both non-monetary deprivation and monetary poverty can be constructed using survey data such as that from the EU-SILC (European Union-Statistics on Income and Living Conditions). In the case of non-monetary deprivation, meaning dimensions or groupings of initial items of deprivation are identified using explanatory and confirmatory factor analyses, and a weighting system is applied to ensure the aggregation of individual items into the dimension they represent. An application of the proposed methodology is conducted using EU-SILC 2007 data from Turkey: estimates are provided at national level and also disaggregated at regional (NUTS1) level.

4.1 Introduction

In order to understand poverty and social exclusion, it is necessary to consider the multiple dimensions of deprivation simultaneously—low income as well as different non-monetary aspects of deprivation. Using fuzzy set representations of different individual risks accruing from poverty, this chapter proposes a multidimensional

G. Betti (✉)
Department of Economics and Statistics,
University of Siena, Siena, Italy
e-mail: gianni.betti@unisi.it

G. Çalik · M. Karakaş
TURKSTAT, Ankara, Turkey
e-mail: gullu.calik@tuik.gov.tr

M. Karakaş
e-mail: murat.karakas@tuik.gov.tr

V. Bérenger and F. Bresson (eds.), *Poverty and Social Exclusion Around the Mediterranean Sea*, Economic Studies in Inequality, Social Exclusion and Well-Being 9,
DOI: 10.1007/978-1-4614-5263-8_4,

methodology for poverty and deprivation which can be applied in a multicountry comparative context.

The need to adopt a multidimensional approach has been noted, among others, by Kolm (1977), Atkinson and Bourguignon (1982), Tsui (1985), Maasoumi (1986), Sen (1999). In the literature, the many attempts to study the poverty phenomenon from a multidimensional perspective appear to fall into two categories: non-axiomatic versus axiomatic approaches. A non-axiomatic approach implicitly considers indicators to be perfectly substitutable. In this approach, different indicators are combined in order to obtain a multidimensional index; this combination can be achieved at macro level (Anand and Sen 1997) or at individual level (Smeeding et al. 1993; Townsend 1979; Klasen 2000). By contrast, Bourguignon and Chakravarty (2003) and Chakravarty et al. (1998) have recently contributed to the development of an axiomatic approach.

In this chapter, we go beyond the conventional study of poverty based simply on the poor/non-poor dichotomy as defined in relation to some chosen poverty line. Instead, poverty and multidimensional deprivation are treated as matters of degree, determined according to an individual's position in the distribution of income and other aspects of their living conditions. The state of deprivation is thus seen in the form of 'fuzzy sets' to which all members of the population belong, but to varying degrees. In this way, we are able to clarify and propose a solution for one of the problems raised in the literature (Atkinson et al. 2002; Duclos et al. 2001; and especially Atkinson 2003): 'How can different attributes be aggregated? [...] A distinction may be drawn between those who adopt a union approach and those who use an intersection measure'. This is due to the fact that the fuzzy supplementary measure proposed in Sect. 4.3 is decomposable by dimension.

A number of authors have evoked the concepts of fuzzy sets in the analysis of poverty and living conditions, for instance (Chiappero 1994), Vero and Werquin (1997). In more specific terms, the present contribution represents a continuation and further development of the work of Cerioli and Zani (1990), Cheli and Lemmi (1995), Cheli (1995), Betti and Verma (1999) and Betti et al. (2006).

Moreover, the multidimensional nature of poverty is a widely recognised fact, not only by the international scientific community, but also by many official statistical agencies (e.g. Eurostat, Istat) as well as by international institutions (United Nations, World Bank). This fact implies a more complete and realistic vision of the phenomenon and also an increased complexity at both the conceptual and analytical levels. Such a complexity determines the need for adequate tools of analysis and the availability of statistical data that also have to be adequate, complete and reliable.

This chapter applies the proposed methodology to the data from the European Survey on Income and Living Conditions (EU-SILC) collected in Turkey in 2007. This is a particularly interesting data set to be analysed because, in contrast to the other Euro–Mediterranean countries, the SILC includes Turkey as a candidate member of the European Union. Collecting data on multidimensional aspects of living conditions is the result of the commitment undertaken by the European Commission in the 2000 Lisbon Summit to promote social inclusion. The list from Laeken and its subsequent refinements has been successful in grouping national diversities into a set of common indicators for monitoring social performance while implying different policy strategies.

Moreover, Turkey, with its peculiar socio–economic characteristics, is a very interesting country for analysis; it is an upper-middle-income country and a member of the OECD. For most of the 2000s, Turkey has experienced high growth rates, stable macroeconomic conditions (decreasing inflation and interest rates) and high levels of direct foreign investment. Turkey is the 16th largest economy in the world and the 6th largest economy relative to the 27 EU countries in 2010. GDP and GDP per capita figures highlight the strength and stability of the national economy as well as its integration into macroeconomic global trends. In the last six years, the Turkish economy has sustained a steady real GDP growth rate of approximately 6 %. The inflation rate decreased to 6.4 % at the end of 2010 (in 2002, it was 30 %).

Despite some positive economic developments, the economy's ability to generate jobs has been limited. Unemployment has been the most visible and perplexing impact of the economic crisis in the country. Unemployment rates stood at 6.5, 10.6 and 10.3 %, respectively, in 2002, 2005 and 2007. Afterwards, there was an increase in these rates to: 11 % in 2008, 14 % in 2009 and 11.9 % in 2010.

With regard to poverty, Turkey's population is slightly less at risk from food poverty, which was 1.35 % in 2002 and 0.48 % in 2009 according to the Household Budget Survey results. By using Income and Living Conditions Survey results, which were first applied in 2006, the poverty rates were 25.4 % in 2006 and 23.5 % in 2007.

The present chapter is organised in the following way. After the Introduction, Sect. 4.2 introduces the theoretical background to poverty measurement and proposes new multidimensional and fuzzy measures (this represents the original nature of the chapter's contribution in terms of theory). Section. 4.3 presents the ES–SILC and shows how the new multidimensional fuzzy approach has been implemented step-by-step. Section 4.4 presents the main results of the methodology as applied to the 2007 wave of SILC data for Turkey, at both country and NUTS1 level; since the study is based on a single cross-section, the largest part of the empirical work is dedicated to regional comparisons. Finally, Sect. 4.5 concludes the chapter.

4.2 Theoretical Background to Poverty Measurement and New Proposals

In the so-called traditional approach, poverty is characterised by a simple dichotomisation of the population into poor and non-poor defined in relation to some chosen poverty line that represents a certain percentage (generally 50 %, 60 % or 70 %) of the mean or the median of the equivalised income[1] distribution.

[1] The equivalised income of a household is obtained by dividing its total disposable income by the household's equivalent size computed using an equivalence scale which takes into account the actual size and composition of the household.

This approach presents two main limitations: first, it is unidimensional, i.e. it refers to only one proxy for poverty, namely low income or consumption expenditure, and secondly it divides the population into a simple dichotomy. However, poverty is a complex phenomenon that cannot be reduced solely to a monetary dimension, but it must also take into account non-monetary indicators of living conditions; moreover, it is not an attribute that characterises an individual in terms of presence or absence, rather a vague predicate that manifests itself in different shades and degrees.

The fuzzy approach considers poverty as a matter of degree rather than an attribute that is simply present or absent for individuals in the population. In this case, two additional aspects have to be introduced:

i. The choice of *membership functions* (m.f.), i.e. the quantitative specification of individuals' or households' degrees of poverty and deprivation;
ii. The choice of rules for the manipulation of the resulting fuzzy sets.

The traditional approach can be seen as a special case of the fuzzy approach, where the membership function may be considered as $\mu_i^H = 1$ if $y_i < z$, $\mu_i^H = 0$ if $y_i \geq z$, where y_i is the equivalised income of individual i and z is the poverty line.

An early attempt to incorporate the concept of poverty as a matter of degree at methodological level was made by Cerioli and Zani (1990) who drew inspiration from the theory of *Fuzzy Sets* initiated by Zadeh (1965). They proposed the introduction of a transition zone $(z_1; z_2)$ between the two states, a zone over which the m.f. declines from 1 to 0 linearly:

$$\mu_i^{CZ} = \begin{cases} 1 & \text{if } y_i \leq z_1 \\ \frac{z_2 - y_i}{z_2 - z_1} & \text{if } z_1 < y_i < z_2 \\ 0 & \text{if } y_i \geq z_2 \end{cases} \tag{4.1}$$

Subsequently, Cheli and Lemmi (1995) proposed the so-called *Totally Fuzzy and Relative* (TFR) approach in which the m.f. is defined as the distribution function $F(y_i)$ of income, and normalised (linearly transformed) so as to equal 1 for the poorest and 0 for the richest person in the population. In order to make this mean equal to some specified value (for instance, 0.1 = 10 % of poor individuals) and therefore facilitate comparison with the conventional poverty rate, Cheli (1995) takes the m.f. as a normalised distribution function, raised to some power $\alpha \geq 1$. Given that w_γ and y_γ are respectively the sample weight and the equivalised income of individual of rank γ in the ascending income distribution, we formally define:

$$\mu_i^{CL} = (1 - F_{(M),i})^\alpha = \left(\frac{\sum_\gamma w_\gamma | y_\gamma > y_i}{\sum_\gamma w_\gamma | y_\gamma > y_1} \right)^\alpha, \quad i = 1, 2, \ldots, n; \ \mu_n = 0 \tag{4.2}$$

where y_i is the equivalised income of the generic ith individual, $F_{(M),i}$ is the value of the income distribution function $F(y_i)$ for the ith individual, $(1 - F_{(M),i})$ is the proportion of individuals who are poorer than the ith individual with a mean of ½ by definition, and α is a parameter. The value of α is arbitrary, but Cheli and Betti (1999) chose the parameter α so that the mean of the m.f. was equal to the Head Count Ratio (HCR) computed for the official poverty line. Increasing the value of this exponent implies giving more weight to the poorest end of the income distribution.

Betti and Verma (1999) used a somewhat refined version of expression (4.2) in order to define what they refer to as a Fuzzy Monetary indicator (FM):

$$\mu_i^{BV} = \mathrm{FM}_i^{BV} = \left(1 - L_{(M)i}\right)^{\alpha} = \left(\frac{\sum\limits_{\gamma} w_\gamma y_\gamma | y_\gamma > y_i}{\sum_\gamma w_\gamma y_\gamma | y_\gamma > y_1}\right)^{\alpha}, \quad i = 1, 2, \ldots, n; \quad \mu_n = 0 \tag{4.3}$$

where y_γ is the equivalised income and $L_{(M),i}$ represents the value of the Lorenz curve of income distribution for individual i; then $1 - L_{(M),i}$ represents the share of the total equivalised income received by all individuals who are poorer than a particular person. It varies from 1 for the poorest individual to 0 for the richest. The mean of the $1 - L_{(M),i}$ values equals (1 + G)/2, where G is the Gini coefficient of the distribution.

4.2.1 Proposal of a New Fuzzy Monetary indicator

In this chapter, we first aim to improve the definition of the FM indicator. The proposed FM indicator is defined as a combination of the $\left(1 - F_{(M),i}\right)$ indicator, i.e. the proportion of individuals who are poorer than a particular person, proposed by Cheli and Lemmi (1995), and of the $\left(1 - L_{(M),i}\right)$ indicator, i.e. the share of the total equivalised income received by all individuals who are poorer than a particular person, proposed by Betti and Verma (1999). Formally:

$$\begin{aligned}\mu_i = \mathrm{FM}_i &= \left(1 - F_{(M),i}\right)^{\alpha-1}\left(1 - L_{(M),i}\right) \\ &= \left(\frac{\sum_{\gamma=i+1}^{n} w_\gamma | y_\gamma > y_i}{\sum_{\gamma=2}^{n} w_\gamma | y_\gamma > y_1}\right)^{\alpha-1}\left(\frac{\sum_{\gamma=i+1}^{n} w_\gamma y_\gamma | y_\gamma > y_i}{\sum_{\gamma=2}^{n} w_\gamma y_\gamma | y_\gamma > y_1}\right)\end{aligned} \tag{4.4}$$

where y_γ is the equivalised income, $F_{(M),i}$ is the income distribution function, w_γ is the sample weight of individual of rank γ ($\gamma = 1, \ldots, n$) in the ascending income distribution, $L_{(M),i}$ represents the value of the Lorenz curve of income distribution for individual i. The parameter α is estimated so that the overall FM indicator (which is calculated simply as the weighted mean of the individual FM_i), is equal to the HCR computed for the official poverty line (60 % of the median).

4.2.2 Proposal of a New Fuzzy Supplementary indicator

In addition to the level of monetary income, the standard of living of households and individuals can be described by a host of indicators, such as housing conditions, possession of durable goods, perception of hardship, expectations, norms and values.

Several steps are necessary in order to quantify and assemble diverse indicators of deprivation. In particular, decisions are required for assigning numerical values to the ordered categories, weighting the score to construct composite indicators, choosing their appropriate distributional form and scaling the resulting measures in a meaningful way.

First, from the large set of indicators which may be available, a selection has to be made of indicators which are substantively meaningful and useful for a given analysis. Secondly, it is helpful to identify the underlying dimensions and to group the indicators, accordingly (these steps will be described in detail in the next sections).

As the first step in an analysis of lifestyle deprivation, Whelan et al. (2001) suggest systematically examining the range of deprivation items to see whether the items cluster into distinct groups. Factor analysis can be used to identify such clusters of interrelated variables. Several steps are necessary in order to quantify and put together diverse indicators:

1. Identification of items;
2. Transformation of the items into the [0, 1] interval;
3. Exploratory and confirmatory factor analyses;
4. Calculation of weights within each dimension (each group);
5. Calculation of scores for each dimension;
6. Calculation of an overall score and the parameter α;
7. Construction of a fuzzy deprivation measure for each dimension (and an overall measure).

Aggregation over a group of items in a particular dimension h ($h = 1, 2, \ldots, m$) is given by a weighted mean taken over j items: $s_{hi} = \sum w_{hj} \cdot s_{hj,i} / \sum w_{hj}$, where w_{hj} is the weight of the jth deprivation variable in the hth dimension.

An overall score for the ith individual is calculated as the unweighted mean:

$$s_i = \frac{\sum_{h=1}^{m} s_{hi}}{m} \tag{4.5}$$

Then we calculate the FS indicator for the ith individual over all dimensions as:

$$\mathrm{FS}_i = \left(1 - F_{(S),i}\right)^{\alpha-1}\left(1 - L_{(S),i}\right) \tag{4.6}$$

As for the FM indicator, the parameter α is determined so as to make the overall non-monetary deprivation rate (which is calculated simply as the weighted mean of the individual FS_i) numerically identical to the HCR computed for the official poverty line (60 % of the median). The estimated parameter α is used to calculate the FS indicator for each dimension of deprivation separately.

The FS indicator for the hth deprivation dimension for the ith individual is defined as a combination of the $\left(1 - F_{(S),hi}\right)$ indicator and the $\left(1 - L_{(S),hi}\right)$ indicator.

$$\begin{aligned}\mu_{hi} =& \mathrm{FS}_{hi} = \left(1 - F_{(S),hi}\right)^{\alpha-1}\left(1 - L_{(S),hi}\right)\\ =& \left[\frac{\sum_{\gamma=i+1}^{n} w_{h\gamma}|s_{h\gamma} > s_{hi}}{\sum_{\gamma=2}^{n} w_{h\gamma}|s_{h\gamma} > s_{h1}}\right]^{\alpha-1}\left[\frac{\sum_{\gamma=i+1}^{n} w_{h\gamma}s_{h\gamma}|s_{h\gamma} > s_{hi}}{\sum_{\gamma=2}^{n} w_{h\gamma}s_{h\gamma}|s_{h\gamma} > s_{h1}}\right],\\ h =& 1, 2, \ldots, m; \quad i = 1, 2, \ldots, n; \quad \mu_{hn} = 0\end{aligned} \tag{4.7}$$

The $\left(1 - F_{(S),hi}\right)$ indicator for the ith individual is the proportion of individuals who are less deprived in the hth dimension than a particular individual. $F_{(S),hi}$ is the value of the score distribution function evaluated for individual i in dimension h and $w_{h\gamma}$ is the sample weight of the ith individual of rank γ in the ascending score distribution in the hth dimension.

The $\left(1 - L_{(S),hi}\right)$ indicator is the share of the total lack of deprivation score assigned to all individuals who are less deprived than a particular person. $L_{(S),hi}$ is the value of the Lorenz curve from the score in the hth dimension for the ith individual.

As for the Fuzzy Monetary and the Fuzzy Supplementary (FS) indicators, the overall index corresponding to each dimension FS_h is calculated simply as the weighted mean of the individual FS_{hi}.

Here, it is interesting to note that the overall ranking of the *FS* indicator cannot be obtained directly from the rankings in each dimension; however, the ranking obtained from FS_i is consistent with the ranking obtained from FS_{hi}. A possible alternative way to define the overall FS_i indicator could simply be as the average of the corresponding FS_{hi} indicators; the advantage would be that the overall indicators would fulfil consistency properties with respect to decomposition (Chakravarty et al. 1998, among others); the drawback would be that the weighted mean of the individual FS_i would not be equal to either FM or HCR indicators.

4.3 EU-SILC Data Set, Identification of Items and Dimensions

In the present section, we use data from the EU-SILC, distributed by Eurostat. The EU-SILC survey was designed to collect detailed information on the income of each household member, and on various aspects of the material and demographic situation of a household. A representative random sample of households throughout the country provides the required information. Data are available at a cross-sectional level for the years 2004, 2005, 2006 and 2007. Only EU 15 countries were present for the 2004 survey round; 26 countries were surveyed for the 2005 and 2006 rounds, and 27 countries for the 2007 round. Firstly, from the large set of EU-SILC variables, a selection was made of indicators which are substantively meaningful and useful for the construction of FS indicators. For our purpose, we have identified a set of items which could serve as indicators of the concept of lifestyle deprivation. All these items are considered at household level, even if some of them are taken from the individual data set and then aggregated to household level.

The first set of items concerns the lack of possession of a widely desired item. These are: telephones including mobile phones (a1); colour TVs (a2); computers (a3); washing machines (a4); cars (a5). In all these cases, we consider a household to be deprived only if the lack of the item is 'enforced', in the sense that the household would like to have the item but cannot afford it.

A second set of items relates to the lack of ability to afford items that are considered as basic: keeping the home adequately warm (b1); paying for a 1-week annual holiday away from home (b2); eating a meal with meat, chicken, fish (or vegetarian equivalent) every second day (b3); being able to meet unexpected financial expenses (b4).

A third set relates to the absence of housing facilities; this set is considered so basic that one can presume all households wish to have them: bath or shower in the dwelling (c1); indoor flushing toilet for sole use of the household (c2).

The fourth set of items relates to problems with accommodation and the environment, with the implicit assumption that households wish to avoid such difficulties: leaking roof, damp walls/floors/foundation or rot in window frames or floor (d1); too dark, not enough light in the dwelling (d2); noise from neighbours or from the street (d3); pollution, grime or other environmental problems (d4); crime violence or vandalism in the area (d5).

The fifth set relates to financial problems and arrears in paying bills that a household has experienced in the last 12 months: arrears on mortgage or rent payments (e1); arrears on utility bills (e2); arrears on hire purchase instalments or other loan payments (e3); the household's ability to make ends meet (e4).

The sixth set relates to the level of health of household members. These items are from individual variables that have been aggregated at household level. We have taken this dimension into consideration because we think that, in dealing with lifestyle deprivation, a lack of good health is also important. The items considered are: general health (f1); suffering from any chronic (long-standing) illness or condition (f2); limitation in activities because of health problems (f3); unmet need for medical examination or treatment (f4); unmet need for dental examination or treatment (f5).

The seventh set relates to education. For this set we have constructed two composite indicators: the first covers households with early school leavers not in education or training (g1), i.e. households with at least one person aged 18–24 years who has only benefitted from lower secondary education or below (ISCED level currently attained: value 2 or less), and who at the same time is not in education or training leading to a qualification to at least upper secondary level. The second covers households with persons having low educational attainment (g2), i.e. households with at least one person aged 25–64 years who has only benefitted from a lower secondary education or below.

The last dimension concerns the labour market. We have also constructed two composite indicators for this set: jobless households (h1); intensity or duration of unemployment at household level (h2). See Appendix for details on the construction of these two indicators.

4.3.1 Transformation of the Items into the [0, 1] Interval

When the item consists of a fixed number of categories, it is then transformed using the procedure as set out below. For each item we determine a deprivation score as follows:

$$d_{hj,i} = \frac{1 - F(c_{hj,i})}{1 - F(1)}; \quad h = 1, 2, \ldots, m; \quad j = 1, 2, \ldots, k; \quad i = 1, 2, \ldots, n \quad (4.8)$$

where $c_{hj,i}$ is the value of the category of the *j*th item for the *i*th individual and $F(c_{hj,i})$ is the value of the *j*th item cumulative distribution function for the *i*th individual. We transform the deprivation score into a positive score as follows:

$$s_{hj,i} = 1 - \frac{1 - F(c_{hj,i})}{1 - F(1)} = \frac{F(c_{hj,i}) - F(1)}{1 - F(1)}; \quad h = 1, 2, \ldots, m; \quad j = 1, 2, \ldots, k; \quad i = 1, 2, \ldots, n \quad (4.9)$$

In both the special and the common cases, where the variable is a dichotomous variable, the deprivation index d is 1 for deprivation and 0 for otherwise; while the positive score s is 0 for deprivation and 1 for otherwise. For a polychotomous item, instead of the real value of the category, we assign each household a value corresponding to the percentage of households that are 'better off' than that household.

In the few cases in which the indicator is a composite one (a set of dichotomies indicating the presence or absence of an experience by household members), the score s represents the proportion of people in the household that experienced it.

4.3.2 Factor Analysis

In order to investigate lifestyle deprivation, we have followed the procedure from Economic and Social Research Ireland (ESRI), as described in Whelan et al. (2001).

In proceeding to construct a summary index of deprivation through the use of different items, we begin by identifying and investigating the latent dimension of deprivation. By 'dimension' we mean a distinct group of individual items of deprivation. Exploratory and confirmatory factor analyses allow us to achieve this objective. In the literature on multidimensional poverty analysis, these approaches usually deal with the problems of measurement errors, of redundancy (and double counting attribute) and of the arbitrariness of assigning weights to indicators. We mainly adopt factor analysis for reaching the first two goals, while we prefer to construct an ad hoc weighting system as described and justified in Sect. 4.3, the next section.

The procedure begins with an exploratory factor analysis to give a preliminary framework of the dimensions.[2] In order to obtain a unique set of dimensions, the analysis has been conducted using the whole EU-SILC data set at European level; this may enable researchers to compare results among countries. We then proceed to rearrange some factors in the dimensions found so as to create more meaningful groups. Finally, we conduct a confirmatory factor analysis to test the goodness of the model hypothesised.

The confirmatory analysis was conducted on the dimensions reported in Table 4.1 in order to achieve substantially more meaningful groupings. The results of the analysis are very good; in fact, all the indicators of goodness of the model are significant.[3]

In conclusion, seven dimensions of well-being were finally obtained from the eight sets of original variables.

4.3.3 Calculation of Weights Within Each Dimension

The weights given to items are determined separately within each dimension and the set of weights is taken to be item specific, i.e. for a given item, weights are common to all individuals in the population.

In multidimensional analyses, there are several methods that can be used to calculate weights for aggregating different items in all the dimensions. Filippone et al. (2001) report a very detailed overview of the most adopted approaches,

---

[2] The exploratory factor analysis identified nine dimensions as follows: items b1, b2, b3, b4 and e4 in dimension 1; items f1, f2 and f3 in dimension 2; items d3, d4 and d5 in dimension 3; items c1 and c2 in dimension 4, items a1, a2, a3, a4 and a5 in dimension 5, items g1, g2, h1 and h2 in dimension 6, items e1, e2 and e3 in dimension 7, items f4 and f5 in dimension 8, and finally items d1 and d2 in dimension 9.

[3] We have considered the Goodness-of-Fit Index (GFI), which is equal to 0.94 and the Adjusted Goodness-of-Fit Index (AGFI), which is equal to 0.93 (they both range between 0 and 1); moreover, we have estimated a value of 0.06 for the Root Mean Square Residual (RMR), which seems to indicate a sufficiently good fit. Finally, the Root Mean Squared Error of Approximation (RMSEA) equal to 0.0475 means a very good fit for the model.

Table 4.1 Dimensions after rearrangement and confirmatory factor analysis

	Dimension	Indicators	Weights
1	Basic lifestyle	Meals with meat, fish or chicken	0.1302
		Household adequately warm	0.7058
		Holiday away from home	0.0592
		Ability to make ends meet	0.1048
2	Consumer durables	Car	0.0512
		PC	0.0695
		Telephone	0.1311
		Washing Machine	0.1838
		TV	0.5644
3	Housing amenities	Bath or shower	0.3968
		Indoor flushing toilet	0.2462
		Leaking roof and damp	0.1300
		Rooms too dark	0.2269
4	Financial situation	Inability to cope with unexpected expenses	0.1058
		Arrears on mortgage or rent payments	0.5060
		Arrears on utility bills	0.1811
		Arrears on hire purchase instalments	0.2071
5	Environment	Crime, violence, vandalism	0.2938
		Pollution	0.3862
		Noise	0.3200
6	Work & Education	Early school leavers	0.4045
		Low education	0.0942
		Worklessness	0.0801
		Duration of unemployment	0.4212
7	Health related	General health	0.2037
		Chronic illness	0.1246
		Mobility restriction	0.1430
		Unmet need for medical examination	0.2149
		Unmet need for dental examination	0.3138

describing the advantages and drawbacks. In the present chapter, we have adopted the approach proposed by Betti and Verma (1999) instead of, for instance, the weights derived from a confirmatory factor analysis. There are three main reasons behind this choice: (i) they solve the problems of measurement errors and of redundancy; (ii) they give less importance to items that are not widespread among the population; (iii) they have been officially adopted by Eurostat (Eurostat 2002) in the 2nd Social Report on Income, Poverty and Social Exclusion, and therefore facilitate comparability of results between Turkey and other EU-SILC countries.

In fact, such weights comprise two factors: the dispersion of deprivation indicators and their correlation with other deprivation indicators from within a given dimension $w_{hj} = w_{hj}^a \cdot w_{hj}^b, \quad h = 1, 2 \ldots, m; \quad j = 1, 2, \ldots, k_h$, where h is a particular dimension and j is a particular deprivation indicator.

The first factor is the coefficient of variation of one minus the deprivation scores s as follows:

$$w_{hj}^{a} = \frac{\mathrm{std}_{hj}}{1 - \mathrm{mean}_{hj}} \tag{4.10}$$

The second factor, as a measure of the correlation, can be computed in the following form:

$$w_{hj}^{b} = \left(\frac{1}{1 + \sum_{j'=1}^{k_h} r_{e_{hj,hj'}} \middle| r_{e_{hj,hj'}} < r^{*}_{e_{hj}}} \right) * \left(\frac{1}{\sum_{j'=1}^{k_h} r_{e_{hj,hj'}} \middle| r_{e_{hj,hj'}} \geq r^{*}_{e_{hj}}} \right) \tag{4.11}$$

where $r_{e_{hj,hj'}}$ is the correlation coefficient between deprivation indicators corresponding to items j and j' in the h-dimension and $r^{*}_{e_{hj}}$ is the particular value determined by dividing the ordered set of correlation values at the point of the largest gap (Betti and Verma 1999).

The last column of Table 4.1 reports the results from our data for w_{hj}, where, as noted earlier, h refers to the dimension and j to a particular item of deprivation in it.

4.4 Empirical Analysis: Estimates at National and NUTS1 Levels

Our empirical analysis has been conducted on the basis of the 2007 cross-sectional wave of the EU-SILC UDB for Turkey (TurkStat 2007). This survey consists of 10,796 households and 42,454 individuals, representing the entire country and the 12 NUTS1 administrative regions in which Turkey is divided. Table 4.2 reports the fuzzy measures introduced in Sect. 4.3 for the whole of Turkey and for the 12 administrative regions. In order to calculate the (HCR and the FM indicator, we consider the distribution of household equivalised disposal income assigned to each individual. First of all, both the values of the FM and of the FS measure calculated for Turkey correspond to the official HCR (at-risk-of-poverty rate) calculated with the poverty line defined as 60 % of the total equivalised income median (23.45 %).

At country level, there are no large differences among the aggregated fuzzy measures in the seven dimensions FS1-FS7 as identified and defined in Sect. 4.2. The relevance and usefulness of such a multidimensional and fuzzy approach is highly evident when analysing disparities at regional level.

First of all, the region of Istanbul, the region of East Marmara and the region of West Anatolia (where the Turkish capital Ankara is located) are better off in terms of FM poverty. This is also clearly consistent with official results based on the traditional approach to poverty measurement adopted by the Turkish National Statistical Office (TurkStat 2007). Conversely, the two regions of Central East Anatolia and South East Anatolia seem to be worse off, with measures of the FM

Table 4.2 Multidimensional and Fuzzy Poverty Measures at National and NUTS1 Levels

	HCR	FM	FS	Basic lifestyle	Consumer durables	Housing amenities	Financial situation	Environment	Work & Education	Health related
Turkey	**0.2345**	**0.2345**	**0.2345**	**0.2441**	**0.1497**	**0.1708**	**0.1844**	**0.2056**	**0.1893**	**0.2344**
Istanbul	0.0190	0.0679	0.2005	0.2564	0.0534	0.0678	0.1848	0.3080	0.1559	0.1577
West Marmara	0.1618	0.1886	0.1808	0.2687	0.1483	0.1776	0.1882	0.1282	0.1299	0.1864
Aegean	0.1674	0.1807	0.2517	0.2666	0.1221	0.1600	0.1785	0.2598	0.1546	0.2650
East Marmara	0.0996	0.1323	0.1634	0.2407	0.0795	0.0954	0.1548	0.1879	0.1429	0.1812
West Anatolia	0.1126	0.1494	0.1741	0.2751	0.0840	0.1211	0.1565	0.1957	0.1586	0.1724
Mediterranean	0.3319	0.3072	0.2850	0.2617	0.1664	0.1975	0.2189	0.2216	0.1726	0.3023
Central Anatolia	0.2254	0.2327	0.2067	0.2765	0.1240	0.1817	0.1616	0.1234	0.2196	0.2206
West Black Sea	0.2369	0.2448	0.1367	0.1984	0.1439	0.1062	0.1312	0.0726	0.1590	0.2433
East Black Sea	0.1580	0.1939	0.1475	0.2021	0.1014	0.1082	0.1453	0.0844	0.1299	0.3046
North East Anatolia	0.3929	0.3573	0.2438	0.1966	0.2877	0.4065	0.1595	0.0550	0.2186	0.1900
Central East Anatolia	0.4985	0.4350	0.3272	0.1669	0.2985	0.3519	0.1910	0.1756	0.2653	0.3161
South East Anatolia	0.7015	0.5733	0.4120	0.2147	0.3796	0.3415	0.2634	0.2229	0.3877	0.3220
α values[a]		**3.95**	**3.47**							

[a] Note that the α value for the Fuzzy Monetary measure is more than 10 % higher than its FS counterpart; this implies that the original equivalised income distribution (y_i) is more unequal than the distribution of the overall score s_i in Eq. (4.5)

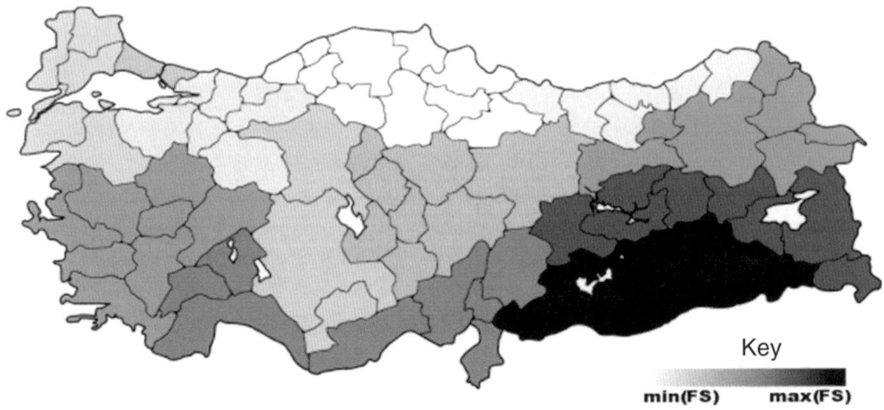

Fig. 4.1 FS poverty measure at NUTS1 level

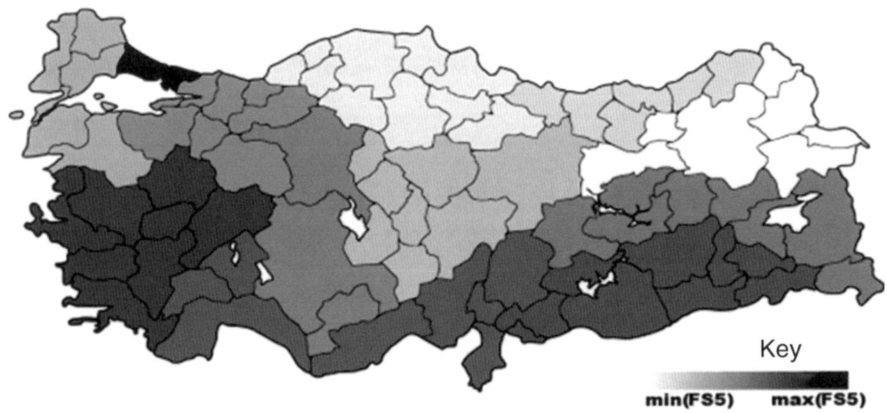

Fig. 4.2 FS5 'environmental problem' poverty measure at NUTS1 level

indicator around 0.5. This confirms the high inequality present in Turkish income distribution, both within and between regions.

On the one hand, the differences are clearly much more marked in comparison to the traditional approach, where the dichotomisation into poor and non-poor individuals allows only the *incidence* of poverty to be taken into account. On the other hand, the FM indicator is capable of revealing poverty *intensity* as well as some aspects of *inequality* (the FM indicator being a transformation of the generalised Gini index; Betti et al. 2006).

Table 4.2 shows less inequality among regions as far as the overall multidimensional and fuzzy measure FS is concerned. When analysing results for each single dimension, differences among regions are considerably marked.

The region of Istanbul performs very well as far as the 'consumer durables' and 'housing amenities' dimensions are concerned, although the indicator referring to

Table 4.3 Correlation coefficients among the ten poverty indicators

	HCR	FM	FS	Basic lifestyle	Consumer durables	Housing amenities	Financial situation	Environment	Work & Education	Health related
HCR	1	0.8860	0.3074	−0.0260	0.4800	0.4002	0.1161	−0.0832	0.2526	0.2299
FM		1	0.3590	0.0023	0.5449	0.4586	0.1578	−0.1123	0.3083	0.2656
FS			1	0.2571	0.3969	0.4204	0.5186	0.4033	0.4036	0.5728
Basic lifestyle				1	−0.0147	−0.1518	0.0640	−0.0345	0.0167	−0.0441
Consumer durables					1	0.4649	0.1397	−0.0823	0.1911	0.2265
Housing amenities						1	0.0838	−0.0214	0.1772	0.2321
Financial situation							1	0.1282	0.1193	0.1877
Environment								1	−0.0231	0.0656
Work & Education									1	0.1703
Health related										1

the 'environmental' dimension shows the worst situation in the country, highlighting problems in crime, violence, vandalism, pollution and noise, which are well above the average for the country. The region of South East Anatolia also performs below par according to the multidimensional fuzzy dimensions—it has the worst indicators for almost all the measures reported in Table 4.2.

In order to demonstrate the differences and inequalities among the 12 administrative regions, Figs. 4.1 and 4.2 below show the fuzzy poverty mapping of the overall FS measure and of the 'environmental dimension' measure (FS5).

According to the overall FS indicators (Fig. 4.1), the regions in black and dark grey show that South East Anatolia and Central East Anatolia are the worst off regions in the country. However, when taking into account the 'environmental dimension' measure (FS5 in Fig. 4.2), the region of Istanbul shows the higher (darker) deprivation measure, followed by the Aegean Coast tourist region.

Another characteristic of the fuzzy approach to poverty measurement is the possibility of analysing the overlap between monetary and non-monetary fuzzy measures of poverty. In the literature on fuzzy measurement of poverty, however, there are several contributions which propose methods for the manipulation of the resulting fuzzy sets, as complements, intersections, unions and aggregations (see, for instance, Chiappero 2006). In Table 4.3, we report the correlation coefficients between each pair from the set of 10 indicators. As can be seen in the table, there is an expected high correlation between the HCR and the FM indicator (almost 90 %); the basic lifestyle dimension seems to be a very clear and distinctive dimension compared to all the others, showing very low correlations (apart from the overall FS measure). Finally, it is interesting to note how the environmental problems dimension is negatively correlated with monetary indicators.

4.5 Concluding remarks

In the present chapter, we have proposed a step-by-step methodology for estimating fuzzy measures of non-monetary deprivation and monetary poverty based on the EU-SILC survey (European Union—Statistics on Income and Living Conditions). We have applied such a methodology to the 2007 wave of Turkish SILC data provided by the Turkish Statistical Institute (TurkStat 2007).

We have shown how the multidimensional nature of poverty can be better revealed by a set of items properly aggregated into several dimensions. This implies a more complete and realistic vision of this phenomenon, especially where the single monetary variable used in the traditional approach (consumption expenditure or income) is affected by measurement errors, under-reporting or tax evasion. Of course, some indicators used in the multidimensional approach are affected by measurement errors and under-reporting; however, this is much less problematic compared to the case of income or consumption expenditure, at least in Turkey.

This is why the proposed method is particularly pertinent for Turkey. In fact, according to Turkstat (TurkStat 2007) and Ballini et al. (2009), tax evasion and the under-reporting of income are relevant in the country, especially in the commercial sector and for the self-employment income component. The proposed multidimensional and fuzzy approach can avoid this problem and give a better picture of the living conditions in the country.

According to the empirical analysis conducted on the SILC data set for 2007, 23.45 % of the Turkish population was deemed to be poor. The HCR varies noticeably among the NUTS1 regions, from less than 2 % in the region of Istanbul to about 70 % in the region of South East Anatolia. However, the differences are less marked when we observe the FM indicator and considerably less marked when observing the FS indicator.

In conclusion, it is interesting to analyse the 'environmental dimension' measure; the region of Istanbul shows the higher deprivation measure, followed by the Aegean Coast tourist region.

Acknowledgments This paper has been partly funded by the Italian PRIN research project n. 2007 HEWTBE-003, and partially funded by the European Commission 7th Framework Project FP7-SSH-2007-1 n. 217565–SAMPLE. The authors are grateful to Dr. Enver Tasti, Dr. Francesca Gagliardi, Dr. Giulio Tarditi and Dr. Janet Donovan for their support and helpful comments.

Appendix

A.1 4.6 List of EU-SILC Variables[4]

HS050: Capacity to afford a meal with meat, chicken, fish (or vegetarian equivalent) every second day
HH050: Ability to keep home adequately warm
HS040: Capacity to afford paying for 1 week annual holiday away from home
HS120: Ability to make ends meet
HS110: Do you have a car?
HS090: Do you have a computer?
HS070: Do you have a telephone (including mobile phone)?
HS100: Do you have a washing machine?
HS080: Do you have a colour TV?
HH080: Bath or shower in dwelling
HH090: Indoor flushing toilet for sole use of household
HH040: Leaking Roof, Damp Walls/Floors/Foundation, Or Rot In Window Frames Or Floor
HS160: Problems with the dwelling: too dark, not enough light

[4] For a detailed description of the coding of the EU-SILC variables, see Eurostat (2009).

HS060: Capacity to face unexpected financial expenses
HS010: Arrears on mortgage or rent payments
HS020: Arrears on utility bills
HS030: Arrears on hire purchase instalments or other loan payments
HS190: Crime violence or vandalism in the area
HS180: Pollution, grime or other environmental problems
HS170: Noise from neighbours or from the street
PE010: Current education activity
PE040: Highest ISCED level attained
PL030: Self-defined current economic status
PL070: Number of months spent at full-time work
PL072: Number of months spent at part-time work
PL080: Number of months spent in unemployment
PL085: Number of months spent in retirement
PL087: Number of months spent studying
PL090: Number of months spent in inactivity
PH010: General health
PH020: Suffer from a chronic (long-standing) illness or condition
PH030: Limitation in activities because of health problems
PH040: Unmet need for medical examination or treatment
PH060: Unmet need for dental examination or treatment

The indicator relating to the worklessness of the household is constructed in the following way. Firstly, we exclude households consisting only of persons who are aged 18–24 in full-time education or who are older than a country-specific retirement age. In order to choose an appropriate retirement age, we have proceeded as follows: among the people who have worked, we consider the distribution of the ones who are retired (PL030 = 5) by age and gender. By examining the ratio of people from a particular age group who are retired, we look for the specific age at which a large jump in this proportion is found to occur. Once this point has been found, we confirm it by examining its relationship to the legal age of retirement for a specific country.

Among the remaining households, we classify the people as employed or not employed using variable PL030. We, thus identify the degree of worklessness of a household by constructing a ratio where the numerator contains all the people in the household for which the variable PL030 takes a value of 1, 2 or 7, and the denominator is the sum of the people of the household for which PL030 takes a value of 1, 2, 3, 6, 7, 9, and values of 5 and 8 only if the age of the person is less than the retirement age as chosen above. So at household level we construct an index reflecting the degree to which eligible household members are engaged in work: a zero indicates a workless household with some of its members at working age; a one indicates that all the working age members are working.

To construct the indicator relating to the duration of unemployment, we calculate the following ratio at household level:

$$1-\frac{\sum_{ind=1}^{HH-size}PL080_{ind}}{\sum_{ind=1}^{HH-size}(\text{ PL070 }+\text{PL072}+\text{ PL080}+\text{ PL085}+\text{ PL087}+\text{ PL090})_{ind}} \tag{A.1}$$

The variable for general health, PH010, is aggregated as follows: to the categories 1–2–3 is assigned the value 1 and to categories 4–5 the value 0. Next, this variable is aggregated at household level, so that a household is considered deprived for that indicator if at least one person in the household is deprived for the item. So the score *s* assumes the value 1 if no one in the household is deprived in relation to that item, and it assumes the value 0 if at least one person is deprived.

The same kind of household aggregation is carried out for all the personal variables relating to health and educational status.

References

Anand S. and Sen A.K. (1997), *Concepts of human development and poverty: a multidimensional perspective.* Human Development Papers, United Nations Development Programme (UNDP), New York

Atkinson, A. B. (2003). Multidimensional deprivation: contrasting social welfare and counting approaches. *Journal of Economic Inequality, 1*, 51–65.

Atkinson, A. B., & Bourguignon, F. (1982). The comparison of multidimensional distributions of economic status. *Review of Economic Studies, 49*, 183–201.

Atkinson, A. B., Cantillon, B., Marlier, E., & Nolan, B. (2002). *Social indicators: The EU and social inclusion.* Oxford: Oxford University Press.

Ballini, F., Betti, G., Çalik, G., Daşkiran, Z., Karakaş, M. (2009). *The Siena micro-simulation model (SM2) for the tax-benefit system in Turkey,* Working Paper n. 81, Dipartimento di Metodi Quantitativi, Università di Siena.

Betti, G., Cheli, B., Lemmi, A., & Verma, V. (2006). Multidimensional and longitudinal poverty: an integrated fuzzy approach. In A. Lemmi & G. Betti (Eds.), *Fuzzy set approach to multidimensional poverty measurement* (pp. 115–137). New York: Springer.

Betti, G., Verma, V., (1999). *Measuring the degree of poverty in a dynamic and comparative context: a multi-dimensional approach using fuzzy set theory, Proceedings, ICCS-VI* (Vol 11, pp 289–301) Lahore, Pakistan.

Bourguignon, F., & Chakravarty, S. R. (2003). The measurement of multidimensional poverty. *Journal of Economic Inequality, 1*, 25–49.

Cerioli, A., & Zani, S. (1990). A fuzzy approach to the measurement of poverty. In C. Dagum & M. Zenga (Eds.), *Income and wealth distribution, inequality and poverty* (pp. 272–284). Berlin: Springer Verlag.

Chakravarty, S. R., Mukherjee, D., & Ranade, R. (1998). On the family of subgroup and factor decomposable measures of multidimensional poverty. *Research on Economic Inequality, 8*, 175–194.

Cheli, B. (1995). Totally Fuzzy and Relative Measures of Poverty in Dynamics Context. *Metron, 53*, 183–205.

Cheli, B., & Betti, G. (1999). Totally fuzzy and relative measures of poverty dynamics in an Italian pseudo panel, 1985–1994. *Metron, 57*, 83–104.

Cheli, B., & Lemmi, A. (1995). A totally fuzzy and relative approach to the multidimensional analysis of poverty. *Economic Notes, 24*, 115–134.

Chiappero, Martinetti E. (1994). A new approach to evaluation of well-being and poverty by fuzzy set theory. *Giornale degli Economisti e Annali di Economia, 53*, 367–388.

Chiappero, Martinetti E. (2006). Capability approach and fuzzy set theory: Description, aggregation and inference issues. In A. Lemmi & G. Betti (Eds.), *Fuzzy set approach to multidimensional poverty measurement* (pp. 93–113). New York: Springer.

Duclos, J.-Y., Sahn, D., & Younger, S. D. (2001). *Robust multidimensional poverty comparisons*. Canada: Université Laval.

Eurostat (2002). *Income, poverty and social exclusion: 2nd Report*. Luxembourg: Office for Official Publications of the European Communities.

Eurostat (2009). *Description of SILC user database variables: cross-sectional and longitudinal*. Version 2007. Luxembourg: Office for Official Publications of the European Communities.

Filippone, A., Cheli, B., and D'Agostino A. (2001). Addressing the interpretation and the aggregation problems in totally fuzzy and relative poverty measures, ISER Working Paper Series number 2001–22, University of Essex.

Klasen, S. (2000). Measuring poverty and deprivation in South Africa. *Review of Income and Wealth, 46*, 33–58.

Kolm, S. C. (1977). Multidimensional egalitarianism. *Quarterly Journal of Economics, 91*, 1–13.

Maasoumi, E. (1986). The measurement and decomposition of multidimensional inequality. *Econometrica, 54*, 771–779.

Sen, A. K. (1999). *Development as freedom*. Oxford: Oxford University Press.

Smeeding, T., Saunders, P., Coder, J., Jenkins, S. P., Fritzell, J., Hagenaars, A., et al. (1993). Poverty, inequality and family living standards impacts across seven nations: The effect of non-cash subsidies for health, education and housing. *Review of Income and Wealth, 39*, 229–256.

Townsend, P. (1979). *Poverty in the United Kingdom*. Harmondsworth: Allen Lane.

Tsui, K. (1985). Multidimensional generalisation of the relative and absolute inequality indices: the Atkinson-Kolm-Sen approach. *Journal of Economic Theory, 67*, 251–265.

TurkStat (2007). *Income and living conditions survey 2007*. Ankara:Turkish Statistical Institute.

Vero, J., & Werquin, P. (1997). Reexamining the measurement of poverty: How do young people in the stage of being integrated in the Labour Force Manage (in French). *Economie et Statistique, 8–10*, 143–156.

Whelan, C. T., Layte, R., Maitre, B., & Nolan, B. (2001). Income, deprivation and economic strain: an analysis of the European community household panel. *European Sociological Review, 17*, 357–372.

Zadeh, L. A. (1965). Fuzzy sets. *Information and Control, 8*, 338–353.

Chapter 5
Regional and Sectoral Distributions of Poverty in Lebanon, 2004

Valérie Bérenger, Florent Bresson, Paul Makdissi and Myra Yazbeck

Abstract In this chapter, we investigate the geographical and sectoral profiles of poverty in Lebanon using data from the country's 2004 National Survey of Households Living Conditions (NSHLC). With this objective in mind, we have adopted both unidimensional monetary approaches and multidimensional approaches to poverty. In the case of multidimensional approaches we focus on four dimensions of poverty: expenditure, education, housing conditions, and access to basic services. The poverty measures are estimated according to standard monetary FGT indices and their extension, based on Alkire and Foster's method, in the multidimensional case. The robustness of rankings (by mohafaza and by occupational sector) resulting from these measures is then tested using stochastic dominance procedures. Our findings suggest that caution should be exercised when the conclusions drawn from typical poverty profiles depend on the researchers' arbitrary decisions. Also, they shed light on the limitations of conducting an analysis that is solely based on the monetary dimension of poverty, as it may not necessarily corroborate the results provided by multidimensional analysis.

V. Bérenger (✉)
University of Nice-Sophia Antipolis, Nice, France
e-mail: valerie.berenger@unice.fr

F. Bresson
University of Orléans, Orléans, France
e-mail: florent.bresson@univ-orleans.fr

P. Makdissi
University of Ottawa, Ottawa, Canada
e-mail: paul.makdissi@uottawa.ca

M. Yazbeck
McGill University, Montreal, Canada
e-mail: myra.yazbeck@mcgill.ca

V. Bérenger and F. Bresson (eds.), *Poverty and Social Exclusion Around the Mediterranean Sea*, Economic Studies in Inequality, Social Exclusion and Well-Being 9,
DOI: 10.1007/978-1-4614-5263-8_5,

5.1 Introduction

Compared to other Southern Mediterranean countries, Lebanon has a regime that can be viewed as being closer to a democracy. However, since its independence, the country has experienced continuous political instability (including 15 years of civil war) that has affected its population's wellbeing. Presently, Lebanon is considered as a middle income country and is ranked in the high human development category (ranked 70 out of 187 countries). Its economic development is primarily based on the service sector, which makes it less productive than other Mediterranean countries (Femise 2011). Although the Lebanese state is known for "a laissez-faire approach", where much is left to market forces, a number of investments in the social sector have been made. However, these social programs have had few targeting mechanisms, as the eligibility criteria are not clearly defined and access to them has been discretionary. The absence of a formalized social sector may be due to the sectarian composition of the society as well as political instability (UNDP: NHDR Lebanon 2009). Since the early 1990s there have been serious attempts to improve this sector. With the support of the UNDP and the Central Administration of Statistics, the Ministry of Social Affairs produced three major studies derived from Household Budget Surveys. The first one is "The Mapping of Living Conditions"; it provides evidence on the geographical distribution of poverty in 1997 (Mosa and UNDP 1998). The other two reports: the "Changes in the Mapping of Living conditions between 1995 and 2004" (Mosa and UNDP 2007) and "The Mapping of Living Conditions in 2004" (Mosa and UNDP 2006) are more recent and are based on the National Survey of Households Living Conditions (NSHLC) (2004). These reports represent a first attempt to offer an accurate and comprehensive review of the situation for poor individuals in the country. The aim was to come up with a solid base for the elaboration of policies that could achieve a set objective.

Recently, the NSHLC (2004) was made available to academic researchers.[1] This survey was first used by Laity et al. (2008), who produced a poverty profile for Lebanon. The authors adopted a money metric approach to poverty based on households' expenditure. The main results of their study indicate the presence of striking regional disparities. It also reveals that poverty is higher among the unemployed and unskilled workers in the agricultural and construction sectors. In addition to regional and sectoral money metric poverty comparisons, the authors supplemented their work with comparisons made using measures drawn from the Unsatisfied Basic Needs (UBN) approach (including housing, water, education, and income indicators). The results indicate close rankings between the two

[1] The authors gratefully thank the Ministry of Social Affairs of Lebanon for providing access to the dataset from the Lebanon Consumption Household Budget Survey: 2004–2005. The authors also wish to thank Rita El Araj for her helpful and efficient assistance in contacting the Ministry of Social Affairs of Lebanon. Lastly, thanks are also due to Adib Yazbeck and Alain Safa for useful information on the Lebanese institutional context.

approaches, which may prove that the results are somehow robust. However, Laity et al. (2008) did not perform global robustness checks for the poverty comparisons they provided. It is well known that the choice of poverty line and poverty measure may alter the poverty rankings. The literature on poverty measurement has pointed out such measurement difficulties associated with this kind of comparisons presented in poverty profiles (see Duclos and Makdissi 2004).

The objective of this chapter is to identify the poverty orderings that are robust to changes in the underlying measurement assumptions. We are particularly interested in comparing poverty between mohafazat and between occupational groups. First, we will use the FGT class of poverty indices (see Foster et al. 1984) in order to obtain monetary measures of poverty. Then, we will perform stochastic dominance tests that will enable us to identify the extent to which the poverty orderings obtained from the FGT indices are robust to a change in the poverty lines or poverty indices. As monetary poverty provides partial information on the population's wellbeing, the Alkire and Foster (2011) multidimensional poverty approach will be used. To this end, four poverty dimensions will be considered, namely, per capita expenditures, housing conditions, education, and access to basic services.[2] The robustness of these multidimensional poverty measures will also be tested using the tools of stochastic dominance.

The remainder of this chapter is organized as follows: the theoretical framework is presented in the next section, Sect. 5.3 presents an empirical investigation of poverty between mohafazat[3] and occupational groups and the last section is the conclusion.

5.2 A Methodological Framework for Poverty Measurement

The theoretical literature on poverty points out two measurement issues (Sen 1976). On the one hand there is an identification problem. Identifying poor individuals requires the selection of an adequate poverty line. If the total expenditure of an individual falls below this poverty line, then the individual is deemed poor. Since many values can be proposed for the poverty line, the researcher is faced with the problem of making an appropriate choice. On the other hand, the selection of an adequate aggregation method is also challenging. An aggregation procedure is necessary in order to summarize individuals' poverty statuses with an aggregate poverty measure. The most commonly used income poverty indices are the FGT poverty measures (Foster et al. 1984), but other measures can be used as well. The rankings produced by these indices may be

[2] It is important to mention that this methodology was a source of inspiration for the recently developed Multidimensional Poverty Index (MPI) (UNDP 2010).

[3] A mohafaza (or mohafazat in the plural form) is an administrative division. We can consider a mohafaza as an administrative region or province.

contingent on the specific choice of poverty line or poverty index. To overcome these measurement issues, we use stochastic dominance procedures that provide robust rankings. When poverty is estimated using a multidimensional approach, the measurement issues remain applicable. In this context, the multidimensional stochastic dominance procedures can be used to obtain robust rankings when comparing two populations.

5.2.1 The Monetary Poverty Index

Let us suppose that we have an additive monetary (or unidimensional) poverty index. Let $F : \Re_+ \rightarrow [0, 1]$ represent the cumulative distribution of expenditures y. Then, total poverty is given by:

$$PF(z) := \int_0^z p(y, z)\, dF(y), \tag{5.1}$$

where

$$\begin{cases} p(y, z) \geq 0 & \text{if } y \leq z \\ p(y, z) = 0 & \text{if } y > z \end{cases} \tag{5.2}$$

Here z is the poverty line. The function $p(y, z)$ represents the contribution to total poverty made by an individual with expenditure y. As mentioned earlier, the most commonly used class of additive poverty indices is the class of FGT indices. It can be defined as follows:

$$FGT_F(\alpha, z) := \int_0^z \left(\frac{z-y}{z}\right)^\alpha dF(y) \tag{5.3}$$

When $\alpha = 0$ the index corresponds to the headcount index. It gives the proportion of poor individuals in the distribution. When $\alpha = 1$, the index corresponds to the poverty gap index. It gives the average poverty gap in the distribution. When $\alpha > 1$, the index is sensitive to inequality in the distribution of this poverty gap. We use $\alpha = 2$ as it is the most widely used value of α in the literature for capturing the inequality dimension of poverty.[4]

In addition to the FGT indices there are many examples of additive indices, for instance, the Chakravarty (1983) poverty indices, the Watts (1968) index, and the Clark et al. (1981) second class of poverty indices.[5] We have chosen the FGT class of poverty indices as it is a good place to start, but our results are not dependent on

[4] We will be using these FGT indices in the empirical section.

[5] The Watts index can be interpreted as a transformation of the Clark, Hemming, and Ulph index.

this choice. In fact, we have performed stochastic dominance tests to ensure that our results remain robust to a change in the poverty index. To this end, we have regrouped these additive indices into classes Π^s, $s = 1, 2, \ldots$, of poverty indices. These classes are defined by:

$$\Pi^s := \left\{ P \left| \begin{array}{c} p(y,z) \in \hat{C}^s(z) \\ (-1)^i p^{(i)}(y,z) \geq 0 \text{ for } i = 0,\ 1, \ldots s \\ p^{(t)}(z,z) = 0 \quad \text{for } i = 0,\ 1, \cdots s-2 \end{array} \right. \right\} \tag{5.4}$$

where $\hat{C}^s(z)$ represents the set of functions that are s times piecewise differentiable on $[0, z)$.[6] The ith derivative of $p(y, z)$ with respect to its first argument is denoted by $p^{(i)}(y, z)$.

At this point, it is useful to supply a normative interpretation of the different classes of indices. When $s = 1$, the indices must be such that poverty decreases weakly when an individual's expenditure increases. These indices are thus of the Pareto type in addition to being symmetric (they obey the anonymity axiom). When $s = 2$, these indices respect the Pigou–Dalton transfer principle. This principle postulates that a mean-preserving transfer of expenditure from an individual with higher level expenditure to an individual with lower level expenditure constitutes a social improvement. When $s = 3$, the indices are also sensitive to favorable composite transfers. These transfers are such that a beneficial Pigou–Dalton transfer within the lower part of the distribution, accompanied by a reverse Pigou–Dalton transfer within the higher part of the distribution, will decrease poverty, provided that the variance of the distribution is not increased.[7] We can use the generalized transfer principles of Fishburn and Willig (1984) to interpret higher orders of dominance. For $s = 4$, for instance, let us consider a combination of composite transfers, the first one being favorable and occurring within the lower part of the distribution, and the second one being unfavorable and occurring within the higher part of the distribution. Because the favorable composite transfer occurs in the lower part of the distribution, indices that are members of the $s = 4$ classes should respond favorably to this combination of composite transfers. Generalized higher order transfer principles essentially postulate that, as s increases, the weight

[6] Note that if the (s–1)th derivative of a function is piecewise differentiable, the (s–1)-th derivative is necessarily continuous and the function itself and its first (s–2) derivatives are continuous and differentiable everywhere. Note that the continuity condition we impose is more restrictive than that in Zheng (1999), which only postulates continuity on the interval $(0, z)$ without any restriction on $p^{(t)}$ $(z, z) = 0$ for $t = 0, 1, \ldots, s - 2$. This difference between his and our assumptions has implications for the analysis developed in this chapter. Specifically, we are able to consider dominance criteria for orders greater than two, even when there is significant uncertainty regarding the value of the lower limits for the ranges of possible poverty lines (for detailed information, see Duclos and Makdissi 2004).

[7] Kolm (1976) was the first to introduce this condition in the inequality literature, and Kakwani (1980) subsequently adapted it to poverty measurement (see also Foster and Shorrocks (1987) for a complete description of this transfer principle).

assigned to the effect of transfers occurring at the bottom of the distribution also increases. Blackorby and Donaldson (1978) consider these indices as becoming more Rawlsian. Finally, when $s \to \infty$ only the lowest level of expenditure counts (Davidson and Duclos 2000).

5.2.2 Stochastic Dominance

In the framework presented in the earlier section, two analysts, using two different poverty lines and/or two different poverty indices may achieve different results in relation to poverty. For this reason, it is important to identify which orderings are robust to a change in measurement choice regarding the poverty line and/or the poverty index. Atkinson (1987) developed stochastic dominance conditions to identify poverty orderings that are valid for all indices in the Π^1 and Π^2 classes. Duclos and Makdissi (2004) generalize these conditions for all Π^s classes.

Let us consider two expenditure distribution functions, F and G. For expositional simplicity, we define stochastic dominance curves $F^1(y) := F(y)$ and $F^s(y) := \int_0^y F^{s-1}(u)\,\mathrm{d}u$ for all integers $s \geq 2$. $G^s(y)$ is defined analogously. When poverty does not increase in a movement from distribution F to distribution G, we have:

$$\Delta P_{\mathrm{FG}}(z) := \int_0^z p(y, z)\,\mathrm{d}G(y) - \int_0^z p(y, z)\,\mathrm{d}F(y) \leq 0 \tag{5.5}$$

If there is a consensus that the deprivation threshold z should not exceed some maximum, denoted by z^+, then it is possible to lay out a necessary and sufficient condition for absolute poverty dominance that is applicable to all orders of stochastic dominance s. In such a situation, Duclos and Makdissi (2004) show that poverty does not increase in a movement from distribution F to distribution G for all poverty indices $P \in \Pi^s$ and all poverty lines $z \in [0, z^+]$ if and only if:

$$F^s(y) - G^s(y) \geq 0 \quad \text{for all } y \leq z^+ \tag{5.6}$$

If the stochastic dominance test fails at order s, two different strategies may be followed. The first involves increasing the order of stochastic dominance until a poverty ordering becomes robust over all of some pre-specified ranges of poverty lines. Davidson and Duclos (2000) indicate that if there is dominance in the lower part of the distribution, then for some higher order of stochastic dominance we will find dominance over all $[0, z^+]$. The second strategy involves inferring a critical bound for a restricted range of poverty lines. This critical poverty line z^s, beyond which Eq. (5.6) does not hold, is then given by:

$$z^s := \sup\{z | F^s(y) \geq G^s(y),\ y \in [0, z]\} \tag{5.7}$$

In our empirical investigation we consider the first two orders of dominance. Also, we refrain from setting a priori any maximum z^+. Instead, we estimate critical values z^1 and z^2 for each comparison when needed. In cases where there is no intersection between the two dominance curves, we conclude that there is social welfare dominance between the two distributions (see Duclos and Makdissi 2004).

5.2.3 The Poverty Index in a Multidimensional Framework

Conducting a multidimensional poverty analysis introduces two additional challenges to the poverty analysis. First, the introduction of many wellbeing attributes makes it difficult to obtain robust rankings. Second, the broadening of the definition of wellbeing beyond the income dimension brings up a new issue: choosing the right set of dimensions for inclusion in the analysis (the horizontal "vagueness" in Quizilbash 2006). Whilst we acknowledge that the choice of the adequate combination of dimensions may raise some discussion, we will not be examining the issue in this chapter. Our main focus is on the measurement problems mentioned earlier; namely, the choice of an adequate poverty line and a satisfactory functional form for the poverty index.

As mentioned earlier, the first problem arises with the identification of poor individuals, (i.e., the definition of the poverty line in the multidimensional wellbeing domain).[8] There is a wide range of approaches to this problem in the literature; some of these are extreme and others are more nuanced. One extreme approach, known as the *intersection approach*, requires the presence of deprivation in each dimension for an individual to be considered as poor. According to this approach, an "old beggar" would be deemed as "non-poor" because the years he or she has lived would exceed the threshold in the longevity dimension (Bourguignon and Chakravarty 2003). Another extreme approach, known as the *union approach*, requires the presence of deprivation in at least one dimension for an individual to be considered as poor. One alternative approach, known as the *intermediate* approach, defines a poverty frontier function that sets the poverty line in each dimension as a function of the level of the attributes in the other dimensions (Duclos et al. 2006). Another *intermediate* approach defines attribute-specific poverty lines in the assessment of shortfalls Alkire and Foster (2011). In each dimension, the threshold is orthogonal to the values of the attribute in the other dimensions. In this situation, an individual is considered to be poor if the number of dimensions for which he or she suffers from deprivation is no less than a given threshold. In this chapter, we address the identification issue using this approach.

Since wellbeing is assessed using a set of $m \geq 2$ attributes, the estimation of poverty for an individual is performed using the m-vector $\mathbf{y}$. The different attributes are defined so that any increase in the value of the attribute corresponds to an increase of wellbeing. Each argument y_j from vector $\mathbf{y}$ (for attribute j) is defined on $\Re_+$ so that the distribution can be fully described by the joint cumulative distribution

---

[8] To describe the reasoning behind multidimensional poverty identification, we will first consider a situation where we measure poverty in relation to a single poverty line. Then, by switching to stochastic dominance we will generalize through the use of a set of poverty lines.

function $F_m : \Re_+^m \to [0, 1]$. Let c be a given threshold such that $c \leq m$[9] and let **z** be the vector obtained from these m attribute-specific poverty lines.[10] For a given population and a given vector **z** the share of the population that is likely to be considered poor cannot increase with c. The set of points in $\Re_+^m$ corresponding to the poverty domain for a given value of c and a given vector **z** of attribute-specific poverty lines is noted as $\Gamma(\mathbf{z}, c)$. Furthermore, we assume that each attribute can be measured using continuous ratio scale variables and we define a subgroup-consistent multidimensional poverty index P_m as follows:

$$P_m(\mathbf{z}, c) := \int \cdots \int_{\Gamma(\mathbf{z}, c)} p_m(\mathbf{y}, \mathbf{z}) \mathrm{d}F_m(\mathbf{y}) \tag{5.8}$$

where:

$$\begin{cases} p_m(\mathbf{y}, \mathbf{z}) \geq 0 & \textit{if} \quad \mathbf{y} \in \Gamma(\mathbf{z}, c) \\ p_m(\mathbf{y}, \mathbf{z}) = 0 & \textit{if} \quad \mathbf{y} \notin \Gamma(\mathbf{z}, c) \end{cases} \tag{5.9}$$

This class of poverty measures includes most of the multidimensional poverty indices provided in the literature, such as the one suggested by Chakravarty et al. (1998); Tsui (2002); Bourguignon and Chakravarty (2003) and Alkire and Foster (2011). Alkire and Foster's (2011) family of poverty measures is an extension of the FGT class to the multidimensional case. It fits the authors' own definition of an intermediate approach to poverty identification and can be defined as follows:

$$AF_F(\boldsymbol{\alpha}, \mathbf{z}, \mathbf{c}) := \int \cdots \int_{\Gamma(\mathbf{z}, c)} \frac{1}{m} \sum_{j=1}^{m} \left(\frac{z_j - y_j}{z_j} \right)^{\alpha_j} \mathrm{d}F_m(\mathbf{y}) \tag{5.10}$$

where $\boldsymbol{\alpha}$ is an m-vector of positive parameters α_j. As in the unidimensional context described in Sect. 5.2.1, the α_j parameters define the sensitivity of the index with respect to inequality within the jth dimension when assessing poverty. If $\alpha_j = 0$

---

[9] Note that the Alkire and Foster (2011) framework includes two limiting cases, namely the "intersection" and the "union" approaches of poverty identification. In the "intersection" approach ($c = m$), a person is considered poor if he or she is deprived in all dimensions. This same person would be regarded as poor if he or she is deprived with respect to any attribute when the "union" approach ($c = 1$) is used.

[10] In fact, the Alkire and Foster (2011) approach is even more general, as the authors also consider the case where different weights are given to each dimension for the comparison with the threshold c. The approach can then be summarized formally by the following identification function Φ:

$$\Phi(\mathbf{y}, \mathbf{z}, \omega, c) := \begin{cases} 1 & \textit{if } \sum_{j=1}^{m} \omega_j \varphi(y_j, z_j) \geq c \\ 0 & \textit{if } \sum_{j=1}^{m} \omega_j \varphi(y_j, z_j) < c \end{cases}$$

With $\boldsymbol{\omega}$ being some m-vector of nonnegative weights ωj such that $\sum_{j=1}^{m} \omega_j = m$ and $\varphi(a, b)$ is a function that returns 1 if a is $<b$ and 0 otherwise.

(for $j = 1,\ldots m$), the resulting index is the adjusted headcount index. It is the product of the proportion of poor individuals in the distribution and the average number of those individuals' deprivations (expressed as a proportion of the number of attributes). When $\alpha_j > 1$, the index is sensitive to the inequality of relative deprivations within dimension j for individuals who are considered as poor, given $\mathbf{z}$ and c.

As mentioned earlier, the AF measures represent only a choice among many for an index of poverty. Therefore, it may be useful to check the robustness of poverty rankings produced by an index. To this end, we regroup all indices in classes that are characterized by ethical principles. For expositional simplicity we follow Duclos et al. (2006) and present these principles in a bidimensional framework. However, in the empirical application four dimensions of wellbeing are analyzed. As in Duclos, Sahn and Younger, we consider only first-and second-order stochastic dominance.[11]

Assuming the poverty domain is $\Gamma(\mathbf{z}, c)$, the largest set of poverty indices that we consider is $\Pi_2^{1,1}$ and is defined as follows:

$$\Pi_2^{1,1} := \left\{ P_2 \middle| \begin{array}{ll} p_2^{(1,0)}(\mathbf{y},\mathbf{z}) \leq 0 \quad \text{and} \quad p_2^{(0,1)}(\mathbf{y},\mathbf{z}) & \forall \mathbf{y} \\ p_2^{(1,1)} \geq 0 & \forall \mathbf{y} \end{array} \right\}, \tag{5.11}$$

where $p_2^{(k,l)}$, the derivative of p_2, is k times with respect to the first element of y and l times with respect to its second element. The first line in Eq. (5.11) stipulates that the indices have to obey the monotonicity axiom. In other words, individual poverty decreases weakly with an increase in the level of one attribute. The second line in Eq. (5.11) stipulates that the benefit of an increase in any attribute decreases in the value of the other attribute. Atkinson and Bourguignon (1982) and Bourguignon and Chakravarty (2002) refer to this property as the axiom of non-decreasing poverty under a correlation increasing switch. As pointed out in Duclos and Makdissi (2005), this weak version of the Pigou–Dalton principle is similar to Sen's weak equity axiom (Sen 1997, p. 18).

A subset of $\Pi_2^{1,1}$ can be defined by imposing the Pigou–Dalton principle on any one of the dimensions. Thus, should the poverty index comply with the Pigou–Dalton principle of transfers for the first attribute, the following class of poverty indices will be of interest:

$$\Pi_2^{2,1} := \left\{ P_2 \in \Pi_2^{1,1} \middle| \begin{array}{c} p_2^{(1,0)}(\mathbf{y},\mathbf{z}) = 0 \quad \forall \mathbf{y} \ s.t.\ p_2(\mathbf{y},\mathbf{z}) = p_2(\mathbf{z},\mathbf{z}), \\ p_2^{(2,0)}(\mathbf{y},\mathbf{z}) \geq 0 \quad \forall \mathbf{y} \\ p_2^{(2,1)}(\mathbf{y},\mathbf{z}) \leq 0 \quad \forall \mathbf{y} \end{array} \right\} \tag{5.12}$$

[11] The generalization of higher order principles to the multidimensional framework is not available in the literature and is beyond the scope of this empirical chapter.

The first line in Eq. (5.12) sets the first order derivative $p_2^{(1,0)}$ to zero at the upper frontier of the poverty domain.[12] The second line in Eq. (5.12) means that a progressive transfer (within the first dimension) between two individuals for whom the other attribute has the same level should decrease poverty weakly. The third condition in Eq. (5.12) simply indicates that the poverty reducing effect of a progressive transfer (within the first dimension) does not increase with the level of the second attribute. Using the analogy in Duclos and Makdissi (2005), this ethical principle constitutes a generalization of Sen's weak equity axiom. Since it is reminiscent of Kolm's (1976) principle, we can refer to this property as weak transfer sensitivity. The subset $\Pi_2^{1,2}$ can be defined analogously.

Finally, it can be argued that sensitivity to within-dimension inequalities should be observed for all dimensions of poverty. To account for this possibility, we define the following class of poverty indices:

$$\Pi_2^{2,2} := \left\{ P_2 \in \Pi_2^{2,1} \cap \Pi_2^{1,2} \left| \begin{array}{l} p_2^{(2,0)}(\mathbf{y},\mathbf{z}) = 0 \text{ and } p_2^{(0,2)}(\mathbf{y},\mathbf{z}) = 0 \ \forall \mathbf{y}\, s.t.\, p_2(\mathbf{y},\mathbf{z}) = p_2(\mathbf{z},\mathbf{z}), \\ p_2^{(2,2)}(\mathbf{y},\mathbf{z}) \geq 0 \ \forall \mathbf{y}. \end{array} \right. \right\} \tag{5.13}$$

Indices belonging to $\Pi_2^{2,2}$ respect the properties of both $\Pi_2^{2,1}$ and $\Pi_2^{1,2}$. In particular, they obey the within-dimension Pigou–Dalton principle and the weak transfer sensitivity axiom. Moreover, as indicated by the second line of Eq. (5.13), second-order derivatives with respect to one attribute are convex with respect to the other attribute. This implies that weak transfer sensitivity is higher when the attribute in the other dimension decreases.

5.2.4 Multidimensional Stochastic Dominance

As in the monetary (unidimensional) case, rankings between different groups in terms of poverty are likely to depend on the choice of identification method and the choice of poverty index. The use of multidimensional stochastic dominance procedures helps in identifying which rankings are robust to changes in those choices. Such procedures were proposed by Duclos et al. (2006), where the following family of stochastic dominance hypersurface functions F_m^s is defined as:

$$F_m^s(\mathbf{z}) := \int \cdots \int_0^{\mathbf{z}} \prod_{j=1}^{m} \left(z_j - y_j\right)^{s_j - 1} dF_m(\mathbf{y}) \tag{5.14}$$

with s being some m-vector of positive integers s_j. It can be easily seen that for $m = 1$, F_m^s collapses toward the functions F^s used for unidimensional poverty

[12] It is a technical requirement.

measurement. As in Sect. 5.2.2, the set of functions $G_m^s(\mathbf{z})$ is defined analogously to obtain the relevant information from the multidimensional distribution G_m.

Assuming that it is possible to agree on setting the maximum values of to be used for each specific poverty line z_j we define $\Gamma(\mathbf{z}^+, c)$ as the set of poverty lines subject to the choice of c, the threshold number of deprived dimensions. By applying the Duclos et al. (2006) theorems, we can state that poverty does not increase in a movement from distribution to distribution for all poverty measures from the class Π_m^s and all poverty lines from the set $\Gamma(\mathbf{z}^+, c)$ if and only if:

$$F_m^s(\mathbf{z}) - G_m^s(\mathbf{z}) \geq 0 \qquad \forall\, \mathbf{z} \in \Gamma(\mathbf{z}^+, c). \tag{5.15}$$

The proposed test is very simple. If $s = \mathbf{1}$, a m-vector of 1, the multidimensional first-order dominance test proposed with Eq. (5.15) consists in verifying if the multidimensional poverty headcount index for distribution F is never higher than for distribution G.[13] The significant benefit of this procedure is that it gives an elegant and definitive solution to the problem of finding an adequate weight for each attribute when computing a multidimensional poverty index, since the result is robust to the choice of weights. In other words, should education and health be the chosen dimensions for poverty analysis, it does not matter if the respective weights for a shortfall in each dimension are 10–90, 50–50, or 90–10 % when measuring poverty, as long as condition (5.15) holds.

If the first-order dominance tests do not yield a robust ranking of distributions F and G, we may restrict the set of poverty indices and test for higher order of dominance. A first step would be to impose the Pigou-Dalton condition in one of the dimensions as in Eq. (5.12). For expositional ease, let us define the set of indices obeying the Pigou-Dalton principle in the kth dimension as Π_m^{1+k}, where $\boldsymbol{k}$ is an m-vector with a 1 for the kth element and 0 otherwise. The corresponding partial second-order dominance test proposed with condition (5.15) involves checking that the difference in the value of the average poverty gap $z_k - y_k$, cumulated over the value of the other attributes, has shrunk when moving from distribution F to distribution G for all poverty line vectors.[14]

It is worth stressing that results obtained for the class Π_m^{1+k} are not instructive in relation to the robustness of rankings when taking into account the class $\Pi_m^{1+k'}$, where $k' \neq k$, since no class is fully included in another. If a robust ranking cannot be obtained using the first-order stochastic dominance check, it is then necessary to bear in mind the m partial second-order stochastic dominance tests in

[13] It is worth noting that F_m^1 corresponds to a multidimensional headcount index consistent with the "intersection" view of poverty identification, though its use for stochastic dominance checks is consistent with any approach of poverty identification that fits the Duclos et al. (2006) framework such as the "intermediate" views of Alkire and Foster (2011) used for the present study.

[14] When the $j \neq k$ other attributes are described by discrete variables, it should be noted that the condition (5.15) is similar to the second-order sequential dominance procedure suggested by Chambaz and Maurin (1998) and applied to multidimensional poverty measurement.

a situation where it makes sense to consider inequalities within each one of the m dimensions of poverty.

A second step would consist in extending the assumption of sensitivity to inequality within a given dimension to pairs of dimensions so as to achieve results with a narrower set of poverty indices $\Pi_m^{1+k+k'}$. When the Pigou-Dalton principle of transfers is applied to each of the m dimensions of poverty, stochastic dominance tests then refer to the class of poverty measures $\Pi_m^{\mathbf{2}}$. Poverty measures from this set are such that sequences of equalizing transfers between two individuals, which can be unambiguously ranked in terms of poverty, have an overall poverty reducing effect if the two individuals can still be unambiguously ranked at the end of the transfer sequence. From a practical point of view, condition (5.15) states that it is necessary to look at the value of the average product of all deprivations cumulated over the poor population for each vector of specific poverty lines inside the maximum poverty domain $\Gamma(\mathbf{z}^+, c)$.

At this point, some remarks on links between unidimensional and multidimensional dominance checks are worth mentioning. It can be easily understood that the power of first-order dominance tests, that is, the probability of yielding a nonambiguous result, may be weaker in the multidimensional case than in the unidimensional case. This depends on the chosen poverty identification approach. For instance, using the "union" view of poverty identification we can see that a direct cause of this shrinking power results from the fact that more people may be deemed poor with this approach than with the monetary approach. However, this effect is also augmented by the necessity to take the cumulative aspects of deprivations into account. Bourguignon and Chakravarty (2002) have hence proved that in the case of two dimensions, multidimensional poverty dominance requires both unidimensional dominance with respect to each specific poverty line and multidimensional poverty dominance for the set of points corresponding to individuals that suffer from deprivations in both dimensions. Conversely, it can be shown that unidimensional dominance within each dimension is not required when moving to other views of poverty identification, as checks are not performed with respect to the same individuals. For instance, when considering the "intersection" view, individuals who are deprived in dimension j as opposed to other dimensions are taken into account when performing the dominance check for the jth dimension, but not when performing the multidimensional dominance test.

Finally, with regard to the search for robust rankings when comparing distributions F and G, as in the unidimensional case, two different strategies may be followed: (1) extending the order of dominance until we obtain conclusive results, taking the maximum poverty domain $\Gamma(\mathbf{z}^+, c)$ into consideration or (2) looking for a critical poverty frontier. This critical poverty frontier Ξ^s may be defined as:

$$\Xi^s := \sup\left\{ \mathbf{z} \,\middle|\, F_m^s(\mathbf{y}) \geq G_m^s(\mathbf{y}),\ \mathbf{y} \in \Gamma(\mathbf{z}, m) \right\}, \tag{5.16}$$

that is, the upper contour of the set of points z such that a dominance relationship holds. While this approach is very appealing, its results cannot be easily presented for our setting. Indeed, while the critical poverty line is a scalar in the monetary case, the critical poverty frontier in our empirical application is a 4-dimensional hypersurface that cannot be represented elegantly as a graph. This is why we have not performed the computations in the multidimensional case and why we have only performed our multidimensional stochastic dominance tests on a pre-specified poverty domain.

5.3 Poverty Measures and Poverty Comparisons Across Regions and Across Occupational Sectors

In this section we use the methodological framework described in Sect. 5.2 and provide an extensive investigation into the nature and extent of poverty in Lebanon. We first analyse the regional aspects of poverty in Lebanon before augmenting this analysis by considering the sectoral dimension of poverty. We test for the robustness of the rankings provided with regard to the choice of the poverty line and the poverty index.

5.3.1 Description of the Data and Dimensions of Poverty Used

Data are from the NSHLC, 2004. The NSHLC is a cross-sectional survey that is representative at both national and mohafaza level. This survey is the result of a joint effort from the Central Administration for Statistics and Capacity Building for Poverty Reduction: a MoSA-UNDP project. It monitors housing conditions, access to basic services and income-related indicators at the household level. It also includes individual information on education attainment, employment characteristics, and some health variables. The survey is composed of a sample of 13,003 households which equates to 56,000 individuals. Expenditure data were collected from a sub-sample of 7,431 observations that can be assumed to be fairly representative at national and mohafaza level. It is worth noting that the data represent the status of households in Lebanon as of 2004. Since then, two important events may have had a considerable impact on household welfare: on 14 February 2005 the former Prime Minister Rafik al Hariri was assassinated and in July 2006 there was a war between Israel and Hezbollah.

This poverty analysis will examine four dimensions of wellbeing. The first dimension is a household's per capita expenditure; this has an instrumental role in determining individual wellbeing. It can be considered as a proxy for control over resources, even if the intra-household distribution is not observable. The poverty line is set at 50 % of the median of the distribution. It will be extended to 150 % of the median value to assess the robustness of the results in a multidimensional setting. The second is educational attainment, which is considered as a core dimension in multidimensional poverty. The threshold is fixed at 6 years and will be varied up to 12 years. The third is the housing conditions of households. This dimension of wellbeing is likely to be an overriding factor as it can affect both the mental and physical health of the individuals involved. It may be considered as a good indicator of their social status and is particularly instructive regarding poverty assessment in Lebanon as shown in Bibi et al. (2012). Indeed, the impact of the civil war, as well as the presence of rent control laws, indicates that housing expenditure may not be a good indicator of the housing services received by the household. In this chapter we gain an insight into this dimension via two indicators. The first captures the overcrowding aspect and the number of rooms (per capita) adjusted by taking into account the nature of the household composition.[15] The second draws on construction materials that are prevalent in the dwelling. By combining this information, it is possible to compute an equivalence scale for the number of rooms while taking the quality of a dwelling into account. Housing deprivations are finally estimated with a threshold set at 0.33 rooms per individual which varies up to 0.7. As for access to basic services, the fourth dimension, it is proxied using five indicators that have been dichotomized, including access to a primary source of water, to electricity, to the means for disposing waste and to water drainage. We also include living in a secure environment. For basic services, we have considered that having access to at least four is a minimum requirement for not being deprived in that dimension.

Table 5.1 presents the mean values of a household's expenditure per capita (in thousands of Lebanese pounds) and non-monetary dimensions of poverty by mohafaza.

This table highlights significant discrepancies between Beirut and the northern and southern regions in terms of expenditure per capita. In the non-monetary dimensions of wellbeing, the population living in Beirut experiences the highest level of wellbeing in all dimensions. However, Nabatieh has the lowest levels of schooling, the worst housing conditions, and the lowest access to basic services when compared to other regions.

5.3.2 Regional and Sectoral Monetary Poverty Comparisons

We first examine the results obtained from monetary measures of poverty. It is worth remembering that our poverty estimates have been computed with a poverty line fixed at 50 % of the median per capita expenditure. The exact value of this poverty line is not of crucial importance since robustness checks for poverty orderings are performed in the next pages. The estimated values for the FGT indices provide an initial benchmark that may be corroborated (or not) by the stochastic dominance tests. For the benefit of non-Lebanese readers, we have normalized income by the value of the poverty line. This means that the critical

[15] More specifically, children under 5 years old are supposed to be less demanding with respect to housing, and therefore represent the equivalent of half an older household member.

Table 5.1 Summary statistics for monetary and nonmonetary attributes

Mohafaza	Mean	Stand. dev.	Min	Max
Income				
Beirut	7,650	6,165	944	61,138
Nabatieh	4,423	3,039	1,191	39,181
Mount Lebanon	5,137	4,447	548	117,723
Bekaa	3,770	3,216	673	63,957
South Lebanon	3,385	2,866	598	30,871
North Lebanon	2,884	2,534	283	25,207
Schooling				
Beirut	8.64	4.42	0	16
Nabatieh	6.69	3.96	0	16
Mount Lebanon	7.82	4.1	0	16
Bekaa	7.12	3.98	0	16
South Lebanon	7.08	3.86	0	16
North Lebanon	6.82	3.52	0	16
Housing				
Beirut	0.418	0.286	0.1	3.2
Nabatieh	0.316	0.203	0	3.2
Mount Lebanon	0.359	0.217	0	2.4
Bekaa	0.326	0.214	0	2.4
South Lebanon	0.334	0.199	0	2.4
North Lebanon	0.382	0.228	0	2.4
Services				
Beirut	4.27	0.715	2	5
Nabatieh	3.19	0.654	1	5
Mount Lebanon	3.73	0.908	1	5
Bekaa	3.7	0.858	1	5
South Lebanon	3.87	0.945	1	5
North Lebanon	3.48	1.07	0	5

poverty lines we have chosen to present are displayed as a proportion of this benchmark poverty line.

Table 5.2 shows the FGT estimates for each mohafaza using per capita household expenditure. For all poverty indices, the ranking of the different mohafazat is the same. Starting from the mohafaza with the lowest poverty level and continuing through to the mohafaza with the highest poverty level, the mohafaza ranking is: Beirut, Nabatieh, Mount Lebanon, the Bekaa, the South, and the North. It is worth mentioning that the difference between Nabatieh and Mount Lebanon is not statistically significant at the 95 % level for FGT(1,0) and FGT(1,1). Also, the difference between the Bekaa and the South is not statistically significant at the 95 % level for FGT(1,1).and FGT(2,1).

Having provided a study of poverty by region we will now focus on poverty by economic sector. Table 5.3 displays the estimates of poverty indices by occupational group. A first important result relates in particular to the financial sector. It shows that the level of poverty in this sector is not significantly different than 0 (at

Table 5.2 Estimates of FGT indices for per capita expenditure by mohafaza

Mohafaza	FGT($\alpha = 0$, $z = 1$)	FGT($\alpha = 1$, $z = 1$)	FGT($\alpha = 2$, $z = 1$)
Beirut	0.0160 (0.0068)	0.0011 (0.0007)	0.0003 (0.0003)
Nabatieh	0.0602 (0.0115)	0.0080 (0.0017)	0.0014 (0.0003)
Mount Lebanon	0.0605 (0.0056)	0.0136 (0.0017)	0.0048 (0.0009)
Bekaa	0.1493 (0.0135)	0.0314 (0.0036)	0.0094 (0.0014)
South	0.2253 (0.0160)	0.0470 (0.0043)	0.0143 (0.0017)
North	0.3447 (0.0171)	0.0902 (0.0058)	0.0331 (0.0028)
National	0.1434 (0.0052)	0.0337 (0.0016)	0.0116 (0.0007)

the 95 % level). Furthermore, it appears that there is significantly less poverty in the financial sector than in any other economic sector in Lebanon. As for the service sector, results show that poverty is significantly less (at the 95 % level) when compared with poverty in the agricultural sector, the construction sector, the trade sector, the inactive individuals category, as well as in the transport, postal services and telecommunications sector (hereafter referred to as the TPC sector).[16]

The difference between the service sector and the manufacturing sector is not statistically significant. The manufacturing sector has significantly less poverty (at the 95 % level) than inactive individuals and those in the agricultural sector. There is significantly less poverty (at the 95 % level) in the TPC sector and in the trade sector than in the agricultural sector.

While the use of poverty indices provides a complete ordering, these rankings do not come without a cost. Comparisons of poverty are highly dependent on the choice of the poverty index and the poverty line. To check the robustness of the comparisons between mohafazat and between occupational sectors, we use stochastic dominance conditions as defined in Sect. 5.2.2. Figure 5.1 shows the first-order dominance curves by mohafaza and Table 5.4 displays the estimated values for the corresponding critical poverty line z^1.

The results suggest that the ranking between Beirut and Nabatieh is not robust, as the curves intersect with each other at many points in the lower part of the income scale.[17]

However, Beirut dominates all other mohafazat for all poverty indices and all poverty lines. This means that social welfare is higher in Beirut than in any other mohafaza and that this is true for any social welfare index. As shown in Table 5.4,

[16] The service sector is vital to the Lebanese economy.

[17] It is worth emphasizing that with respect to this result, though not as rich as Beirut, Nabatieh exhibits a greater value for the minimum income with our dataset (see Table 5.1).

Table 5.3 Estimates of FGT indices for per capita income by occupational group

Sector	FGT($\alpha = 0, z = 1$)	FGT($\alpha = 1, z = 1$)	FGT($\alpha = 2, z = 1$)
Financial sector (FS)	0.0108 (0.0108)	0.0009 (0.0009)	0.00007 (0.00007)
Services (SE)	0.0779 (0.0090)	0.0169 (0.0024)	0.0051 (0.0009)
Manufacturing (MA)	0.1111 (0.0140)	0.0256 (0.0038)	0.0077 (0.0013)
Inactive (IN)	0.1234 (0.0095)	0.0320 (0.0033)	0.0120 (0.0017)
Trans.,post and comm. (TPC)	0.1462 (0.0190)	0.0290 (0.0049)	0.0088 (0.0019)
Trade (TR)	0.1540 (0.0133)	0.0361 (0.0039)	0.0121 (0.0017)
Construction (CO)	0.2221 (0.0220)	0.0455 (0.0060)	0.0144 (0.0024)
Agriculture (AG)	0.3427 (0.0254)	0.0877 (0.0094)	0.0339 (0.0051)
National	0.1434 (0.0052)	0.0337 (0.0016)	0.0116 (0.0007)

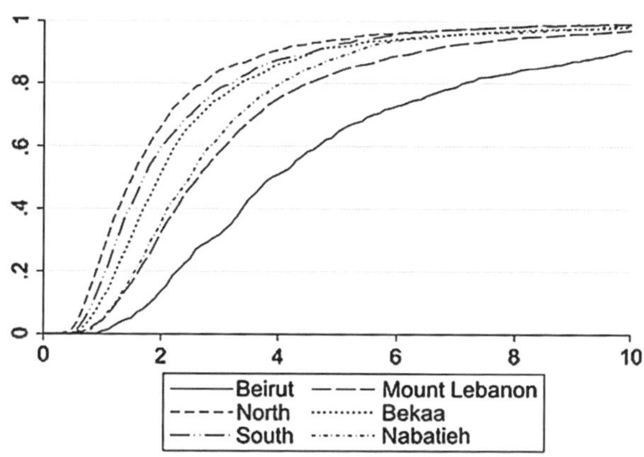

Fig. 5.1 First-order dominance curves by mohafaza

the ranking between Nabatieh and Mount Lebanon is not robust. Also, Nabatieh dominates the other mohafazat for all poverty indices and for all poverty lines that are below 6.464 for the Bekaa, below 11.240 for the South and below 11.094 for the North. The rankings between Mount Lebanon and the Bekaa (or the South) are not robust, yet the ranking between Mount Lebanon and the North is. Mount Lebanon dominates the North for all poverty indices and all poverty lines. This means that social welfare is higher in Mount Lebanon for all social welfare indices (compared to the North). In addition, the Bekaa dominates the South and the North for all poverty indices and all poverty lines. Thus, social welfare is considered higher in the Bekaa than in the South and North for all social welfare indices. Finally, the South dominates the North for all poverty indices and all poverty lines.

Table 5.4 Robust rankings between mohafazat at the first-order

Mohafaza	Nabatieh	Mount Lebanon	Bekaa	South	North
Beirut	ND	WD	WD	WD	WD
Nabatieh		ND	$z^1 = 6.464$	$z^1 = 11.24$	$z^1 = 11.094$
Mount Lebanon			ND	ND	WD
Bekaa				ND	$z^1 = 10.517$
South					$z^1 = 10.136$

Note WD stands for welfare dominance and ND for no dominance ($z^1 < 1$)

We can therefore conclude that social welfare is higher in the South than in the North for all social welfare indices.

Table 5.5 displays the values for the critical poverty lines z^2 corresponding to second-order dominance checks.[18] The tabulated results suggest that the ranking between Beirut and Nabatieh is, once again, not robust. The stochastic dominance between Nabatieh and the other mohafazat (except for Mount Lebanon) is now strict. Nabatieh dominates the Bekaa, the South and the North for all poverty lines and all poverty indices that comply with the Pigou-Dalton transfer principle. This implies that social welfare is higher in Nabatieh than in the Bekaa, the South and the North for all social welfare indices that display inequality aversion. In addition, Nabatieh dominates Mount Lebanon for all poverty indices that display inequality aversion and all poverty lines below 1.491. The ranking between Mount Lebanon and the Bekaa or Mount Lebanon and the South remains similar to that provided by the first-order dominance. As for the Bekaa, there is strict second-order dominance. It dominates the South and the North for all monetary poverty indices that display inequality aversion and for all poverty lines. This means that social welfare is higher in the Bekaa than in the South and the North for all social welfare indices that display inequality aversion. Finally, the South dominates the North for all poverty indices that display inequality aversion and for all poverty lines. This implies that social welfare is higher in the South than in the North for all social welfare indices that display inequality aversion.

While we cannot provide a complete ranking of mohafazat, the relative position of Nabatieh comes as a surprise. On saying this, our results are in line with previous results obtained by other academic researchers who have used the same dataset (see for instance Laity et al. 2008). There are three possible explanations for these surprising results. First, prices vary spatially from one mohafaza to another, for instance, prices in the North are lower than prices in Beirut. Theoretically, it is possible to address this issue, but data limitations prevent us from doing so.[19] Second, the Nabatieh region does not benefit from a level of infrastructure that is equivalent to that in Beirut or Mount Lebanon (i.e., in terms of access to water, electricity, etc.). In such cases, a high level of expenditure may not

[18] Since first-order dominance implies second-order dominance, we will not discuss the cases where first-order dominance can be observed.

[19] On the grounds of privacy, unit prices were not made available in the data.

Table 5.5 Robust rankings between mohafazat at the second-order

Mohafaza	Nabateieh	Mount Lebanon	Bekaa	South	North
Beirut	ND	WD	WD	WD	WD
Nabateieh		$z^2 = 1.491$	WD	WD	WD
Mount Lebanon			ND	ND	WD
Bekaa				WD	WD
South					WD

Note WD stands for welfare dominance and ND for no dominance ($z^1 < 1$)

reflect the real level of wellbeing, rather a compensation for the lack of infrastructure, for instance, an individual may be required to pay for a private generator to compensate for long hours of electricity cuts. Also, in the presence of an unreliable electricity source, an individual may need to acquire an alternative heating source and pay for an alternative energy source (such as a gas heater). As a result expenditure may be doubled or tripled, thus increasing households' overall expenditure. A similar argument (in a different context) was made by Foster et al. (2001) and by Makdissi and Wodon (2006). Another possible explanation for these surprising results could be the remittances sent by emigrants from Nabatieh. Emigrants may have been sending money to relatives living in Nabatieh, thereby allowing the residents to have high expenditure levels. In addition, a lot of these residents enrolled in the Lebanon South Army during the South Lebanon conflict that occurred between 1978 and 2000, and were consequently allowed to work on Israeli territory. Higher salaries and better job opportunities in these areas provided them with large flows of cash that can explain the high expenditure levels observed in Nabatieh.

Table 5.6 displays the values for critical z^1 for first-order dominance by occupational sector. These results indicate that monetary poverty is more prominent in the agricultural sector than in any other sector. This conclusion remains valid for all poverty indices and for all poverty lines. This means that, for all social welfare indices, social welfare is lower in the agricultural sector than in all other sectors for all social welfare indices.

The results also suggest that the financial sector dominates the trade sector as well as the TPC sector for all poverty monetary indices and for all poverty lines. Thus for all social welfare indices, social welfare is higher in the financial sector than in the TPC sector. The financial sector dominates (for all monetary poverty indices) the service sector, the manufacturing sector, and the inactive individuals category for all poverty lines below 12.834. It also dominates the construction sector for all poverty indices and for all poverty lines below 32.083. The ranking between the service sector and the manufacturing sector is not robust. In addition, the service sector dominates (for all poverty indices) the inactive individuals category, the TPC sector, the trade sector, and the construction sector for all poverty lines up to 10.206, 22.022, 23.264, and 17.820 respectively, times our initial poverty line. The rankings between the manufacturing sector and the inactive individuals category and the TPC sector are not robust. The manufacturing sector dominates (for all poverty indices) the trade sector and the construction sector for

Table 5.6 Robust rankings between occupations at the first-order

Sector	SE	MA	IN	TPC	TR	CO	AG
FS	$z^1 = 12.834$	$z^1 = 12.834$	$z^1 = 12.834$	WD	WD	$z^1 = 32.083$	WD
SE		ND	$z^1 = 10.206$	$z^1 = 22.022$	$z^1 = 23.164$	$z^1 = 23.164$	WD
MA			ND	ND	$z^1 = 1.655$	$z^1 = 18.300$	WD
IN				ND	ND	ND	WD
TPC					ND	$z^1 = 3.959$	WD
TR						ND	WD
CO							WD

Note WD stands for welfare dominance and ND for no dominance ($z^1 < 1$)

all poverty lines up to 1.655 and 18.300, respectively. The rankings between the inactive individuals category and the TPC sector, the trade sector, and the construction sector are not robust. The same applies to the rankings between the TPC sector and the trade sector. The TPC sector dominates the construction sector for all poverty indices and for all poverty lines below 3.959 times the initial poverty line. Finally, the ranking between the trade sector and the construction sector is not robust.

Table 5.7 displays the values for critical z^2 for second-order dominance. This table focuses on indices showing aversion to monetary inequalities and offers additional dominance results. The financial sector dominates all other sectors for all poverty indices that display inequality aversion and for all poverty lines. This implies that, for all social welfare indices displaying inequality aversion, social welfare is higher in this sector than in all other sectors. In addition, the service sector dominates the inactive individuals category, the TPC sector, the trade sector, and the construction sector for all poverty indices displaying inequality aversion and for all poverty lines. This means that for all social welfare indices that display inequality aversion social welfare is higher in the service sector in relation to these other sectors. The manufacturing sector dominates the TPC sector as well as the construction sector for all poverty lines and for all poverty indices that convey aversion to inequalities.

Therefore, social welfare is higher in the manufacturing sector in relation to these other sectors for all social welfare indices. The manufacturing sector dominates the inactive individuals category and the trade sector for all poverty lines up to 1.751 and 3.043, respectively. The distribution of income for the inactive individuals category dominates the distribution corresponding to the TPC sector for all poverty indices displaying inequality aversion and for all poverty lines below 1.088. The TPC sector dominates the construction sector for all poverty lines and for all poverty indices obeying the Pigou-Dalton transfer principle. This implies that social welfare is higher in the TPC sector than in the construction sector for all social welfare indices displaying inequality aversion. Finally, the TPC sector dominates the trade sector for all poverty indices exhibiting inequality aversion and for all poverty lines below 1.625 times our benchmark monetary poverty line.

Table 5.7 Robust rankings between occupations at the second-order

Sector	SE	MA	IN	TPC	TR	CO	AG
FS	WD	WD	WD	WD	WD	WD	WD
SE		ND	WD	WD	WD	WD	WD
MA			$z^1 = 1.751$	WD	$z^1 = 3.043$	WD	WD
IN				$z^1 = 1.088$	ND	ND	WD
TPC					$z^1 = 1.625$	WD	WD
TR						ND	WD
CO							WD

Note WD stands for welfare dominance and ND for no dominance ($z^1 < 1$)

5.3.3 Multidimensional Poverty Comparisons

Let us now consider a broader approach to poverty using multidimensional measures in order to see whether or not they provide new insights into the well-being of the Lebanese population. For this purpose, multidimensional FGT-type poverty measures are estimated using the dual cutoff and counting approach developed by Alkire and Foster (2011) as defined in Eq. (5.10). Hence, Tables 5.8 and 5.9 present our estimates of these FGT-type measures for different approaches to the definition of the poverty domain. More specifically, with $c = 1$ the poverty domain corresponds to the traditional "union" approach; $c = 2$ and $c = 3$ refer to "intermediate" approaches and $c = 4$ relates to the "intersection" approach. The latter can be seen as a way to analyze the "core" of poverty. For each value of c, H is the multidimensional headcount. It indicates the percentage of people who are deprived in at least one, two, three, and all dimensions. The measure associated with $\alpha = 0$ is the adjusted headcount ratio that is sensitive to the breadth of poverty. Unlike H, this measure takes into account the average share of deprivations among poor individuals. The columns corresponding to values of $\alpha = 1$ and $\alpha = 2$ are the counterpart of the poverty gap and square poverty gap in the multidimensional setting.

As expected, we observe that values of H are higher than the corresponding values of the adjusted headcount ratio ($\alpha = 0$) because poor individuals are rarely deprived in all dimensions.[20] Moreover, the values of H and those of adjusted headcount ratios decrease with values of c. The difference between the two measures narrows as c increases. This can be due to the high number of individuals that are deprived in only one dimension.

The results provided by Table 5.8 show that Beirut has the lowest percentage of poor individuals (H) and is the least deprived ($\alpha = 0$) regardless of the choice made for the cutoff c. Also, the population of the Bekaa ranks among the poorest regardless of the value chosen for c. The rankings of mohafazat change slightly

[20] Of course, the two measures are equal with the intersection view since poor individuals are then by definition systematically deprived with respect to all attributes.

with the values of c. However, there is evidence of significant re-arrangements in the rankings of Nabatieh and the North when it comes to the headcount ratios. Thus, if we consider poor individuals as being those who are deprived in at least one ($c = 1$) or two dimensions ($c = 2$), then people living in Nabatieh appear to be experiencing the highest level of poverty. Yet, Nabatieh is one of the mohafazat with the least deprived population and ranks just after Beirut and Mount Lebanon when $c = 3$. It also ranks just after Beirut when $c = 4$. This means that when we restrict the definition of poverty to individuals who are deprived in all four dimensions, Nabatieh's ranking is the same as the one resulting from a monetary measure (see Table 5.2). A similar result can be observed once the headcount is adjusted by the average share of deprivations among the poor population. To a lesser extent, a reverse scenario is observed in the South. Using an extensive view of poverty, we observe that the percentage of poor individuals in the South is among the lowest (it ranks just after Beirut). Once we move toward more restricted views of poverty, we observe that the South records the highest level of deprivation. It ranks after the North (when $c = 3$) and after the Bekaa and the North using the intersection approach (i.e., $c = 4$). Other noticeable changes in the rankings can be observed once the definition of the poverty domain is limited to individuals with a higher number of deprived dimensions. While the North has a lower number of poor individuals than the Bekaa for $c \leq 2$, an opposite ranking is obtained when $c \geq 3$.

Generally speaking, accounting for the breadth of deprivations does not change the overall picture obtained with the multidimensional headcount ratio. However, there are differences that deserve our interest, for instance, results obtained using the "union" view of poverty identification show that the South performs better than Mount Lebanon. Nevertheless, once the multidimensional headcount is adjusted ($\alpha = 0$), the South shows more poverty than Mount Lebanon and it becomes evident that the average share of deprivations among poor individuals is higher in the South. Also, the Bekaa and the North are characterized by showing the same value of adjusted headcount index, but a larger value of headcount index for the Bekaa when $c = 1$. Therefore, it can be concluded that (on average) poor individuals in the North experience more deprivations than poor individuals in the Bekaa. It is clear that that the rankings obtained with adjusted headcount ratios using the intersection approach to poverty provide the same ranking among mohafazat as those obtained with FGT monetary measures shown in Table 5.3. This suggests that the ranking provided using a monetary poverty approach is a good predictor of the ranking obtained for "core" poverty, given the values chosen to define the poverty domain in our multidimensional analysis. The variations of the estimates of the adjusted headcount ratio with the value of c provide interesting additional information. The estimates for $\alpha = 0$ only become close to zero when $c = 3$ for Beirut, Mount Lebanon, and Nabatieh. The highest relative decrease in poverty is recorded in Nabatieh (when c varies from 2 to 3). As for the North, the Bekaa and the South (that are worse off notably for $c = 3$ and $c = 4$), values of the adjusted headcount only become closer to zero with the intersection approach. This suggests that in the North, the Bekaa and the South, dimensions of poverty seem to be more correlated than in Beirut, Mount Lebanon, and Nabatieh. All these conclusions seem to be robust when we consider the other AF indices. This means that the poverty rankings may depend on the number of deprivations required to consider an individual as poor.

Here again, these findings lead us to emphasize the puzzling issue raised by the case of the mohafazat of Nabatieh and the South. When adopting a multidimensional view of poverty, it is clear that policy interventions are sensitive to the decisions linked to the dimension cutoff c. Furthermore, significant differences emerge from the rankings obtained by monetary poverty measures and multidimensional measures using the union approach for Nabatieh, the North and the South that would deserve further in-depth investigations of the dimensions of poverty.

Similarly to monetary poverty, it is worth analyzing the multidimensional poverty measures by occupational group. Table 5.9 shows multidimensional poverty measures by occupational group. The multidimensional headcount ratio and the adjusted headcount ratio are lowest in the financial sector and service sector regardless of the approach used to define poor individuals. Thus, households in which the head of the household is employed in these two sectors are at a lower risk of poverty and experience fewer deprivations than those employed in other sectors. The reverse scenario is true for the agricultural sector, regardless of the choice of dimension cutoff c. The TPC and construction sectors have the same percentage of poor individuals (for $c = 1$), yet poor individuals in the construction sector suffer from more deprivations (on average). The same situation is observed between the manufacturing sector and the trade sector when an intermediate approach ($c = 2$) is used. However, once the headcount ratio is adjusted, poor individuals in the trade sector are slightly more deprived. In contrast to the poverty estimates by region of residence, the rankings of multidimensional poverty based on the headcount ratio and the adjusted headcount ratio among sectors are very close. The rankings do not seem to be sensitive to the value chosen for c. Overall, the least deprived are individuals who are employed in the financial sector, followed by those who are in the service sector, the manufacturing sector, and the trade sector. We can note that the financial sector shows zero poverty when poor individuals are defined as those who cumulate deprivations in all dimensions ($c = 4$). Individuals who work in the agricultural sector, the construction sector, the transportation sector, and those who are inactive are more likely to suffer from multiple deprivations. It is also interesting to mention that the rankings between monetary poverty and multidimensional poverty are almost the same for these occupational groups.

While the use of poverty indices provides a complete ordering, these rankings do not come without a cost. Comparisons of poverty are highly dependent on several decisions relative to the choice of poverty lines, the identification of poor individuals according to the number of dimensions, the weights assigned to each dimension and the choice of a particular measure. To check the robustness of the comparisons between mohafazat and between occupational sectors, we use stochastic dominance conditions as defined in Sect. 2.4. We focus on comparisons of

Table 5.8 Multidimensional FGT indices by mohafaza using the Alkire and Foster (2011) approach

Years	$c = 1$				$c = 2$			
	H	$\alpha = 0$	$\alpha = 1$	$\alpha = 2$	H	$\alpha = 0$	$\alpha = 1$	$\alpha = 2$
Beirut	0.8 (0.0085)	0.3 (0.0043)	0.073 (0.0021)	0.046 (0.0018)	0.34 (0.01)	0.18 (0.0056)	0.048 (0.002)	0.031 (0.0016)
Nabatieh	0.98 (0.0022)	0.46 (0.003)	0.16 (0.0022)	0.088 (0.0019)	0.72 (0.0081)	0.39 (0.0047)	0.14 (0.0024)	0.083 (0.0019)
Mount Lebanon	0.92 (0.0027)	0.38 (0.002)	0.11 (0.0011)	0.064 (0.00095)	0.51 (0.0049)	0.28 (0.0028)	0.086 (0.0012)	0.054 (0.00095)
Bekaa	0.95 (0.003)	0.44 (0.0028)	0.13 (0.0017)	0.08 (0.0014)	0.64 (0.0064)	0.37 (0.0039)	0.12 (0.0018)	0.073 (0.0014)
South	0.89 (0.0051)	0.41 (0.0037)	0.12 (0.002)	0.072 (0.0016)	0.56 (0.008)	0.33 (0.0049)	0.11 (0.0021)	0.063 (0.0016)
North	0.91 (0.0042)	0.44 (0.0037)	0.14 (0.0019)	0.072 (0.0015)	0.54 (0.0072)	0.35 (0.0049)	0.11 (0.0021)	0.064 (0.0015)
National	0.92 (0.0016)	0.41 (0.0013)	0.12 (0.00071)	0.071 (0.00058)	0.55 (0.0029)	0.32 (0.0017)	0.1 (0.00075)	0.062 (0.00058)
	$c = 3$				$c = 4$			
	H	$\alpha = 0$	$\alpha = 1$	$\alpha = 2$	H	$\alpha = 0$	$\alpha = 1$	$\alpha = 2$
Beirut	0.043 (0.0043)	0.034 (0.0034)	0.013 (0.0014)	0.011 (0.0011)	0.005 (0.0015)	0.005 (0.0015)	0.002 (0.00061)	0.0016 (0.0005)
Nabatieh	0.12 (0.0059)	0.094 (0.0046)	0.043 (0.0022)	0.031 (0.0017)	0.0099 (0.0018)	0.0099 (0.0018)	0.0044 (0.00081)	0.0032 (0.00059)
Mount Lebanon	0.098 (0.0029)	0.077 (0.0023)	0.031 (0.001)	0.023 (0.00078)	0.014 (0.0011)	0.014 (0.0011)	0.0066 (0.00055)	0.0052 (0.00044)
Bekaa	0.16 (0.048)	0.12 (0.0038)	0.051 (0.0017)	0.035 (0.0013)	0.021 (0.0019)	0.021 (0.0019)	0.011 (0.00097)	0.008 (0.00073)
South	0.17 (0.006)	0.13 (0.0047)	0.048 (0.0019)	0.03 (0.0013)	0.019 (0.0022)	0.019 (0.0022)	0.0086 (0.001)	0.0064 (0.00075)
North	0.26 (0.0063)	0.2 (0.005)	0.073 (0.002)	0.04 (0.013)	0.035 (0.0027)	0.035 (0.0027)	0.018 (0.0014)	0.013 (0.00099)
National	0.14 (0.002)	0.11 (0.0016)	0.044 (0.00068)	0.029 (0.00049)	0.018 (0.00077)	0.018 (0.00077)	0.0088 (0.00038)	0.0066 (0.00029)

Table 5.9 Multidimensional FGT indices by occupational sector using the Alkire and Foster (2011) approach

Year	$c = 1$				$c = 2$			
	H	$\alpha = 0$	$\alpha = 1$	$\alpha = 2$	H	$\alpha = 0$	$\alpha = 1$	$\alpha = 2$
Financial sector (FS)	0.082	0.29	0.068	0.04	0.32	0.17	0.047	0.029
	(0.023)	(0.011)	(0.0055)	(0.0046)	(0.028)	(0.015)	(0.0054)	(0.0044)
Services (SE)	0.9	0.37	0.096	0.052	0.48	0.26	0.075	0.043
	(0.0039)	(0.0026)	(0.0014)	(0.0011)	(0.0065)	(0.0037)	(0.0015)	(0.0011)
Manufacturing (MA)	0.091	0.38	0.11	0.056	0.51	0.28	0.082	0.045
	(0.0051)	(0.0037)	(0.0019)	(0.0015)	(0.0088)	(0.0037)	(0.0021)	(0.0015)
Inactive (IN)	0.9	0.4	0.13	0.081	0.53	0.31	0.11	0.071
	(0.0032)	(0.0024)	(0.0014)	(0.0012)	(0.0054)	(0.0033)	(0.0015)	(0.0012)
Trans.,post and comm. (TPC)	0.95	0.44	0.13	0.07	0.63	0.36	0.11	0.063
	(0.0043)	(0.004)	(0.0023)	(0.0019)	(0.0096)	(0.0058)	(0.0025)	(0.0019)
Trade (TR)	0.9	0.39	0.11	0.06	0.51	0.29	0.087	0.051
	(0.0044)	(0.0031)	(0.0016)	(0.0013)	(0.0072)	(0.0042)	(0.0017)	(0.0013)
Construction (CO)	0.95	0.45	0.14	0.08	0.65	0.38	0.12	0.073
	(0.0045)	(0.0045)	(0.0026)	(0.0021)	(0.0099)	(0.0063)	(0.0028)	(0.0022)
Agriculture (AG)	0.98	0.53	0.19	0.11	0.79	0.49	0.18	0.1
	(0.0029)	(0.0041)	(0.0027)	(0.0022)	(0.0079)	(0.0055)	(0.0029)	(0.0023)
National	0.92	0.41	0.12	0.0071	0.55	0.32	0.1	0.062
	(0.0016)	(0.0013)	(0.00071)	(0.00058)	(0.0029)	(0.0017)	(0.00075)	(0.00058)
	$c = 3$				$c = 4$			
	H	$\alpha = 0$	$\alpha = 1$	$\alpha = 2$	H	$\alpha = 0$	$\alpha = 1$	$\alpha = 2$
Financial sector (FS)	0.25	0.019	0.0061	0.005	0	0	0	0
	(0.0095)	(0.0071)	(0.0027)	(0.0022)	(0)	(0)	(0)	(0)
Services (SE)	0.086	0.066	0.025	0.017	0.0084	0.0084	0.0039	0.0029
	(0.0037)	(0.0029)	(0.0012)	(0.00086)	(0.0012)	(0.0012)	(0.0056)	(0.00042)

(continued)

Table 5.9 (continued)

	$c = 3$				$c = 4$			
	H	$\alpha = 0$	$\alpha = 1$	$\alpha = 2$	H	$\alpha = 0$	$\alpha = 1$	$\alpha = 2$
Manufacturing (MA)	0.11 (0.0055)	0.084 (0.0043)	0.031 (0.0017)	0.019 (0.0012)	0.01 (0.0018)	0.01 (0.0018)	0.0047 (0.00083)	0.0034 (0.0036)
Inactive (IN)	0.14 (0.0038)	0.11 (0.003)	0.044 (0.0013)	0.03 (0.00099)	0.022 (0.0016)	0.022 (0.0016)	0.01 (0.00078)	0.008 (0.0006)
Trans.,post and comm. (TPC)	0.15 (0.0071)	0.12 (0.0056)	0.047 (0.0024)	0.032 (0.0018)	0.019 (0.0027)	0.019 (0.0027)	0.0089 (0.0013)	0.0068 (0.001)
Trade (TR)	0.12 (0.0046)	0.091 (0.0036)	0.034 (0.0015)	0.022 (0.0011)	0.012 (0.0016)	0.012 (0.0016)	0.0059 (0.00077)	0.0044 (0.00057)
Construction (CO)	0.19 (0.0082)	0.15 (0.0065)	0.059 (0.0028)	0.039 (0.002)	0.027 (0.0034)	0.027 (0.0034)	0.014 (0.0017)	0.01 (0.0012)
Agriculture (AG)	0.33 (0.0091)	0.26 (0.0072)	0.1 (0.0032)	0.065 (0.0023)	0.044 (0.004)	0.044(0.)	0.022 (0.002)	0.016 (0.0015)
National	0.14 (0.002)	0.11 (0.16)	0.044 (0.00068)	0.029 (0.00049)	0.018 (0.00077)	0.018 (0.00077)	0.0088 (0.00038)	0.0066 (0.0029)

the joint distributions of expenditure, education, housing, and access to services. Also, we consider the classes of poverty measures and intermediate sets such that sensitivity to extreme deprivations is imposed on either one dimension or on pairs of dimensions. Therefore, in each case the parameter s_k to be used for the function F^s_m can then be equal to 1 or 2 for each dimension. Let s_{inc}, s_{ed}, s_{hous}, s_{serv} be the coefficients referring to expenditure, education, housing conditions, and services, it is then possible to consider 16 possible combinations including {1, 1, 1, 1} (or **1** using the notations from 2.4) and {2, 2, 2, 2} or equivalently **2.**[21] As dominance at a given order systematically implies dominance to hold at higher orders, the next tables only provide the lowest orders of dominance observed for each comparison. Many vectors can be given for each pair of mohafazat since the different classes of poverty measures used for the dominance checks are not systematically nested. For instance, if dominance does not hold with **1** but can be observed with (2, 1, 1, 1), nothing can be said at the order (1, 1, 2, 2) until the corresponding test has been performed.

Dominance tests have also been carried out on poverty domains by adopting the union, intermediate, and intersection views of poverty. As mentioned earlier, robust multidimensional poverty comparisons are more demanding when using an extensive view of poverty, since dominance in each dimension is required. A higher number of robust rankings are expected in the "intersection approach" that cumulates deprivation in all dimensions. Thus, dominance with the intersection approach to poverty identification is a necessary condition for dominance to be observed with the intermediate and union approaches. This is why our discussion of the results mainly concentrates on the intersection approach.

Results of multidimensional poverty dominance tests are presented in Tables 5.10 and 5.11 by region and occupational sector.

Although monetary dominance can be generally observed, it is difficult to depict unambiguous rankings of multidimensional poverty among mohafazat and occupational sectors. This is regardless of the approach used for poverty identification. The only mohafaza for which robust comparisons can be detected is Beirut. For the largest class of multidimensional poverty indices Π^1_m, the population living in Beirut benefits from better living conditions than the population living in Mount Lebanon and the North. When Beirut is compared with the other mohafazat, it is always necessary to turn to more restrictive classes of poverty indices (i.e., to suppose inequality aversion with respect to at least one attribute). For instance, higher poverty in the South compared to Beirut can only be observed when emphasis is put on severe deprivations in housing or access to services. It is clear that the monetary dimension plays a crucial role in obtaining a dominance relationship between the mohafazat. In our unidimensional analysis, we have seen that Beirut exhibits robustly less monetary poverty than the other mohafazat. Once other dimensions of wellbeing are included, inequality aversion with respect to these attributes often has to be supposed in order to obtain unambiguous results in

[21] These correspond to the first-and second-order dominance tests, respectively.

Table 5.10 Robust multidimensional rankings by mohafaza

Mohafaza	Nabateieh	Mount Lebanon	Bekaa	South	North
Intersection approach					
Beirut	(1,1,1,2)	(1,1,1,1)	(2,1,1,1) (1,2,1,1) (1,1,2,1) (1,1,1,2)	(1,1,2,1) (1,1,1,2)	(1,1,1,1)
Nabatieh		ND	ND	ND	ND
Mount Lebanon			ND	ND	ND
Bekaa				ND	ND
South					ND
Intermediate approach ($c = 3$)					
Beirut	ND	(1,1,1,1)	(1,2,1,1) (1,1,2,1) (1,1,1,2)	(1,1,2,1) (1,1,1,2)	(1,1,1,1)
Nabatieh		ND	ND	ND	ND
Mount Lebanon			ND	ND	ND
Bekaa				ND	ND
South					ND
Intermediate approach ($c = 2$)					
Beirut	ND	(1,1,1,1)	(1,2,1,1) (1,1,2,1) (1,1,1,2)	(1,1,2,1) (1,1,1,2)	(1,1,1,1)
Nabatieh		ND	ND	ND	ND
Mount Lebanon			ND	ND	ND
Bekaa				ND	ND
South					ND
Union approach					
Beirut	ND	(1,1,1,1)	(1,2,1,1) (1,1,2,1) (1,1,1,2)	(1,1,2,1) (1,1,1,2)	(1,1,1,1)
Nabatieh		ND	ND	ND	ND
Mount Lebanon			ND	ND	ND
Bekaa				ND	ND
South					ND

Note (s_{inc}, s_{ed}, s_{hous}, s_{serv}) indicates the lowest dominance orders and ND denotes no dominance

terms of poverty while s_{inc} can remain equal to 1. This does not mean that the contribution of income to poverty is more significant, but that it is a more powerful attribute to differentiate between mohafazat in terms of poverty within our multidimensional framework.

Widening the poverty domain to include broader definitions changes our results marginally, except for Nabatieh. Moving away from the intersection approach makes it impossible to draw robust conclusions regarding the situation for Nabatieh, even when restricted classes of poverty measures are considered. This indicates the existence of multiple crossings of the dominance surfaces which correspond to the joint distribution of the selected attributes outside the intersection poverty domain.

Table 5.11 Robust multidimensional rankings between occupational sectors

Sector	SE	MA	IN	TPC	TR	CO	AG
Intersection approach							
FS	(1,1,1,1)	(2,1,1,1) (1,1,2,1) (1,1,1,2)	(1,1,1,1)	(1,1,1,1)	(1,1,1,1)	(1,1,1,1)	(1,1,1,1)
SE		ND	ND	ND	ND	ND	ND
MA			(2,2,2,2)	ND	ND	ND	ND
IN				ND	ND	ND	ND
TPC						ND	ND
TR						ND	ND
CO							ND
Intermediate approach (c = 3)							
FS	(2,1,1,1) (1,1,1,2)	(2,1,1,1) (1,1,1,2) (1,2,2,1)	(2,1,1,1) (1,2,1,1) (1,1,1,2)	(2,1,1,1) (1,2,1,1) (1,1,1,2)	(2,1,1,1) (1,1,1,2)	(1,1,1,1)	(2,1,1,1) (1,2,1,1) (1,1,1,2)
SE		ND	ND	ND	ND	ND	ND
MA			ND	ND	ND	ND	ND
IN				ND	ND	ND	ND
TPC					ND	ND	ND
TR						ND	ND
CO							ND
Intermediate approach (c = 2)							
FS	(2,1,1,1) (1,1,1,2)	(2,1,1,1) (1,1,1,2) (1,2,2,1)	(2,1,1,1) (1,2,1,1) (1,1,1,2)	(2,1,1,1) (1,2,1,1) (1,1,1,2)	(2,1,1,1) (1,1,1,2)	(1,1,1,1)	(2,1,1,1) (1,2,1,1) (1,1,1,2)
SE		ND	ND	ND	ND	ND	ND
MA			ND	ND	ND	ND	ND
IN				ND	ND	ND	ND
TPC					ND	ND	ND
TR						ND	ND
CO							ND
Union approach							
FS	(2,1,1,1) (1,1,1,2)	(2,1,1,1) (1,1,1,2)	(2,1,1,1) (1,2,1,1) (1,1,1,2)	(2,1,1,1) (1,2,1,1) (1,1,1,2)	(2,1,1,1) (1,1,1,2)	(1,1,1,1)	(2,1,1,1) (1,2,1,1) (1,1,1,2)
SE		ND	ND	ND	ND	ND	ND
MA			ND	ND	ND	ND	ND
IN				ND	ND	ND	ND
TPC					ND	ND	ND
TR						ND	ND
CO							ND

Note (s_{inc}, s_{ed}, s_{hous}, s_{serv}) indicates the lowest dominance orders and ND denotes no dominance

For comparisons by occupational sector (Table 5.11), it appears that the financial sector is the only occupational sector that can be unambiguously ranked in terms of poverty, regardless of the value used for the dimension cutoff c. More precisely, the dominance checks confirm that poverty is lower for households whose head works in the financial sector. With the intersection view of poverty identification, dominance frequently holds at the first-order. However, with looser definitions of poverty, holding the conclusion of lower poverty levels is often to the cost of introducing inequality aversion with respect to some attributes. For instance, poverty is robustly greater in the service sector than in the financial sector when we consider poverty indices that display inequality aversion with respect to income or access to services.

When the financial sector is compared with the other occupational sectors, the discriminatory variable becomes housing conditions, since s_{hous} rarely has to be set to equal 2 to achieve robust results. Apart from the financial sector, the sole comparison that yields robust results is the one between individuals from the manufacturing sector and individuals who are in the inactive group ($c = 4$). For poverty indices that exhibit inequality aversion simultaneously within the four dimensions, poverty is lower when the head of the household works in the manufacturing sector compared to poverty when the head of the household is in the inactive group. Unlike the dominance checks performed for the monetary approach, individuals working in the agricultural sector do not seem to be at a higher risk of poverty.

While our findings may be seen as disappointing at first glance, the loss of robust poverty comparisons in the multidimensional approach to poverty mitigates previous findings based on money metric measures. They also highlight how results from poverty quantification are fragile and debatable as soon as we challenge the sensitivity of poverty estimates with respect to identification and aggregation choices.

5.4 Conclusion

In this chapter, we analyzed the geographical and sectoral dimensions of poverty in Lebanon resulting from both unidimensional and multidimensional approaches to poverty. Our findings enrich previous studies on poverty in Lebanon and stress that considerable caution should be taken when the conclusions drawn from poverty rely on the researcher's arbitrary decisions. They also shed light on the limitations of conducting an analysis that is solely based on the monetary dimension of poverty, as it may not necessarily corroborate the results provided by multidimensional analysis. The latter makes it possible to focus on sub-groups of individuals that cumulate different forms of deprivations.

In the unidimensional case, using stochastic dominance techniques, we were able to identify the extent to which the orderings of poverty between mohafazat and occupational groups are robust. Two principal results need to be highlighted. First, Beirut is the region with the lowest level of poverty. This conclusion seems to be robust. Second, the North shows the highest monetary poverty. While this result might seem surprising, it has to be borne in mind that the use of an expenditure approach to compare the North with the South and Nabatieh may raise some issues. The mohafazat of Nabatieh and the South, having been the regions that suffered the most from the conflict between Lebanon and Israel, have a level of public infrastructure and services that may be less developed than in the other mohafazat. In this context, higher levels of expenditure may not directly imply a higher level of wellbeing. This is why the reader has to be cautious when interpreting these results. The occupational sector of the head of the household also provides relevant information regarding the likelihood of being monetarily deprived. Indeed, monetary poverty is marginal in the financial sector; poverty estimates are not significantly greater than zero. The agricultural sector is the occupational group with the highest level of poverty. This result seems to be robust to changes in measurement assumptions. The construction sector follows the agricultural sector. Thus, it is reasonable to believe that labor market policies focussing on these two sectors may produce high returns in terms of poverty reduction.

The introduction of a multidimensional analysis of poverty to some extent mitigates these findings. By considering information on education, housing conditions, and access to basic services alongside expenditure, our multidimensional poverty measures uncovered particularities in the profiles of Nabatieh and the South. For Nabatieh, poverty comparisons were highly dependent on the definition of the set of poor individuals. In particular, the relatively high rankings of Nabatieh in monetary poverty could hide deprivations in combined dimensions such as education, housing, and access to services. To a lesser extent the same is true for the South. Our results suggest that monetary poverty seems highly correlated with an accumulation of deprivations for ultra-poor individuals. The investigation of poverty by occupational group reveals comparisons that are consistent with the monetary approach. The impression may be given that being educated translates into greater opportunities for finding highly paid jobs located in areas endowed with basic services and better housing conditions. The reverse scenario would be particularly true for individuals employed in the agricultural sector. However, multidimensional poverty dominance tests challenge this simplistic view. Indeed, they indicate that it is not easy to extract a unique sector exhibiting the highest level of poverty. The only exception is the dominance of the financial sector. The same caution should be taken for poverty comparisons at a regional level. Robust comparisons seem to indicate that individuals living in Beirut are at the lowest risk of poverty while it is not possible to provide a clear ranking between the other mohafazat. This supports the argument for investigating observable characteristics other than expenditure when the objective is to improve the targeting of anti-poverty policy measures.

References

Alkire, S., & Foster, J. (2011). Counting and multidimensional poverty measurement. *Journal of Public Economics, 95*(7–8), 476–487.

Atkinson, A. (1987). On the measurement of poverty. *Econometrica, 55*(4), 749–764.

Atkinson, A., & Bourguignon, F. R., (1982). The comparison of multi-dimensioned distributions of economic status. *Review of Economic Studies,* 49(2), 183–201.

Bibi, S., Makdissi, P., & Yazbeck, M. (2012). Equivalence scales and housing deprivation orderings: An example using lebanese data. *Applied Economics*, 44(7), 853–866.

Blackorby, C., & Donaldson, D. (1978). Measures of relative equality and their meaning in terms of social welfare. *Journal of Economic Theory, 18*, 59–80.

Bourguignon, F., & Chakravarty, S. (2002). Multi-dimensional poverty orderings, Working Paper 2002–22, DELTA.

Bourguignon, F., & Chakravarty, S. (2003). The measurement of multidimensional poverty. *Journal of Economic Inequality* 1(1), 25–49.

Chakravarty, S. (1983). A new index of poverty. *Mathematical Social Sciences, 6*, 307–313.

Chakravarty, S., Mukherjee, D., & Ranade, R. (1998). On the family of subgroup and factor decomposable measures of multidimensional poverty. In D. Slottje, (ed.), *Research on Economic Inequality*, vol. 8. JAI Press, New York, pp. 175–194.

Chakravarty, S., & Silber, J. (2008). Measuring multidimensional poverty: The axiomatic approach. In N. Kakwani, & J. Silber, (eds.), Quantitative approaches to multidimensional poverty measurement. Palgrave Macmillan, New York, Chapter 11, pp. 192–209.

Chambaz, C., & Maurin, E. (1998). Atkinson and Bourguignon's dominance criteria: Extended and applied to the measurement of poverty in France. *Review of Income and Wealth, 44*(4), 497–513.

Clark, S., Hemming, R., & Ulph, D. (1981). On indices for the measurement of poverty. *The Economic Journal, 91*, 515–526.

Davidson, R., & Duclos, J.-Y. (2000). Statistical inference for stochastic dominance and for the measurement of poverty and inequality. *Econometrica, 68*(6), 1435–1464.

Duclos, J.-Y., & Makdissi, P. (2004). Restricted and unrestricted dominance for welfare. Inequality and poverty orderings. *Journal of Public Economic Theory, 6*, 145–164.

Duclos, J.-Y., & Makdissi, P. (2005). Sequential stochastic dominance and the robustness of poverty orderings, *Review of Income et Wealth*, vol. 51(1), 63–87.

Duclos, J. –. Y., Sahn, D., & Younger, S. (2006). Robust multidimensional poverty comparisons. *The Economic Journal, 116*(514), 943–968.

Laity, H., Abu-Ismail, K., & Hamdam, K. (2008). Poverty, growth and income distribution in lebanon, international poverty center country study no. 13.

Femise (2011). Towards a new Med region: achieving fundamental transitions, annual report, Femise.

Fishburn, P., & Willig, R. (1984). Transfer principles in income redistribution. *Journal of Public Economics, 25*, 323–328.

Foster, J., Greer, J., & Thorbecke, E. (1984). A class of decomposable poverty measures. *Econometrica, 52*(3), 761–766.

Foster, J., & Shorrocks, A. (1987). Transfer sensitive inequality measures. *Review of economic studies, 54*(3), 485–497.

Kakwani, N. (1980). On a class of poverty measures. *Econometrica, 48*(2), 437–446.

Kolm, S.-C. (1976). Unequal inequalities I. *Journal of Economic Theory, 12*, 416–442.

Foster, V., Tre, J. -P., & Wodon, Q. (2001). Energy prices, energy efficiency, and fuel poverty, Mimeo, World Bank, Washington DC

Makdissi, P., Wondon, Q. (2006). Environmental regulation and economic growth under education Externalities. *Journal of Economic Development, 31*(1), 45–51.

Ministry of Social Affairs & the United Nations Development Programme. (1998). *Mapping of living conditions in Lebanon*. Beirut: MOSA & UNDP.

Ministry of Social Affairs, the Central Administration of Statistics, & the United Nations Development Program (2006). Living conditions of households-the national survey of household living conditions, 2004, Beirut.

Ministry of Social Affairs & the United Nations Development Programme. (2007). *Development of mapping of living conditions in Lebanon, 1995–2004: A comparison with the results of "mapping of living conditions in Lebanon, 1998*. Beirut: MOSA & UNDP.

Quizilbash, M. (2006), Philosophical account of vagueness, fuzzy poverty measures and multidimensionality. In G. Betti., & A. Lemmi., (eds.), Fuzzy set approach to multidimensional poverty measurement, Springer, New York, Chapter 1, pp. 9–28.

Sen, A. (1976). Poverty: An ordinal approach to measurement. *Econometrica, 44*(2), 219–231.

Sen, A. K. (1997). On economic inequality, expanded edition, Clarendon Press, Oxford

Tsui, K.-Y. (2002). Multidimensional poverty indices. *Social Choice and Welfare, 19*(1), 69–93.

UNDP (2009). NHDR Lebanon 2008–2009, Toward a citizen's State, Technical report, UNDP.

UNDP (2010). Human Development Report: The Real Wealth of Nations: Pathways to Human Development New york, Palgrave Macmillan.

UNDP and Ministry of Social Affairs of Lebanon (2007). Development of mapping of living conditions in Lebanon, 1995–2004, "Capacity building for poverty reduction" Joint project between MoSA and UNDP.

Watts, H. (1968). An economic definition of poverty. In D. P. Moynihan (Ed.), *On understanding poverty*. New York, NY: Basic Books.

Zheng, B. (1999). On the power of poverty orderings. *Social Choice and Welfare, 16*(3), 349–371.

Chapter 6
A Non-Monetary Multidimensional Poverty Analysis of Tunisia Using Generalized Sen-Shorrocks-Thon Measures

Naouel Chtioui and Mohamed Ayadi

Abstract A monetary approach cannot represent the complex and multidimensional phenomena of poverty, as it only takes economic aspects into account. Many attempts have been made to propose multidimensional approaches to poverty using basic needs or capabilities approaches. For our study we adopted a non-monetary approach, using multivariate correspondence analysis to construct a composite welfare indicator as an aggregation index of the various well-being attributes. In order to measure poverty within a composite indicator distribution we first developed new classes of poverty measures. Indeed, we developed classes of ethical generalized Sen-Shorrocks-Thon (SST) poverty measures. These are a generalization of the Shorrocks (Econometrica 63:1225–1230, 1995) poverty measure which itself is a modified version of the Sen (Econometrica 44:219–231, 1976) poverty measure. We analyzed the non-monetary poverty trend in Tunisia between 1994 and 2006 using the composite welfare indicator and the generalized SST poverty measures. In order to achieve this, we constructed confidence intervals and tested hypotheses based on the bootstrap method. We found that poverty decreased between 1994 and 2006, but that there were inequalities within the poorer population. We also found that poverty was essentially rural poverty and that it was concentrated in the North West, Center West, and South East of Tunisia.

N. Chtioui (✉)
Département d'Economie, Méthodes Quantitatives et Informatique, Institut Supérieur d'Administration des Entreprises de Gafsa, et UAQUAP, Institut Supérieur de Gestion, Tunis, UAQUAP, I.S.G, 41 Cité Bouchoucha, 2000 Le Bardo, Tunisie
e-mail: naouel.tunisie@gmail.com

M. Ayadi
Département d'Economie et Méthodes, Institut Supérieur de Gestion et UAQUAP, Université de Tunis, I.S.G, 41 Cité Bouchoucha, 2000 Le Bardo, Tunisie
e-mail: med.ayadi@gnet.tn

V. Bérenger and F. Bresson (eds.), *Poverty and Social Exclusion Around the Mediterranean Sea*, Economic Studies in Inequality, Social Exclusion and Well-Being 9, DOI: 10.1007/978-1-4614-5263-8_6,

6.1 Introduction

For many years, poverty measurement has been based strictly on a monetary approach which views poverty as the consequence of unequal income distribution. Many measures have been proposed in the literature to measure monetary poverty.[1] However, currently there is a conviction among scientists, and also empirical evidence (see, e.g. Ravallion 1992) that income cannot be the only welfare indicator since it cannot reflect many other non-monetary aspects such as health, literacy, and social position and life perceptions. New approaches to welfare and poverty analysis have been proposed, for instance, the basic needs approach and the capability approach, which provide a new framework and a wide-ranging definition of welfare and poverty. The basic needs approach considers the well being of a person as encompassing minimum basic needs like education and health. Consequently, a person is poor if she is deprived of these basic needs (for further information about this approach see Streeten 1981). The capability approach, first developed by Sen (1985),[2] considers that the well being of a person is not her utility or her fulfillment of basic needs but is primarily the capabilities she has. They include her ability to transform the set of opportunities available to her into functionings achieved, as well as the freedom to choose these functionings—being educated, being in good health, and having a respectable position in society. A monetary approach has been used to measure the extent of economic poverty in Tunisia since about 1970. Ayadi et al. (2007)[3] were the first to analyze non-monetary multidimensional poverty in Tunisia between 1988 and 2001 by constructing a composite welfare indicator from a set of non-monetary primary factors. Following the methodology set out in Ayadi et al. (2007), we also propose to construct a composite welfare indicator using multivariate correspondence analysis.

In this article, we initially propose to develop new classes of poverty measures in order to measure poverty within a composite indicator distribution. To achieve this, we will construct classes of both relative and absolute generalized

---

[1] There currently exist different literature reviews on one-dimensional poverty measures (see for example, Foster 1984; Seidl 1988; Chakravarty 1990; Foster and Sen 1997; Zheng 1997).

[2] There exist different literature reviews of the capabilities approach, see e.g. Robeyns (2005).

[3] In an unpublished master's thesis Chtioui (2004) constructed a composite welfare indicator using factor analysis as developed by Stifel et al. (1999) and analyzed poverty in Tunisia. Bibi (2004) only applied multidimensional poverty measures developed in the literature on Tunisia. In relation to Tunisia both applied bidimensional stochastic dominance as developed by Duclos et al. (2006).

Sen-Shorrocks-Thon (SST) measures[4] using a normative approach.[5] We have chosen to generalize the SST index because it is a rank-based poverty measure and has the advantage of simultaneously measuring the Incidence, Intensity, and Inequality[6] among those who are poor. Also, the SST index is related to the Gini coefficient of inequality and is derived from a Gini social evaluation function. This relation will allow us to develop generalized SST poverty measures in the same manner as the generalized Gini inequality indices developed by Weymark (1981) and by Donaldson and Weymark (1980). And the widespread use of the Gini index will ensure that a class of generalized SST measures is applied as well.

At the beginning of 2011, we assisted to the Tunisian revolution against Ben Ali's 23 years long regime. This popular uprising can be explained in many ways e.g., as a result of the increase in unemployment and corruption, the absence of regional development in some poorer areas, and a high level of deprivation. We will apply the poverty measures we develop to see how poverty evolved between 1994 and 2006, and highlight regional disparities in Tunisia using the bootstrap method.

The remainder of the paper is organized as follows. Section 6.2 develops classes of generalized SST poverty measures. In Sect. 6.3, we present the methodology with the construction of a composite welfare indicator and statistical inference. The results are presented in Sect. 6.4. The final section is the conclusion.

6.2 Poverty Measurement

In order to measure poverty within a composite indicator distribution, we propose to develop classes of poverty indices.

According to Sen (1976), constructing a poverty measure invokes two types of problems: (i) the identification of those who are poor within the whole population

---

[4] A measure that satisfies the scale invariance axiom is called a relative poverty measure. This axiom requires that the value of a poverty measure should remain unchanged after a relative change in both real incomes and the real poverty line. A relative change consists for example of doubling real incomes and the real poverty line or dividing them by the poverty line. A measure that satisfies the translation invariance axiom is called an absolute poverty measure. This axiom stipulates that the value of a poverty measure should remain unchanged after an absolute change of both real incomes and the real poverty line. It requires that adding a real value to all real incomes and the real poverty line should leave the poverty value unchanged.

[5] The normative approach was proposed by Dalton (1920) for the measurement of income inequality. He recommended that an inequality index should have a normative foundation in terms of social evaluation and should incorporate society's judgments regarding inequality. Two well-known procedures were developed for the construction of a normative inequality measure. The first procedure, the Atkinson-Kolm-Sen (AKS) approach, was proposed separately by Atkinson (1970) and Kolm (1969) and popularized by Sen (1973) for developing relative indices. The second procedure, the Kolm-Pollak (KP) approach, was proposed separately by Kolm (1969) in income inequality literature and by Pollak (1971) in the consumer theory for developing absolute indices.

[6] The three 'I's of Poverty Curve were first introduced by Jenkins and Lambert (1997).

and (ii) the construction of an index that contains the information available on this poorer section of the population.

In order to solve the first problem, the literature adopted the poverty line method as proposed by Rowntree (1901) and Booth (1902). A poverty line (or threshold) is defined as the necessary income needed to attain the minimum of well being. Consequently, a person is poor if her income is below the poverty line. A poverty measure is then an aggregation of individual privations suffered by poorer people.

The first two poverty measures proposed in the literature were the headcount ratio and the poverty gap ratio. The headcount ratio represents the proportion of poor people in a population and the poverty gap ratio is the mean distance between the poverty line and the mean income of those who are poor. The simplicity of these ratios has meant that they have been used accurately in the field. But they have been criticized by Sen (1976) because they violate the following two basic axioms: the Monotonicity Axiom, which requires that a poverty measure should increase if the income of a poor person decreases; the Transfer Axiom, which requires that a transfer of income from a poor person to a less poor person should increase the poverty measure. At this point, according to Sen (1976), a poverty measure should have an axiomatic foundation. Following Sen's reasoning (1976) scholars have developed several poverty measures and proposed different intuitive and reasonable axioms. Zhen (1993) delimited the basic axioms for an effective poverty measure i.e., focus, symmetry, monotonicity, continuity, increasing poverty line, replication invariance, scale invariance, transfer, transfer sensitivity, and subgroup consistency.

The Sen (1976) poverty measure violates the more robust version of the transfer axiom; it is not continuous and is not a replication invariant. For that, Shorrocks proposed a modified version of the Sen's measure which is continuous, replication invariant and satisfies the transfer axiom. Zheng (1997) showed that the limits of the Shorrock (1995)' poverty measure are identical to the Thon (1979, 1983) measure. The Shorrocks measure is otherwise known as the Sen-Shorrocks-Thon measure.

Duclos and Gregoire (2002) proposed a class of generalized SST poverty measures[7] that "depend upon an ethical parameter which captures the sensitivity of poverty measurement to "exclusion" or "relative deprivation" aversion" and that are functions of the cumulative poverty gap curve.[8]

Using the normative approach, we have chosen to develop classes of ethical relative and absolute generalized SST poverty measures derived from the same generalized SST social evaluation function and based on axiomatic and ethical foundations. The relative and absolute generalized SST poverty measures that we propose to develop will be functions of a weighting system. This system is a generalization of the weighting system for the SST poverty measure. The advantage of this weighting system generalization is that it will transform the

---

[7] The functional form of the members of this class was first proposed by Chakravarty (1983) as an example of an ethical poverty measure.

[8] Jenkins and Lambert (1997), who introduced this curve in the measurement of poverty, called it a "TIP" curve (TIP or the three "I"s of poverty: Incidence, Intensity and Inequality among poorer people).

measures from rank-based poverty measures to more general poverty measures that can even be additively separable.

6.2.1 Preliminary Work and Notations

We have chosen to consider income distributions $\underline{x}(x_1, \ldots, x_i, \ldots, x_n)$ represented by vectors drawn from the set $\mathrm{N} = \bigcup_{n=1}^{\infty} S_n$; where $S = R^+$ or R^{++} and S_n is the set of all n-tuples of elements from S.

Let, $z \in S$ be the poverty threshold and $n = n\ (\underline{x})$ be the population size corresponding to $\underline{x}$;

Let, S_n^* representing the set of the censored distributions $\underline{x}^*\ (x_1^*, \ldots, x_i^*, \ldots, x_n^*)$; where $\underline{x}^*$ is censored at the poverty line z, $x_i^* = \left\{ \begin{matrix} x_i\ , x_i < z \\ z,\ \ x_i \geq z \end{matrix} \right\}$.

Let, $\underline{\tilde{x}}^*$ be a permutation of $\underline{x}^*$ such that $\underline{\tilde{x}}_1^* \geq \underline{\tilde{x}}_2^* \geq \ldots \geq \underline{\tilde{x}}_n^*$ and let $\underline{\hat{x}}^*$ be a permutation of $\underline{x}^*$ such that $\underline{\hat{x}}_1^* \leq \underline{\hat{x}}_2^* \leq \ldots \leq \underline{\hat{x}}_n^*$;

Let, $W(\underline{x})$ be a social evaluation function, the equally distributed equivalent (EDE) income[9] $\Xi(\underline{x})$ is given by:

$$W(\Xi(\underline{x})) = W(\mathrm{x}_1, \ldots, \mathrm{x}_\mathrm{n}). \tag{6.1}$$

6.2.2 Development of Generalized SST Poverty Measures

Using the normative approach, we have chosen to develop classes of both relative and absolute generalized Sen-Shorrocks-Thon poverty measures.

The Sen-Shorrocks-Thon (SST) poverty measure has the following expression:

$$\mathrm{SST}(\underline{x}^*; z) = \frac{1}{n^2} \sum_{i=1}^{n} (2n - 2i + 1) \frac{(z - \hat{x}_i^*)}{z}. \tag{6.2}$$

It can be rewritten as:

$$\begin{aligned} \mathrm{SST}(\underline{x}^*; z) = \mathrm{SST}^R(\underline{x}^*; z) &= 1 - \frac{\sum_{i=1}^{n} (2n - 2i + 1) \hat{x}_i^*}{n^2 z} \\ &= 1 - \frac{\sum_{i=1}^{n} (2i - 1) \tilde{x}_i^*}{n^2 z} = 1 - \frac{\sum_{i=1}^{n} (2i - 1) \tilde{x}_i^*}{\left(\sum_{i=1}^{n} (2i - 1)\right) z} \\ &= 1 - \frac{\Xi_G^n(\underline{x}^*)}{z}. \end{aligned} \tag{6.3}$$

[9] If each person receives an equally distributed equivalent income then we obtain a distribution which is socially equivalent to the initial distribution, but which is less unequal.

The function:

$$\Xi_G^n(\underline{x}^*) = \frac{\sum_{i=1}^n (2i-1)\tilde{x}_i^n}{\sum_{i=1}^n (2i-1)}. \tag{6.4}$$

represents the expression of the equally distributed equivalent income[10], and is a particular representation of the underlying social evaluation function of the Gini index in the case of a censored distribution function ($\underline{x}^*$) given by:

$$W_G^n(\underline{x}^*) = f\left(\Xi_G^n(\underline{x}^*)\right), \tag{6.5}$$

where f is an arbitrary increasing transformation function.

Hence, the SST poverty measure is a function of a particular Gini social evaluation function. This justifies our approach in developing generalizations of the SST poverty measure similar to those of the Gini index of inequality developed by Donaldson and Weymark (1980).

The function $\Xi_G^n(\underline{x}^*)$ is also positively linearly homogeneous and unit-translatable[11] and then can be used to construct relative SST poverty measures (given by the last line of the Eq. (6.3)) and also to construct absolute SST poverty measures given by.

$$\begin{aligned} \mathrm{SST}^A(\underline{x}^*; z) &= z - \Xi_G^n(\underline{x}^*) \\ &= z - \frac{\sum_{i=1}^n (2i-1)\tilde{x}_i^*}{n^2}. \end{aligned} \tag{6.6}$$

6.2.2.1 Class of Welfare Poverty Measures

Following the work of Donaldson and Weymark (1980), we propose to generalize the weights of the function $\Xi_G^n(\underline{x}^*)$ which will enable us to construct a class of generalized SST poverty measures whose EDE income function has the following expression.

$$\Xi_{\mathrm{GSST}}^n(\tilde{\underline{x}}^*) = \frac{\sum_{i=1}^n w_i^n \tilde{x}_i^*}{\sum_{i=1}^n w_i^n}. \tag{6.7}$$

We have chosen to assume that $w_i^n > 0$ and $\{w_i^n\}$ constitute a non-decreasing sequence of positive integers and are a function of n.

$$w_1^n \leq w_2^n \leq \ldots \leq w_n^n. \tag{6.8}$$

Similarly to Donaldson and Weymark (1980) and without any loss of generality, we set $w_1^n = 1$.

---

[10] From this point onwards and until the end of the paper, we will use the term "income" to signify a welfare indicator, which is not necessarily money income.

[11] The function $\Xi_G^n(\underline{x}^*)$ is positively linearly homogeneous if and only if W is homothetic and it is unit-translatable if and only if W is translatable.

The corresponding relative generalized SST poverty measures are derived as:

$$\mathrm{GSST}^R(\underline{x}^*; z) = 1 - \frac{\Xi^n_{\mathrm{GSST}}(\underline{\tilde{x}}^*)}{z}. \tag{6.9}$$

And the corresponding absolute generalized SST poverty measures are:

$$\mathrm{GSST}^A(\underline{x}^*; z) = z - \Xi^n_{\mathrm{GSST}}(\underline{\tilde{x}}^*). \tag{6.10}$$

The class of generalized SST poverty measures developed is a collection of families of indices, each of which corresponds to a population size. When the weighting function becomes independent of the population size, we obtain a subclass of generalized SST poverty measures that we have referred to as a class of single-series SST poverty measures and which is characterized by the following expression for its EDE income function:

$$\Xi^n_{\mathrm{SS}}(\underline{\tilde{x}}^*) = \frac{\sum_{i=1}^n w_i \tilde{x}_i^*}{\sum_{i=1}^n w_i}. \tag{6.11}$$

with

$$w_1 \le w_2 \le \ldots \le w_n. \tag{6.12}$$

Their corresponding relative and absolute poverty measures are respectively given by $1 - \frac{\Xi^n_{\mathrm{SS}}(\underline{\tilde{x}}^*)}{z}$ and $z - \Xi^n_{\mathrm{SS}}(\underline{\tilde{x}}^*)$.

The SST poverty measure is a member of this class with $w_i = 2i - 1$.

An inequality measure (as well as a poverty measure) must satisfy basic axioms for inequality measurement (poverty measurement). Donaldson and Weymark (1980) noted that the various generalized Ginis which have been developed fail to satisfy the principle of population. To achieve this, they identify families of welfare (and illfare) generalized Ginis and families of welfare (and illfare) single-series Ginis that respect the principle of population. Based on the relation of the SST with the Gini, we can identify the families of welfare generalized SST poverty measures and those for single-series SST poverty measures that satisfy the replication invariance axiom. Also, we can identify the illfare families that respect the principle of population.

First, we have chosen to begin with the class of welfare single-series SST poverty measures and define a function φ.

Définition 1. The function $\varphi : \{0, I\} \to R$ has the form:

$$\varphi(m) = \left\{ \begin{array}{c} 0 \; , m = 0 \\ \sum_{i=1}^m w_i \; , m \in I \end{array} \right\}, \tag{6.13}$$

Thus,

$$w_i = \varphi(i) - \varphi(i-1), \tag{6.14}$$

for all $i \in I / I = (1,\ 2,\ 3,\ 4, \ldots)$..

When the function φ takes a particular form, the EDE income function $\Xi^n_{SS}(\underline{\tilde{x}}^*)$ and its corresponding single-series SST measures satisfy the principle of population.

Proposition 1. *The family { Ξ^n_{SS}} given by:*

$$\begin{aligned}\Xi^n_{SS}(\underline{\tilde{x}}^*) &= \frac{\sum_{i=1}^n w_i \tilde{x}_i^*}{\sum_{i=1}^n w_i} \\ &= \frac{\sum_{i=1}^n [\varphi(i) - \varphi(i-1)]\tilde{x}_i^*}{\varphi(n)},\end{aligned} \tag{6.15}$$

$\varphi(0) = 0, \varphi(1) = 1$, *satisfies the PP if and only if:*

$$\varphi(qn) = \varphi(q)\varphi(n), \tag{6.16}$$

for all $q, n \in I$.

Proof. Similar to the proof provided by theorem 1 in Donaldson and Weymark (1980).

The form of the function φ, given by the Eq. (6.16), associated with the restriction on the weighting function $\{w_i\}$ (Eq. 6.12) leads to a particular class that we have referred to as the class of single-parameter SST (or simply S-SST) poverty measures.

Proposition 2. *The function* φ *with* $\varphi(0) = 0, \varphi(1) = 1$ *and*

$$\varphi(n+1) - \varphi(n) \geq \varphi(n) - \varphi(n-1), \tag{6.17}$$

for all $n \in I$,[12] satisfies the Eq. (6.16), if and only if:

$$\varphi(n) = n^v, \tag{6.18}$$

where $v \geq 1$, *is an arbitrary constant.*

Proof. Similar to the proof provided by theorem 2 in Donaldson and Weymark (1980).

This leads to the following form of the EDE income function:

$$\Xi^n_{S-SST}(\underline{\tilde{x}}^*) = \frac{\sum_{i=1}^n [i^v - (i-1)^v]\tilde{x}_i^*}{n^v}. \tag{6.19}$$

[12] This inequality is obtained from Eq. (6.12) where $w_{n+1} \geq w_n$..

Consequently, a relative S-SST poverty measure is obtained as[13]:

$$\mathrm{S-SST}^R(\underline{x}^*; z) = z - \frac{\sum_{i=1}^{n} [i^v - (i-1)^v]\tilde{x}_i^*}{n^v z}. \qquad (6.20)$$

And an absolute S-SST poverty measure is obtained as:

$$\mathrm{S-SST}^A(\underline{x}^*; z) = z - \frac{\sum_{i=1}^{n} [i^v - (i-1)^v]\tilde{x}_i^*}{n^v}. \qquad (6.21)$$

The parameter v can be considered as a parameter that represents the social judgments of the population and that reinforces the ethical nature of the measures which incorporate it, where different values of v represent different judgments. In line with Duclos and Grégoire (2002), we consider the ethical parameter v to represent aversion to relative deprivation.

- When $v = 1$, the S-SST poverty measure does not incorporate relative deprivation;
- $v = 2$, the S-SST poverty measure is the classical SST poverty measure;
- Increasing the values of v (due to an increase in the relative deprivation aversion) has as consequence an increase in the values of the S-SST poverty measure—a considerable weight is given to relative deprivation in computing poverty;
- $v \to \infty$ represents the maxmin rule; total poverty is assessed according to the relative deprivation of the most deprived people in the population.

6.2.2.2 Class of Illfare Poverty Measures

Most one-dimensional and multidimensional poverty measures have been based on illfare-ranked incomes ($\hat{x}_1^* \leq \hat{x}_2^* \leq \ldots \leq \hat{x}_n^*$). This is why we propose to construct a class of illfare poverty measures that depend on a weighting function.

The weighting system must respect our principle of according higher weights to most deprived people:

$$\pi_1 \geq \ldots \geq \pi_j \geq \ldots \geq \pi_n; \quad \hat{x}_1^* \leq \ldots \leq \hat{x}_j^* \leq \ldots \leq \hat{x}_n^*; \quad j = 1, \ldots, n. \qquad (6.22)$$

We begin by developing a class of illfare single-series SST poverty measures. For simplicity, we have chosen a weighting function $\{\pi_j\}$ as a particular permutation of the weighting function $\{w_i\}$.

Définition 2. We define the weighting function $\{\pi_j\}$ by:

$$\pi_j = w_{n-j+1}; \quad \text{for all} \quad j \in I, \qquad (6.23)$$

where $\pi_j > 0$.

---

[13] This expression was first proposed by Chakravarty (1983) as noted earlier.

Using this definition and the inequality relation on $\{w_i\}$ (Eq. 6.12), we obtain:

$$\pi_1 = w_{n-1+1} = w_n \geq \pi_2 = w_{n-2+1} = w_{n-1} \geq \ldots \succ \pi_n = w_{n-n+1} = w_1 = 1. \tag{6.24}$$

The definition also leads to the following equality:

$$\sum_{j=1}^{n} \pi_j = \sum_{j=1}^{n} w_{n-j+1} = \sum_{i=1}^{n} w_i. \tag{6.25}$$

The corresponding EDE income function is then defined by:

$$\hat{\Xi}^n_{\text{SS}}(\underline{\hat{x}}^*) = \frac{\sum_{j=1}^{n} \pi_j \hat{x}^*_j}{\sum_{j=1}^{n} \pi_j}. \tag{6.26}$$

which represents the class of illfare single-series SST poverty measures.

And their corresponding relative and absolute poverty measures are respectively given by $1 - \frac{\hat{\Xi}^n_{\text{SS}}(\underline{\hat{x}}^*)}{z}$ and $z - \hat{\Xi}^n_{\text{SS}}(\underline{\hat{x}}^*)$.

The SST poverty measure is a member of this class with $\pi_j = 2n - 2j + 1$.

Next, we can generalize this class of illfare single-series SST poverty measures in order to incorporate families of indices corresponding to different population sizes. We obtain a more general class of measures that we have referred to as a class of illfare generalized SST poverty measures with the following EDE income function:

$$\hat{\Xi}^n_{\text{GSST}}(\underline{\hat{x}}^*) = \frac{\sum_{j=1}^{n} \pi^n_j \hat{x}^*_j}{\sum_{j=1}^{n} \pi^n_j}, \tag{6.27}$$

where $\pi^n_j > 0$ and $\left\{\pi^n_j\right\}$ consists of a non-increasing sequence of numbers and are a function of n.

$$\pi^n_1 \geq \pi^n_2 \geq \ldots \geq \pi^n_n. \tag{6.28}$$

Without any loss of generality, we set $\pi^n_n = 1$.

As with the welfare class, we have chosen to identify the members of the class of illfare single-series poverty measures that satisfy the principle of population. For the welfare case, we have defined a function φ that can also be used to develop the illfare single-series poverty measures which satisfy the principle.

Proposition 3. *Function* φ*as defined previously can be rewritten as a function of* j *and* π_j.

$$\phi(m) = \left\{ \begin{array}{ll} 0, & m = 0 \\ \sum_{j=1}^{m} \pi_j, & m \in I \end{array} \right\}, \tag{6.29}$$

Thus,

$$\pi_j = \phi(m - j + 1) - \varphi(m - j), \quad \text{for all} \quad j \in I. \tag{6.30}$$

Proof. See Appendix A.

We have proved that the EDE income function (and its corresponding welfare single-series poverty measures) satisfies the principle of population (PP). This finding can be easily translated to the EDE income function that derives the illfare single-series SST measures.

Proposition 4. *The family* $\{\hat{\Xi}^n_{SS}\}$ *given by:*

$$\begin{aligned}\hat{\Xi}^n_{\mathrm{SS}}(\underline{\hat{x}}^*) &= \frac{\sum_{j=1}^n \pi_j \hat{x}^*_j}{\sum_{j=1}^n \pi_j} \\ &= \frac{\sum_{j=1}^n [\phi(n-j+1) - \phi(n-j)]\hat{x}^*_j}{\phi(n)},\end{aligned} \tag{6.31}$$

$\varphi(0) = 0, \varphi(1) = \varphi(n - n + 1) = 1$, *satisfies the PP if and only if* $\varphi(qn) = \varphi(q)\varphi(n)$, for all $q, n \in I$.

Proof. See Appendix A.

The form of the function φ, given by the Eq. (6.16), associated with the restriction on the weighting function $\{\pi_j\}$ (Eq. 6.24) leads to a particular class that we have referred to as a class of illfare single-parameter SST (or simply illfare S-SST) poverty measures, where the function φ has the following expression.

Proposition 5. *The function* φ, *with* $\varphi(0) = 0, \varphi(1) = 1$ *and* [14]

$$\varphi(n) - \varphi(n-1) \geq \varphi(n-1) - \varphi(n-2) \tag{6.32}$$

for all $n \in I$, *satisfies the Eq.* (6.16), *if and only if* $\varphi(n) = n^v$,*where* $v \geq 1$, *is an arbitrary constant.*

Proof. See Appendix A.

This leads to the following form of the EDE income function:

$$\hat{\Xi}^n_{\mathrm{S-SST}}(\underline{\hat{x}}^*) = \frac{\sum_{j=1}^n [(n-j+1)^v - (n-j)^v]\hat{x}^*_j}{n^v}. \tag{6.33}$$

Then, an illfare relative S-SST poverty measure is obtained as:

$$\mathrm{S-S\hat{S}T}^R(\underline{x}^*; z) = 1 - \frac{\sum_{j=1}^n [(n-j+1)^v - (n-j)^v]\hat{x}^*_j}{n^v z}. \tag{6.34}$$

And an illfare absolute S-SST poverty measure is obtained as:

$$\mathrm{S-S\hat{S}T}^A(\underline{x}^*; z) = z - \frac{\sum_{j=1}^n [(n-j+1)^v - (n-j)^v]\hat{x}^*_j}{n^v}. \tag{6.35}$$

---

[14] This inequality is obtained from Eq. (6.24) with $\pi_1 \geq \pi_2$.

$v = 2$ represents the SST poverty measure, which means that the SST measure simultaneously belongs to both the class of welfare single-series poverty measures and the illfare class.

6.2.2.3 Ethical Poverty Measures

Let, $\underline{y}^* = \left(y_1^*, \ldots, y_n^*\right)$ be the censored poverty gap vector of the population with $y_i^* = \max\left\{\left(z - x_i^*\right)/z, 0\right\}$ as the relative censored poverty gap of a person i.

As mentioned before, Dalton (1920) recommended that an inequality index should have a normative foundation in terms of social evaluation. The AKS approach and the KP approach developed ethical inequality measures from social evaluation functions (or from their representative EDE income functions).

Using the normative approach, Chakravarty (1983) developed a class of ethical poverty measures given by:

$$\mathrm{Ch}(\underline{x}^*; z) = \frac{[z - \Xi(\underline{x}^*)]}{z}. \tag{6.36}$$

Xu and Osberg (2002) showed that the SST poverty measure is a particular case of the ethical Chakravarty (1983) poverty measures.

$$\mathrm{SST}(\underline{x}^*; z) = \frac{[z - \Xi_G^n(\underline{x}^*)]}{z} = \Xi_G\left(\underline{y}^*\right). \tag{6.37}$$

The SST measure uses the Gini social evaluation function.

We propose to generalize Xu and Osberg's result to show that the members of the class of relative single-series SST measures and of the class of relative single-parameter (S-SST) measures are particular cases of the ethical Chakravarty (1983) poverty measures.[15]

For our generalization of the Eq. (6.37), we must have the following equality.

Lemma 1. *We have for all* $i \in I, v \geq 0$,

$$\sum\nolimits_{i=1}^{n} [i^v - (i-1)^v] = \sum\nolimits_{j=1}^{n} [(n-j+1)^v - (n-j)^v] = n^v, \tag{6.38}$$

Then, for all $\underline{x}^* \in S_n^*$, we have:

$$\sum\nolimits_{i=1}^{n} [i^v - (i-1)^v]\tilde{x}_i^* = \sum\nolimits_{j=1}^{n} [(n-j+1)^v - (n-j)^v]\hat{x}_j^*, \tag{6.39}$$

and:

$$\Xi_{\text{S-SST}}^n(\underline{\tilde{x}}^*) = \hat{\Xi}_{\text{S-SST}}^n(\underline{\hat{x}}^*). \tag{6.40}$$

[15] Similar results can easily be derived for absolute single-series SST measures and absolute single-parameter measures.

Proof. See Appendix A.

The lemma shows that the class of welfare relative S-SST poverty measures is equivalent to the class of illfare relative S-SST poverty measures and that the SST poverty measure ($v = 2$) belongs to both classes. In fact the Eq. (6.40) leads to have:

$$\begin{aligned} \mathrm{S} - \mathrm{SST}^R(\underline{x}^*; z) &= 1 - \frac{\Xi^n_{\mathrm{S-SST}}(\underline{\tilde{x}}^*)}{z} \\ &= 1 - \frac{\hat{\Xi}^n_{\mathrm{S-SST}}(\underline{\hat{x}}^*)}{z} \\ &= \mathrm{S} - \mathrm{S\hat{S}T}^R(\underline{x}^*; z). \end{aligned} \tag{6.41}$$

Our generalization of Eq. (6.37) is then given by:

Proposition 6. *The single-parameter SST poverty measure is a particular case of Chakravarty (1983)'s class of ethical poverty measures where.*

$$\mathrm{S} - \mathrm{SST}^R(\underline{x}^*; z) = \mathrm{SST}^v(\underline{x}^*; z) = \frac{\left[z - \hat{\Xi}_{G^v}(\underline{\hat{x}}^*)\right]}{z} = \hat{\Xi}_{G^v}\left(\underline{y}\right). \tag{6.42}$$

where $\hat{\Xi}_{G^v}(\underline{\hat{x}}^*)$ *is a particular representation of the underlying social evaluation function of a generalized Gini index in the case of a censored distribution function* $\underline{x}^*$.

Proof. See Appendix A.

The results of Lemma 1 are now generalized to the class of relative single-series SST poverty measures. In fact, the welfare relative single-series class is equivalent to the illfare relative single-series class where $\pi_j = w_i$.

Lemma 2. *We have for all* $i \in I, j \in I,$.

$$\sum\nolimits_{i=1}^{n} w_i = \sum\nolimits_{j=1}^{n} \pi_j, \tag{6.43}$$

Then, for all $\underline{x}^* \in S^*_n$, we have.

$$\sum\nolimits_{i=1}^{n} w_i \tilde{x}^*_i = \sum\nolimits_{j=1}^{n} \pi_j \hat{x}^*_j, \tag{6.44}$$

and

$$\Xi^n_{\mathrm{SS}}(\underline{\tilde{x}}^*) = \hat{\Xi}^n_{\mathrm{SS}}(\underline{\hat{x}}^*). \tag{6.45}$$

Proof. See Appendix A.

The generalization of Proposition 6 is given by:

Proposition 7. *The relative single-series SST poverty measure is a particular case of Chakravarty (1983)'s class of ethical poverty measures where:*

$$\text{SS} - \text{SST}^R(\underline{x}^*; z) = \text{SS} - \hat{\text{SST}}^R(\underline{x}^*; z) = \frac{\left[z - \hat{\Xi}_G^{\text{SS}}(\underline{x}^*)\right]}{z} = \hat{\Xi}_G^{\text{SS}}\left(\underline{y}\right). \tag{6.46}$$

where $\hat{\Xi}_G^{\text{SS}}(\underline{x}^*) = \frac{\sum_{j=1}^{n} \pi_j \hat{x}_j^*}{\sum_{j=1}^{n} \pi_j}$ *is a particular representation of the underlying social evaluation function of a generalized Gini index in the case of a censored distribution function* $\underline{x}^*$.

Proof. See Appendix A.

Here, we have developed an ethical structure based on a function of the social evaluation function (and on its corresponding EDE income function). It has enabled us to derive both relative and absolute poverty measures and this for different functions of weights.

6.3 Method

Unlike the monetary approach, which considers income as the only indicator of welfare, the non-monetary approach includes many primary indicators of welfare such as education, housing conditions, and household durables. It represents a multidimensional approach to welfare and consequently to poverty.

We adopted a non-monetary approach and constructed a composite welfare indicator as an aggregation of different attributes of well being. Hence, we developed the statistical inference necessary to conduct a multidimensional analysis of welfare and poverty in Tunisia.

6.3.1 Estimation of the Composite Welfare Indicator

We constructed an index of the assets owned by a household based on two categories of assets: household durables (in possession of a TV, bicycle, etc.) and household conditions (source of drinking water, floor material, etc.). We also considered human capital to be represented by education.

The index of assets has the following expression:

$$A_i = \sum_{p=1}^{P} \gamma_p a_{ip} \tag{6.47}$$

where A_i is the index of assets owned by the household $i (i = 1, \ldots, n)$, a_{ip} are the different assets considered $(p = 1, \ldots, P)$, and γsare the coefficients to be estimated.

Different methods have been proposed for the estimation. Hammer (1998) and Filmer and Pritchett (1998) used principal component analysis to construct an asset index and estimate the coefficients. Stifel, Sahn and Younger (1999) replaced it with factor analysis. Ayadi et al.(2007) adopted multivariate correspondence analysis (MCA)[16] as developed by Asselin (2002). We also use MCA to estimate the coefficients because our surveys contain categorical data.

In order to analyze categorical variables, we consider that each primary indicator (p) has K_p response categories so that the index of assets becomes a function of the categories as follows:

$$A_i = \frac{\sum_{p=1}^{P} \sum_{k=1}^{K_p} W_k^p I_{ipk}}{P} \tag{6.48}$$

where I_{ipk} are binary variables (take value 1 if the household i owns the category k and 0 if not) and W_k^p are the coefficients to be estimated.

6.3.2 Description of Data and Selection of the Attributes

6.3.2.1 Data

Our empirical analysis was carried out using primary microdata from three demographic and health surveys conducted by the 'Office National de la Famille et de la Population' (ONFP) as a part of the Pan Arab Project for the development of children (PAPCHILD project) in collaboration with the League of Arab Countries' Pan Arab Project for Family Health. The surveys contain demographic information, housing conditions, health status, social-cultural characteristics and the level of education for each member of the family. The 1994 survey covered 6,085 households (with 33,750 individuals) reduced to 6,037 households, the 2001 survey covered 6,083 households (with 31,505 individuals) reduced to 6,051 households, and the 2006 survey covered 8,682 households (with 42,677 individuals) reduced to 8,186 households.

6.3.2.2 Choice of the attributes

We selected 16 variables from the three sets of data; these variables represent a range of primary indicators of well being or of achieved functionings. We classified them into five categories as shown in Table 6.1. Then, in order to harmonize the data in the three surveys, we recoded and renamed the modalities of the selected variables. Next, we merged the three sets of data and applied MCA.[17] The estimated weights (W_k^p in Eq. 6.48) for the attributes were used to compute the values of the composite welfare for the three sets of data (Eq. 6.48).

[16] This method is more appropriate when using and retaining only categorical variables for the analysis.

[17] Ayadi et al. (2007) conducted MCA using Stata. We opted to use the *R* language of programing because it gives more appropriate weights for the attributes.

Table 6.1 Preliminary list of 16 variables

Dimensions (primary indicators)	Attributes (variables)
1. Housing conditions	Floor material
	Type of house
	Statute of occupation
	Number of persons per bedroom
	Source of energy
2. Water and toilet availability	Water inside the house
	Toilet inside the house
3. Human capital	Level of education of the household head
4. Ownership of household durables	Refrigerator
	Gas cooker
	Water heating
	Washing machine
	Conditioner
5. Access to the technologies of information and telecommunication (TIC)	Television
	Radio
	Telephone

6.3.3 Statistical Inference

For poverty measurement, we first needed to choose a poverty threshold. We used a relative poverty threshold as a proportion k of the median of the distribution of the composite welfare indicator. We chose k as equal to 50 % and applied the absolute single-parameter SST poverty indices (Eq. 6.21 with $v = 2, 3, 4$). In order to compare them to the FGT poverty measures, we also compute the headcount ratio and absolute FGT measures (with $\alpha = 1, 2$):

$$FGT\left(\underline{x}_p; z; \alpha\right) = \frac{1}{n}\sum\nolimits_{i=1}^{q} \left(z - x_i\right)^{\alpha}, \alpha \geq O, \tag{6.49}$$

where α is a parameter of poverty aversion.

In order to analyze poverty in Tunisia, we computed:

1. the estimated values from 1994, 2001, and 2006 (k equal to 50 %) to see how poverty evolved;
2. the estimated values for rural and urban areas (same threshold equal to 75 % of median value of 2006) and this for the year 2006, to see which areas were more deprived;
3. the estimated values for the 7 regions (see Table 6.2, same threshold equal to 75 % of median value of 2006) and this for the year 2006, to see which regions had more poverty.

The values could be influenced by sampling variability which can be eliminated by constructing confidence intervals and developing hypothesis tests. For that, we needed to compute the standard errors (or variance) for the estimates, which is

Table 6.2 The seven regions in Tunisia

Tunis and nearby	North East	North West	Center East	Center West	South East	South West
Tunis	Nabeul	Beja	Sousse	Kairouan	Gabes	Gafsa
Ariana	Zaghouan	Jendouba	Monastir	Kasserine	Medenine	Tozeur
Manouba	Bizerte	Kef	Mahdia	Sidi Bouzid	Tataouine	Kebili
Ben Arous		Siliana	Sfax			

complex for poverty measures. Also, the data distribution is unknown and we could not apply parametric tests. The bootstrap procedure[18] resolves these two problems and makes it both possible and easy to compute standard errors, to construct confidence intervals and to test hypotheses.

Xu (1998) was the first to develop the statistical inference of the SST index and used the bootstrap method to compute the standard error of its estimator. Osberg and Xu (1999)[19] used the SST measure and its multiplicative decomposition as developed by Xu and Osberg (2002) to compare poverty intensity between provinces in Canada and he used the bootstrap method to construct the confidence intervals. Osberg (2000) used the SST measure and its multiplicative decomposition to compare poverty between Canada and the United States. Osberg and Xu (2000)[20] proposed an alternative formulation for the SST index which was more appropriate when surveying data with sampling weight. They also used the bootstrap procedure and carried out international comparisons of poverty from the Luxembourg Income Study (LIS) data. In order to analyze poverty in Tunisia, we have chosen to develop the bootstrap inferential procedure for the generalized SST measures.

We developed the generalized SST poverty measures in the same manner as generalized Gini inequality indices. Xu (2000) showed that the generalized Gini estimator is not asymptotically pivotal, which means that we could not use the standard method to construct bootstrap confidence intervals. To achieve this he used the iterated bootstrap method. The generalized SST poverty measure is also not asymptotically pivotal, but rather than using the iterated bootstrap method, we used the percentile method (we constructed intervals using the standard percentile method and an alternative method developed by Hall (1992).[21]

[18] The bootstrap technique was introduced by Efron (1979, 1981, 1982) and further developed by Efron and Tibshirani (1993) among others.

[19] The Osberg and Xu (1999) approach has been adopted by the World Bank in the Handbook of Poverty Analysis.

[20] The Osberg and Xu (2000) approach has been adopted by Luxembourg Income Study.

[21] The Hall (1992) method consists of computing the difference between each bootstrap point estimate and the point estimate of the initial data: $\hat{y}_b^* = \hat{P}_b^* - \hat{P}$. Then, it establishes the $\frac{\alpha}{2}$ and the $(1-\frac{\alpha}{2})$ limits of the new distribution of the $\hat{y}_b^*$s. Finally, the confidence interval limits are obtained as follows

$$(1-\alpha)100\%CI = \left[\hat{P} - \hat{y}^*_{\left(1-\frac{\alpha}{2}\right)}, \hat{P} - \hat{y}^*_{\left(\frac{\alpha}{2}\right)}\right];\ \alpha\text{: significance level.}).$$

Finally, we carried out hypothesis tests. We tested the null hypothesis of poverty equality between A and B against the alternative hypothesis that A has less poverty than B. In order to do this we constructed the following test:

$$\begin{aligned} H_0 &: P(x_A, z_A) = P(x_B, z_B) \\ &\text{versus} \\ H_1 &: P(x_A, z_A) < P(x_B, z_B) \end{aligned} \tag{6.50}$$

where $P(x_A, z_A)$: poverty measure of A and $P(x_B, z_B)$: poverty measure of B and significance level equal 5 %.

In adherence to Moran (2005) work's, we computed the confidence interval of the bootstrap estimates for the difference $P_A - P_B$. If zero did not figure in this interval, we rejected H_0 and concluded that A has less poverty than B.[22]

6.4 Results

6.4.1 *Results of the MCA*

6.4.1.1 Analysis of the MCA

We conducted a final MCA on fifteen variables.[23] The histogram of the eigenvalues (Table B.1, Appendix B) shows the importance of the first factorial axis, which accounts for 18.61 % of total inertia while the remainder of the axes account for less than 8 %. Hence, the first factorial axis can be considered as the axis of welfare and the analysis of the MCA carried out with regard to the relations between the variables and this axis. The wealth in welfare is represented by variables positively correlated to the first factorial axis whereas the poverty in welfare is represented by variables negatively correlated to this axis. Next, the first axis was used to compute the values of the composite welfare indicator for the households in the three sets of data and to compute poverty measures for 1994, 2001, and 2006.

The representation of the first factorial plan in Table B.2 (Appendix B) shows the importance of the first factorial axis in representing a household's welfare. The variables representing wealth in welfare are in the left part (positively correlated to

[22] We developed a program using the *R* language of programing which contains estimates of different poverty measures, statistics to be computed (mean, standard error), estimates of the different bootstrap confidence intervals and estimates of the different hypothesis tests to be conducted.

[23] We conducted a first MCA on the preliminary attributes selected. This technique produced non-intuitive results for the "statute of occupation" variable. They showed a negative weight for its "owner" and "joint owner" modalities and a positive weight for its "Rent the house" modality along with a low contribution to the first factor. It appears evident that these variables should not be included.

the axis) and those representing poverty in welfare are in the right part (negatively correlated to the axis). The first factorial axis distinguishes between the richer population and the poorer population. The second factorial axis distinguishes between two categories inside each population; rich and very rich households/poor and very poor households.

6.4.1.2 Consistency of the MCA

Table B.3 (Appendix B) presents the relations of the variables with the two first factors and contains their scores and contributions. All the variables verify the appropriateness of the first axis ordering consistency (FAOC) which requires that the ordinal structure of each variable is respected by the ordinal structure of its modalities scores. For example, considering the gas cooker variable, its "having it" modality is in the left (positively correlated to the first factorial axis) and its other modality is in the right (negatively correlated to the first factorial axis). This appropriateness ensures that the composite welfare indicator reflects a welfare situation.

The analysis of the weights of the 15 attributes shows that owning durables, possessing high quality floor material, having a popular house, a villa, or an apartment, having fewer than two persons per bedroom, having electricity, having running water and a toilet inside the house and having an education contribute positively to the composite welfare indicator of the household.

On the other hand, having poor quality floor material, not having a house, having more than two persons per bedroom, having no access to electricity, no running water, and no toilet inside the house, no durables, no access to TIC, being without an education all contribute negatively to the composite welfare indicator of the household and represents its poverty.

We note that the most significant contributing attributes to the first factorial axis were durables (refrigerator, washing machine, boiler, and gas cooker), type of house, number of persons per bedroom, source of energy, education of the household head, water and toilet availability, and three means of TIC: television, telephone, and radio. Air conditioning and floor material made the smallest contribution to the first factor.

6.4.1.3 Evolution of the Composite Welfare Indicator Between 1994, 2001, and 2006

Table B.4 (Appendix B) reports the mean values of the composite welfare indicators for the 3 years. It shows an improvement in the indicator value between 1994 and 2006, denoting an improvement in well being of households during this period.

This is confirmed by Fig. 6.1 which represents the cumulative distributions of the composite welfare indicator for 1994, 2001, and 2006, respectively.

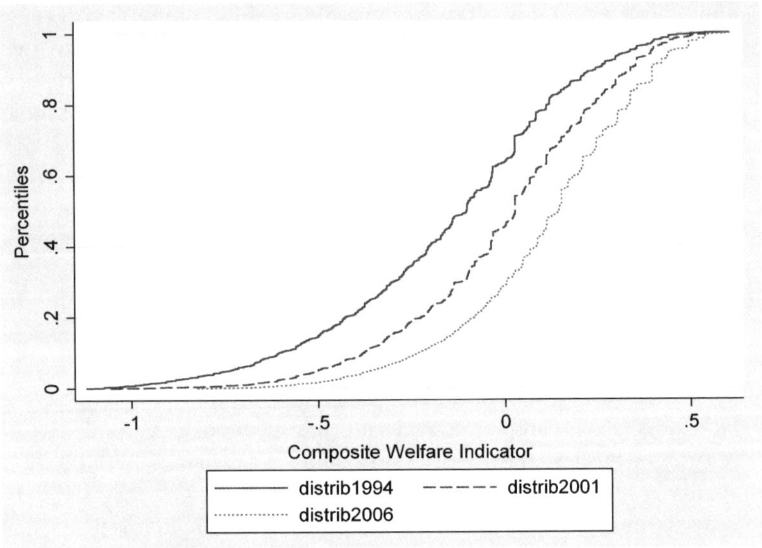

Fig. 6.1 Cumulative distributions of the composite welfare indicator for 1994, 2001 and 2006

6.4.2 Analysis of Poverty in Tunisia

The distribution of the computed composite welfare indicator has some negative values. Hence, we transformed it by adding the absolute value of the lowest score to each household's score in order to obtain a new distribution for the indicator defined on a positive support.[24]

The results of the application of the program developed on data from 1994, 2001, and 2006 are reported in Tables C.1, C.2, C.3, C.4, C.5, C.6, C.7, C.8 (see Appendix C).

6.4.2.1 Evolution of Poverty Between 1994 and 2006

We can now compare poverty between 1994 and 2006 to see if poverty has increased or decreased in Tunisia during this period (Table C.1). All poverty measure values show a decrease in poverty between 1994 and 2001 and between 2001 and 2006. The decrease in poverty over the period 1994–2001 is more important than the decrease in poverty over the period 2001–2006 for all the poverty measures (as shown in Fig. 6.2). Thus, poverty decreased between 1994 and 2006.

We also computed poverty measures by choosing three poverty lines. Table C.2 demonstrates that the poverty measure values increase with the poverty line for each of the poverty measures and for each of the 3 years (1994, 2001, and 2006).

[24] As noted before, this absolute change has no effect in terms of poverty on the values of absolute poverty measures that we apply in this section.

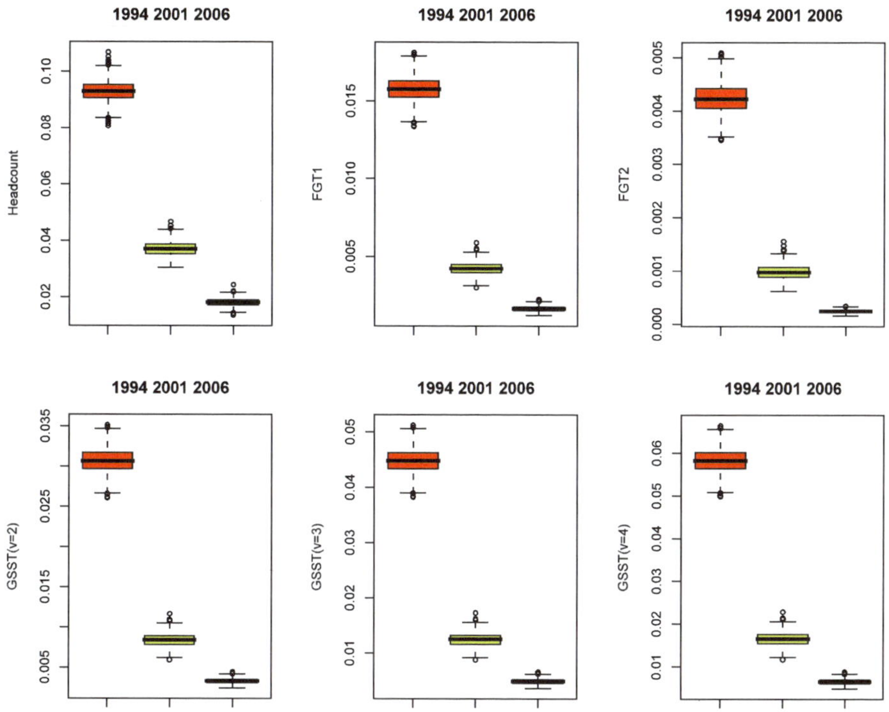

Fig. 6.2 Poverty in the composite welfare indicator distribution in 1994, 2001 and 2006

We note that the value of the generalized SST poverty measure increases when the value of the parameter $v$increases and this is due to the increase in the aversion to relative deprivation.

We cannot, according to the values of the measures, conclude that poverty has decreased in Tunisia. For that, we must construct confidence intervals and conduct hypothesis tests for all the measures. The confidence intervals are an estimation of the real limits of the poverty measures of the population and the hypothesis test is a robust way to validate or invalidate our conclusion concerning the decrease in poverty. Table C.3 contains all the confidence intervals computed for all the years and all the poverty measures.

We also tested the statistical significance of poverty changes in Tunisia (see Table C.4), first between 1991 and 2001 (Case 1 with 2001 being distribution A) and then between 2001 and 2006 (Case 2 with 2006 being distribution A).

We can see that zero does not figure in any confidence interval in both case 1 and case 2, so we can reject the null hypothesis and conclude that poverty in Tunisia decreased significantly between 1994 and 2001 and decreased significantly between 2001 and 2006.

Figure 6.2 represents the trend of the values of the poverty measures as applied to Tunisia between 1994 and 2001. It shows that the value of the headcount ratio, the values of the FGT poverty measures and the values of the generalized SST poverty measures decline in this period, hence confirming our conclusion that poverty did indeed decrease.

6.4.2.2 Comparison Between Rural Poverty and Urban Poverty in 2006

We compared poverty between rural areas and urban areas in 2006 to see which type of area was more deprived (Table C.5). All poverty measure values show that rural areas have more poverty than urban areas.[25]

In order to validate or invalidate this result, we conducted a hypothesis test (Table C.6, A represents an urban area and B represents a rural area). We can see that zero does not figure in any confidence interval, so we can reject the null hypothesis of poverty equality between areas and conclude that urban areas were significantly less deprived than rural areas in 2006.

6.4.2.3 Regional Disparities

We computed poverty measures for the seven regions in Tunisia in 2006 (Table C.7).

All poverty measures values gave the same ranking for the regions: The Capital Tunis and surrounding area was less deprived than the Center East, itself less deprived than the South West. The latter was less deprived than the North East which was less deprived than the South East. While the Center West of Tunisia was the most deprived region and was more deprived than the North West.

In order to validate or invalidate the presence of regions disparities, we conducted hypothesis tests (Table C.8):

Case 1: A represents the Capital Tunis and surrounding area; B represents the North West.

Case 2: A represents the Capital Tunis and surrounding area; B represents the Center West.

Case 3: A represents the Capital Tunis and surrounding area; B represents the South East.[26]

In fact, we compared the Capital and the surrounding area to the North West (Case 1), the Center West (Case 2), and the South East (Case 3) to see if the Capital and the surrounding area had less poverty than the other regions or not.

---

[25] For area comparisons and regional comparisons we chose a poverty line equal to 75 % of the median rather than a poverty line equal to 50 % of the median. This is because for rural areas and some regions, when applying the second threshold we only obtained a small number of poorer people and this prevented us from using the bootstrap technique and computing poverty measures.

[26] We limited our regional comparisons. For each of the three cases we chose to test regional disparities between the most deprived regions and the least deprived regions (the Capital Tunis and its surrounding area, as shown by the poverty values in Table C.8).

We can see that zero does not figure in any confidence interval in Case 1, Case 2, and Case 3, so we can reject the null hypothesis of poverty equality between regions and conclude that in 2006 Tunis and the surrounding area was significantly less deprived than the North West, the Center West, and the South East.

6.5 Discussion

In a similar vein to Ayadi et al. (2007), we proposed a non-monetary multidimensional poverty analysis for Tunisia, but using different methodology and a different period of analysis. In fact, we developed a similar composite welfare indicator, but based on more primary indicators. Also, we developed and applied some generalized SST poverty measures whereas Ayadi et al. (2007) only applied the FGT poverty measures. They analyzed poverty using a DHS survey from 1988 and ONFP surveys from 1994 and 2001. We, however, considered poverty in Tunisia between 1994 and 2006 using the 1994, 2001, and 2006 ONFP surveys.

Both (our results and the Ayadi et al. (2007) results) show that poverty decreased between 1994 and 2001. We also found a decrease in poverty between 2001 and 2006. This decrease can be explained by the economic growth in Tunisia which led to an improvement in the quality of life for Tunisian households, an improvement in household conditions, and an increase in the access of households to public services (water, electricity, health).[27]

The disparities between the areas in 2006 show how this economic growth benefited urban families more. These families had better access to public services than rural families. In addition, they had better household conditions, for example, good quality floor material, toilet and running water inside the house, and generally a better level of education for the household head. This means that poverty was still a rural poverty in 2006 (Ayadi et al. (2007) showed that poverty was a rural poverty in the period 1988–1994).

Similarly to Ayadi et al. (2007), we also tried to compare regions in terms of privation. We found that the least deprived regions were Tunis and its surrounding area, Center East and South West. The first two regions contain the capital and its surrounding area, and Sfax (both Tunis and Sfax are Tunisia's most industrialized cities).

They also contain Sousse, Monastir and Mahdia, which are tourist cities. All of them provide many jobs and contribute to the economic growth of the country. This is why the households in these regions enjoy good conditions, are modernized and experience high levels of education.

Like Ayadi et al. (2007), we found that the more deprived regions were Center West and North West. They are agricultural regions and they face many problems, for example:

---

[27] For more details on the economical and social changes in Tunisia since the independence see Ayadi et al. (2007).

- People and young people do not like working in agriculture and choose to either go to industrialized cities for work or leave the country;
- Until independence in 1956 the Tunisian government did not try to promote the economic development of these regions; it concentrated its efforts on Tunis and Center East. The other regions were marginalized and this has contributed to their privation and poverty.

These reasons clearly explain the changes which occurred in the country at the begining of 2011 when the Tunisian revolution started in Sidi Bouzid and Kasserine, both located in Center West, the most deprived region, because people do not want to be poor; they want to work and to improve their well being.

We have also shown that the South East is among the deprived regions despite its natural resources (gas, mineral oil) in Tatatouine, Zarzis. The free trade zone in Zarzis/Djerba provides many jobs but these jobs are not given to people originating from the region. This accounts for the higher unemployment rate and the recent phenomenon of illegal migration by young people from Zarzis to Europe via Lampadoza in Italy.

6.6 Conclusion

A monetary approach cannot represent the complex and multidimensional phenomena of poverty as it is limited to economic considerations. New approaches to welfare and poverty analysis have been proposed such as the basic needs approach and the capability approach which provide a new framework and a wide-ranging definition of welfare and poverty. We chose to adopt a non-monetary approach which considers the well being of a person to be represented by the fulfillment of some basic needs or the achievement of some functionings. We then constructed a composite welfare indicator as an aggregation for different attributes of well being using MCA.

We also developed an ethical structure (based on the social evaluation function) which allowed us to derive both relative and absolute poverty measures and this for different functions of weights and different parameters representing different social judgments. This structure can be easily implemented and used by decision makers and practitioners to measure (monetary or non monetary) poverty for any country.

The application of the composite welfare indicator and single-parameter SST poverty measures developed for Tunisia gives an idea of how poverty evolved between 1994 and 2006; it shows the disparities between regions in 2006. We found that poverty decreased between 1994 and 2006, but that there was inequality within the poorer population. We also found that during this period poverty was a rural phenomenon and that it was concentrated in the North West, Center West and South East of Tunisia explaining the revolution against Ben Ali's 23-year-long regime.

Appendix A: Proofs

Proof of Proposition 3

Proof 1:$\phi(m) = \sum_{j=1}^{m} \pi_j, m \in I$.

$$\begin{aligned}
\varphi(m) &= \sum_{i=1}^{m} w_i \\
&= w_1 + w_2 + \ldots + w_m \\
&= w_{m-m+1} + w_{m-(m-1)+1} + \ldots + w_{m-1+1} \\
&= \sum_{j=1}^{m} w_{m-j+1} \\
&= \pi_m + \pi_{m-1} + \ldots + \pi_1 \\
&= \sum_{j=1}^{m} \pi_j.
\end{aligned}$$

Proof 2:$\pi_j = \varphi(i) - \varphi(i-1)$.

For $i = m, j = 1$; prove that $\pi_1 = \varphi(m) - \varphi(m-1)$..

By proof 1, we have $\varphi(m) = \sum_{j=1}^{m} \pi_j = \sum_{j=1}^{m} w_{m-j+1} = \sum_{i=1}^{m} w_i$.

We have $j = m - i + 1$, if $i = 1 \rightarrow j = m$,if $i = m \rightarrow j = 1$and if $i = m - 1 \rightarrow j = 2$.

Then, by change of variable in the precedent equation, we obtain

$$\varphi(m-1) = \sum_{j=2}^{m} \pi_j = \sum_{j=2}^{m} w_{m-j+1} = \sum_{i=1}^{m-1} w_i.$$

Verification

$$\begin{aligned}
\varphi(m-1) &= \sum_{i=1}^{m-1} w_i \\
&= w_1 + w_2 + \ldots + w_{m-1} \\
&= w_{m-m+1} + w_{m-(m-1)+1} + \ldots + w_{m-2+1} \\
&= \sum_{j=2}^{m} w_{m-j+1} \\
&= \pi_m + \pi_{m-1} + \ldots + \pi_2 \\
&= \sum_{j=2}^{m} \pi_j.
\end{aligned}$$

Then,

$$\begin{aligned}
\pi_1 &= (\pi_1 + \pi_2 + \ldots + \pi_m) - (\pi_2 + \pi_3 + \ldots + \pi_m) \\
&= \sum_{j=1}^{m} \pi_j - \sum_{j=2}^{m} \pi_j \\
&= \varphi(m) - \varphi(m-1).
\end{aligned}$$

Gradually, we obtain $\pi_j = \varphi(i) - \varphi(i-1)$.

Proof 3: $\pi_j = \varphi(m - j + 1) - \varphi(m - j)$.

We have, $j = m{-}i + 1$, then by change of variable, we obtain:

$\pi_j = \phi i - \phi i - 1 = \phi m - j + 1 - \phi m - j$.

Proof of Proposition 4

Similar to the proof of Proposition 1, we just have to replace i by $n{-}j + 1$.

$$\begin{aligned}\pi_j &= \phi(i) - \phi(i-1)\\ &= \phi(m-j+1) - \phi(m-j)\end{aligned}$$

Proof of Proposition 5

We have $\pi_j = w_{n-j+1}$.

Then $w_1 \leq w_2 \leq \ldots \leq w_n \Leftrightarrow \pi_1 \geq \pi_2 \geq \ldots \geq \pi_n$.

So the function φ with $\pi_1 \geq \pi_2 \geq \ldots \geq \pi_n$.

satisfies the equation $\varphi(kn) = \varphi(k)\varphi(n)$,

if and only if $\varphi(n) = n^v$.

Proof of lemma 1

I/$\sum_{i=1}^n [i^v - (i-1)^v] = \sum_{j=1}^n [(n-j+1)^v - (n-j)^v] = n^v$.

Let, $k = i - 1 \Rightarrow i = 1 \rightarrow k = 0$ and $i = n \rightarrow k = n - 1$.

Proof 1:$\sum_{i=1}^n [i^v - (i-1)^v] = n^v$.

$$\begin{aligned}\sum_{i=1}^n [i^v - (i-1)^v] &= \sum_{i=1}^n i^v - \sum_{i=1}^n (i-1)^v\\ &= \sum_{i=1}^n i^v - \sum_{k=0}^{n-1} k^v\\ &= \sum_{i=1}^n i^v - \left[\sum_{k=1}^n k^v + 0^v - n^v\right]\\ &= \sum_{i=1}^n i^v - \sum_{i=1}^n i^v + n^v\\ &= n^v.\end{aligned}$$

Proof 2: $\sum_{j=1}^n [(n-j+1)^v - (n-j)^v] = n^v$.

We have $j = n{-}i + 1$, then $\sum_{j=1}^n [(n-j+1)^v - (n-j)^v] = \sum_{i=1}^n [i^v - (i-1)^v]$.

By proof 1, we have $\sum_{i=1}^n [i^v - (i-1)^v] = n^v$.

Then $\sum_{j=1}^n [(n-j+1)^v - (n-j)^v] = n^v$.

II/$\sum_{i=1}^{n} [i^v - (i-1)^v]\tilde{x}_i^* = \sum_{j=1}^{n} [(n-j+1)^v - (n-j)^v]\hat{x}_j^*$.

We have $j = n-i+1$, then $i^v - (i-1)^v = (n-j+1)^v - (n-j)^v$.

And we have $\sum_{i=1}^{n} \tilde{x}_i^* = \sum_{j=1}^{n} \hat{x}_j^* . \Rightarrow \sum_{i=1}^{n} [i^v - (i-1)^v]\tilde{x}_i^* = \sum_{j=1}^{n} [i^v - (i-1)^v]\hat{x}_j^* . \Rightarrow \sum_{i=1}^{n} [i^v - (i-1)^v]\tilde{x}_i^* = \sum_{j=1}^{n} [(n-j+1)^v - (n-j)^v]\hat{x}_j^*$.

Dividing by n^v, we obtain $\Xi_{S-SST}^{n}(\tilde{x}^*) = \hat{\Xi}_{S-SST}^{n}(\hat{x}^*)$.

Proof of Proposition 6

Proof 1: $S - SST^R(\underline{x}^*; z) = \frac{\left[z - \hat{\Xi}_{G^v}(\hat{\underline{x}}^*)\right]}{z}$.

By Eq. (6.41), we have

$$\begin{aligned}
S - SST^R(\underline{x}^*; z) &= S - S\hat{S}T^R(\underline{x}^*; z) \\
&= 1 - \frac{\sum_{j=1}^{n} [(n-j+1)^v - (n-j)^v]\hat{x}_j^*}{n^v z} \\
&= SST^v(\underline{x}^*; z) \\
&= 1 - \frac{\hat{\Xi}_{S-SST}^{n}(\hat{\underline{x}}^*)}{z} \\
&= \frac{z - \hat{\Xi}_{S-SST}^{n}(\hat{\underline{x}}^*)}{z} \\
&= \frac{z - \hat{\Xi}_{G^v}(\hat{\underline{x}}^*)}{z}.
\end{aligned}$$

Proof 2: $S - SST^R(\underline{x}^*; z) = \hat{\Xi}_{G^v}\left(\underline{y}^*\right)..$

We have $\sum_{j=1}^{n} [(n-j+1)^v - (n-j)^v] = n^v$.

Hence,

$$\begin{aligned}
S - SST^R(\underline{x}^*; z) &= S - S\hat{S}T^R(\underline{x}^*; z) \\
&= 1 - \frac{\sum_{j=1}^{n} [(n-j+1)^v - (n-j)^v]\hat{x}_j^*}{n^v z} \\
&= \frac{n^v z - \sum_{j=1}^{n} [(n-j+1)^v - (n-j)^v]\hat{x}_j^*}{n^v z} \\
&= \frac{\sum_{j=1}^{n} [(n-j+1)^v - (n-j)^v]}{n^v}\left(\frac{z - \hat{x}_j^*}{z}\right) \\
&= \frac{\sum_{j=1}^{n} [(n-j+1)^v - (n-j)^v]}{n^v} y_j^* \\
&= \hat{\Xi}_{G^v}\left(\underline{y}^*\right).
\end{aligned}$$

where $y_j^* = \frac{z - \hat{x}_j^*}{z}$ so $\hat{x}_j^* \uparrow \Rightarrow y_j^* \downarrow$.

Proof of Lemma 2

Similar to the proof of Lemma 1, we just have to replace $i^v - (i-1)^v$ by w_i and $(n-j+1)^v - (n-j)^v$ by π_j.

Proof of Proposition 7

Similar to the proof of Proposition 6, we just have to replace $i^v - (i-1)^v$by w_i and $(n-j+1)^v - (n-j)^v$ by π_j.

Appendix B: Multivariate Correspondence Analysis

Table B.1–B.4

Table B.1 Histogram of the eigen values

Number	Eigenvalue	Percentage	Cumulative Percentage
1	0.2977	18.61	18.61
2	0.1270	7.93	26.54
3	0.0774	4.84	31.38
4	0.0765	4.78	36.16
5	0.0752	4.70	40.86

Source MCA with code R on ONFP's data of 1994, 2001, and 2006

Table B.2 The first factorial plan

Factor 2 (vertical axis): 7.93 % of inertia Poverty	Factor 1 (vertical axis): 18.61 % of inertia Wealth
Between 2 and 4 persons per bedroom	Floor material of high quality
Having water but outside the house	Popular house or Villa
Having toilet but outside the house	Source of energy: Electricity
No diploma	Professional/Technical diploma
No access to TIC (telephone)	Access to TIC (television, radio)
No durables (water heating, washing machine, Conditioner)	Durables (refrigerator, gas cooker)
Floor material of low quality	Apartment
House of other types	Less than 2 persons per bedroom
More than 4 persons per bedroom	Having water inside the house
Source of energy: Gas and others	Having toilet inside the house
No water	Primary, Secondary and higher diploma
No toilet	Access to TIC (telephone
No access to TIC (television, radio)	Durables (water heating, washing machine,
No durables (refrigerator, gas cooker)	Conditioner)

Source MCA with code R on ONFP's data of 1994, 2001 and 2006

Table B.3 Scores and contributions of the modalities from the final MCA

	Scores		Contributions (%)	
	Factor 1	Factor 2	Factor 1	Factor 2
1. Floor Material			*1.642*	*7.427*
Cemented/"Marble"/"Carrelage"	0.052	0.072	0.058	0.264
Earth or «clay» floor	–1.218	–1.736	0.964	4.589
Others	–1.910	–2.540	0.62	2.574
2. Type of house			*3.368*	*11.267*
Villa/Popular house	0.018	0.096	0.007	0.459
Apartment	0.968	–1.060	0.673	1.891
Others	–2.501	–2.975	2.688	8.917
3. Number of persons per bedroom			*6.029*	*5.782*
Less than 2 persons per bedroom	0.559	–0.406	2.406	2.985
Between 2 and 4 persons per bedroom	–0.043	0.331	0.019	2.720
More than 4 persons per bedroom	–0.935	–0.089	3.604	0.077
4. Source of Energy			*5.922*	*10.116*
Electricity	0.140	0.117	0.408	0.670
Gas	–1.543	–1.147	2.021	2.622
Other sources of Energy	–2.177	–1.987	3.493	6.824
5. Water			*9.614*	*3.982*
Having water inside the house	0.458	–0.024	3.1	0.019
Having water but outside the house	–0.574	0.619	1.006	2.742
No water	–1.098	–0.338	5.508	1.220
6. Toilet			*11.897*	*10.794*

(continued)

Table B.3 (continued)

	Scores		Contributions (%)	
	Factor 1	Factor 2	Factor 1	Factor 2
Having toilet inside the house	0.463	–0.064	3.225	0.145
Having toilet but outside the house	–0.460	0.854	0.831	6.718
No toilet	–1.518	–0.702	7.841	3.931
7. Level of education of the household head			*5.657*	*6.515*
No diploma	–0.357	0.168	1.834	0.951
Primary diploma	0.506	–0.120	1.201	0.159
Professional/Technical	0.468	0.181	0.263	0.092
Secondary diploma	0.920	–0.660	0.882	1.064
Higher diploma	1.182	–1.309	1.477	4.249
8. Refrigerator			*11.77*	*1.259*
Yes	0.434	0.093	3.101	0.332
No	–1.212	–0.259	8.669	0.927
9. Gas Cooker			*3.44*	*6.186*
Yes	0.101	0.088	0.212	0.382
No	–1.529	–0.338	3.228	5.804
10. Water heating			*7.919*	*9.757*
Yes	1.063	–0.771	6.032	7.432
No	–0.333	0.241	1.887	2.325
11. Washing machine			*8.777*	*8.857*
Yes	1.029	–0.675	6.404	6.463
No	–0.381	0.250	2.373	2.394
12. Conditioner			*2.235*	*8.785*
Yes	1.430	–1.851	2.131	8.376
No	–0.070	0.090	0.104	0.409
13. Television			*4.308*	*0.081*
Yes	0.274	0.025	1.21	0.023
No	–0.702	–0.063	3.098	0.058
14. Radio			*8.36*	*6.362*
Yes	0.247	0.140	1.17	0.891
No	–1.515	–0.863	7.19	5.471
15. Telephone			*9.063*	*2.831*
Yes	0.692	–0.253	4.911	1.534
No	–0.585	0.213	4.152	1.297

Source MCA with code R on ONFP's data of 1994, 2001 and 2006

Table B.4 Descriptive statistics of the composite welfare indicator

Variable	Mean	Median	Standard deviation	Minimum	Maximum	Range	N
CWI 1994	–0.149	–0.102	0.321	–1.122	0.606	1.728	6,037
CWI 2001	–0.002	0.028	0.272	–1.056	0.606	1.663	6,051
CWI 2006	0.164	0.178	0.230	–0.823	0.606	1.429	8,186

Appendix C: Analysis of Poverty in Tunisia

Table C.1–C.8

Table C.1 Evolution of poverty in Tunisia between 1994, 2001 and 2006

Years	1994		2001		2006	
Poverty	Value	Bias*	Value	Bias	Value	Bias
Headcount	0.0942	–0.0014	0.0372	–0.0001	0.0182	–0.0002
	(0.0037)		(0.0024)		(0.0015)	
FGT1	0.0159	–0.0001	0.0043	–0.0001	0.0017	0.0000
	(0.0008)		(0.0004)		(0.0002)	
FGT2	0.0043	–0.0001	0.0010	0.0000	0.0002	0.0000
	(0.0003)		(0.0001)		(0.0000)	
GSST,2	0.0309	–0.0002	0.0084	0.0000	0.0033	0.0000
	(0.0015)		(0.0008)		(0.0003)	
GSST,3	0.0451	–0.0003	0.0125	0.0000	0.0050	–0.0001
	(0.0022)		(0.0012)		(0.0005)	
GSST,4	0.0586	–0.0004	0.0166	–0.0001	0.0066	–0.0004
	(0.0028)		(0.0015)		(0.0007)	

Note bootstrap standard errors in parentheses; bias is difference between the mean of the 1000 bootstrapped poverty measure estimates and sample poverty measure estimate

Table C.2 Choice of different poverty lines

Poverty measure	1994 (%)			2001 (%)			2006 (%)		
Threshold (% of median)	30	50	75	30	50	75	30	50	75
Headcount	0.0348	0.0942	0.2504	0.0055	0.0372	0.1803	0.0005	0.0182	0.1305
FGT1	0.0039	0.0159	0.0571	0.0006	0.0043	0.0313	0.0000	0.0017	0.0218
FGT2	0.0007	0.0043	0.0212	0.0001	0.0010	0.0092	0.0000	0.0002	0.0058
GSST,2	0.0077	0.0309	0.1061	0.0013	0.0084	0.0595	0.0000	0.0033	0.0419
GSST,3	0.0115	0.0451	0.1486	0.0019	0.0125	0.0850	0.0001	0.0050	0.0606
GSST,4	0.0151	0.0586	0.1855	0.0026	0.0166	0.1080	0.0001	0.0066	0.0779

Table C.3 Construction of confidence intervals

Years	1994				2001				2006			
Method	Simple percentile				Simple percentile				Simple percentile			
	Standard		Hall's		Standard		Hall's		Standard		Hall's	
	2.5 %	97.5 %	2.5 %	97.5 %	2.5 %	97.5 %	2.5 %	97.5 %	2.5 %	97.5 %	2.5 %	97.5 %
Headcount	0.0853	0.1004	0.0882	0.1032	0.0324	0.0416	0.0327	0.0420	0.015	0.021	0.0154	0.0214
FGT1	0.0142	0.0174	0.0143	0.0175	0.0035	0.0050	0.0035	0.0050	0.0013	0.0020	0.0014	0.0020
FGT2	0.0037	0.0048	0.0038	0.0049	0.0007	0.0013	0.0007	0.0012	0.0002	0.0003	0.0001	0.0003
GSST,2	0.0277	0.0339	0.0279	0.0341	0.007	0.010	0.0069	0.0099	0.0027	0.0039	0.0027	0.0040
GSST,3	0.0405	0.0494	0.0408	0.0496	0.0104	0.0148	0.0102	0.0147	0.0040	0.0058	0.0041	0.0060
GSST,4	0.0527	0.0640	0.0532	0.0644	0.0137	0.0196	0.0136	0.0195	0.0053	0.0077	0.0054	0.0079

Table C.4 Rule of decision for the hypothesis tests of evolution of poverty in Tunisia

Confidence interval				
Poverty Measure	Case 1 (2001 versus 1994)		Case 2 (2006 versus 2001)	
Headcount	–0.065	–0.045	–0.025	–0.014
FGT1	–0.013	–0.010	–0.003	–0.002
FGT2	–0.004	–0.003	–0.001	–0.0005
GSST,2	–0.026	–0.019	–0.007	–0.004
GSST,3	–0.037	–0.027	–0.010	–0.005
GSST,4	–0.048	–0.035	–0.013	–0.007

Table C.5 Comparison between rural and urban poverty in Tunisia

Year	2006			
Poverty Measure	Rural		Urban	
	Value	Bias	Value	Bias
Headcount	0.28854 (0.00810)	0.000070	0.02979 (0.00238)	–0.000054
FGT1	0.05039 (0.00189)	0.000075	0.00358 (0.00038)	–0.000017
FGT2	0.01370 (0.00070)	0.000031	0.00076 (0.00012)	–0.00004
GSST,2	0.09227 (0.00324)	0.000118	0.00710 (0.00074)	–0.000025
GSST,3	0.12738 (0.00422)	0.000141	0.01056 (0.00110)	–0.000046
GSST,4	0.15707 (0.00493)	0.000149	0.01396 (0.00145)	–0.000061

Note bootstrap standard errors in parentheses
*bias: difference between the mean of the 1000 bootstrapped poverty measure estimates and sample poverty measure estimate

Table C.6 Rule of decision for the hypothesis tests of areas disparities in terms of poverty in 2006

Poverty measure	Confidence interval	
Headcount	–0.274	–0.242
FGT1	–0.050	–0.043
FGT2	–0.014	–0.011
GSST, 2	–0.091	–0.078
GSST, 3	–0.125	–0.108
GSST, 4	–0.153	–0.133

Table C.7 Regions disparities in terms of poverty in 2006

Years	Tunis and nearby		North East		North West		Center East	
Poverty measure	Value	Bias*	Value	Bias	Value	Bias	Value	Bias*
Headcount	0.02219	–0.000106	0.07339	9.4e-05	0.13659	6.5e-05	0.03367	–0.000093
	(0.00433)		(0.00850)		(0.01063)		(0.00559)	
FGT1	0.00261	–0.000003	0.00954	–3.0e-06	0.01964	3.1e-05	0.00541	–0.000048
	(0.00066)		(0.00145)		(0.00201)		(0.00115)	
FGT2	0.00055	–0.000001	0.00211	–1.0e-06	0.00458	3.0e-06	0.00133	–0.000023
	(0.00020)		(0.00044)		(0.00074)		(0.00039)	
GSST, 2	0.00519	–0.000004	0.01870	–1.0e-05	0.03769	4.1e-05	0.01071	–0.000107
	(0.00130)		(0.00280)		(0.00376)		(0.00225)	
GSST, 3	0.00774	–0.000005	0.02749	–3.3e-05	0.05431	3.7e-05	0.01590	–0.000169
	(0.00193)		(0.00406)		(0.00530)		(0.00332)	
GSST, 4	0.01026	–0.000015	0.03594	-5.9e-05	0.06965	5.0e-06	0.02099	–0.000227
	(0.00254)		(0.00524)		(0.00665)		(0.00435)	

Years	Center West		South East		South West	
Poverty measure	Value	Bias	Value	Bias	Value	Bias
Headcount	0.29035	0.268174	0.08701	0.000032	0.06187	–0.000209
	(0.00936)		(0.00995)		(0.00878)	
FGT1	0.05338	0.050747	0.01162	–0.000018	0.00875	–0.000010
	(0.00222)		(0.00181)		(0.00147)	
FGT2	0.01508	0.014519	0.00287	–0.000009	0.00173	–0.000002
	(0.00085)		(0.00068)		(0.00038)	
GSST, 2	0.09762	0.092386	0.02270	–0.000054	0.01715	–0.000031
	(0.00380)		(0.00350)		(0.00283)	
GSST, 3	0.13461	0.126785	0.03329	–0.000092	0.02522	–0.000058
	(0.00493)		(0.00506)		(0.00410)	
GSST, 4	0.16580	0.155415	0.04341	–0.000141	0.03297	–0.000087
	(0.00574)		(0.00652)		(0.00528)	

Table C.8 Rule of decision for the hypothesis tests of regions disparities in terms of poverty in 2006

	Confidence interval					
Poverty Measure	Case 1 (Tunis and nearby versus North West)		Case 2 (Tunis and nearby versus the Center West)		Case 3 (Tunis and nearby versus the South East)	
Headcount	–0.139	–0.092	–0.288	–0.248	–0.086	–0.044
FGT1	–0.021	–0.013	–0.055	–0.046	–0.013	–0.005
FGT2	–0.005	–0.002	–0.016	–0.012	–0.003	–0.001
GSST, 2	–0.040	–0.025	–0.100	–0.084	–0.025	–0.010
GSST, 3	–0.057	–0.035	–0.137	–0.116	–0.036	–0.015
GSST, 4	–0.073	–0.045	–0.167	–0.143	–0.047	–0.020

References

Asselin, L. M. (2002). *Multidimensional poverty: composite indicator of multidimensional poverty*. Lévis: Institut de Mathématique Gauss.

Atkinson, A. B. (1970). On the measurement of inequality. *Journal of Economic Theory, 2*, 244–263.

Ayadi, M., El Lahga, A., Chtioui, N. (2007). *Poverty and inequality in Tunisia: a non-monetary approach*. PMMA Working Paper No. 2007-05.

Bibi, S. (2004). *Comparing multidimensional poverty between Egypt and Tunisia*. CIRPEE Working Paper No. 04-16.

Booth, C. (1902). *A case of town life*. London: Macmillan.

Chakravarty, S. R. (1983). Ethically flexible measures of poverty. *The Canadian Journal of Economics, 16*, 74–85.

Chakravarty, S. R. (1990). *Ethical social index numbers*. New York: Springer-Verlag.

Chtioui, N. (2004). *Les Mesures de la Pauvreté Multidimensionnelle. Mémoire de Mastère de Modélisation non publiée*. Tunis: Institut Supérieur de Gestion.

Dalton, H. (1920). The measurement of the inequality incomes. *Economic Journal, 30*, 348–362.

Donaldson, D., & Weymark, J. A. (1980). A single-parameter generalization of the Gini indices of inequality. *Journal of Economic Theory, 22*, 67–86.

Duclos, J.-Y., & Gregoire, P. (2002), Absolute and Relative Deprivation and the Measurement of Poverty, Review of Income and Wealth, series 48(4).

Duclos, J.-Y., Sahn, D., & Younger, S.D. (2006). Robust multidimensional poverty comparisons. *Economic Journal, 116*(514), 943–968.

Efron, B. (1979). Bootstrap methods: another look at the jackknife. *Annals of Statistics, 7*, 1–26.

Efron, B. (1981). Nonparametric estimates of standard error: The jackknife, the bootstrap and other methods. *Biometrika, 68*, 589–599.

Efron, B. (1982). *The Jackknife, the Bootstrap, and other resampling plans*. Philadelphia: Society for Industrial and Applied Mathematics.

Efron, B., & Tibshirani, R. J. (1993). *An introduction to the Bootstrap*. New York: Chapman and Hall.

Filmer, D., & Pritchett, L. (1998). *Estimating wealth effects without expenditure data or tears: an application of educational enrollment in states of India, Mimeo*. Washington: The World Bank.

Foster, J. E. (1984). On economic poverty: a survey a aggregate measures. In R. L. Basmann, G. F. Rhodes (Eds.), *Advances in Econometrics, 3.Connecticut* (pp. 215–251). Bingley: JAI Press.

Foster, J. E. & Sen, A. (1997). Economic inequality after a quarter century. In *On Economic Inequality (expanded edition)*. Oxford: Clarendon Press.

Hall, P. (1992). *The Bootstrap and Edgeworth expansion*. New York: Springer.

Hammer, J. (1998). *Health outcomes across wealth groups in Brazil and India. Mimeo. DECRG.* Washington: The World Bank.

Jenkins, S. P., & Lambert, P. J. (1997). Three is of poverty curves, with an analysis of UK poverty trends. *Oxford Economic Papers, 49*, 317–327

Kolm, S. C. (1969). The optimal production of social justice. In J. Magolis & H. Guittor (Eds.), *Public economics* (pp. 145–200). London: Macmillan.

Moran, T. (2005). *Bootstrapping the LIS: statistical inference and patterns of inequality in the Global North*. LIS Working Paper No. 378.

Osberg, L. (2000). Poverty in Canada and the United States: measurement, trends, and implications. *The Canadian Journal of Economics/Revue Canadienne d'Economique*, *33*(4), 847–877.

Osberg, L. & Xu, K. (1999). Poverty intensity-How well do Canadian provinces compare? *Canadian Public Policy*, *25*(2), 179–195.

Osberg, L. & Xu, K. (2000). International comparison of poverty intensity: Index decomposition and bootstrap inference. *Journal of Human Resources*, *35*(1), 51–81.

Pollak, R. A. (1971). Additive utility functions and linear Engel curves. *Review of Economic Studies, 38*, 401–414.

Ravallion, M. (1992). On hunger and public action: a review article on the book by Jean Dreze and Amartya Sen. *The World Bank Research Observer, 7*, 1–16.

R Development Core Team (2008). *R: A language and environment for statistical computing*. Vienna: R Foundation for Statistical Computing. ISBN 3-900051-07-0, URL http://www.R-project.org

Robeyns, I. (2005). The capability approach: a theoretical survey. *Journal of Human Development, 6*(1), 93–117.

Rowntree, B. S. (1901). *Poverty: a study of town life*. London: Macmillan.

Seidl, C. (1988). Poverty measurement: a survey. In D. Bös, M. Rose & C. Seidl (Eds.), *Welfare and efficiency in public economics* (pp. 219–231). Heidelberg: Springer-Verlag.

Sen, A. (1973). *On Economic Inequality*. Oxford: Clarendon Press.

Sen, A. (1976). Poverty: an ordinal approach to measurement. *Econometrica, 44*, 219–231.

Sen, A. (1985). *Commodities and capabilities* (p. 89). New Delhi: Oxford India paper backs.

Shorrocks, A. F. (1995). Revisiting the Sen Poverty index. *Econometrica, 63*, 1225–1230.

Stifel, D., Sahn, D. & Younger, S. (1999). Inter-temporal changes in welfare: preliminary results from nine African countries. *CFNPP WP*, 6–94.

Streeten, P. (1981). *First things first: meeting basic human needs in developing countries*. New York: Oxford University Press.

Thon, D. (1979). On measuring poverty. *Review of Income and Wealth, 25*, 429–439.

Thon, D. (1983). A note on a troublesome axiom for poverty indices. *Economic Journal, 93*, 199–200.

Weymark, J. A. (1981). Generalized Gini inequality indices. *Mathematical Social Sciences, 1*, 409–430.

Xu, K. (1998). The statistical inference for the Sen-Shorrocks-Thon index of poverty intensity. *Journal of Income Distribution*, *8*(1), 143–152.

Xu, K. (2000). Inference for generalized Gini indices using the iterated bootstrap method. *Journal of Business and Economics Statistics*, *18*(2), 223–227.

Xu, K., & Osberg, L. (2002). On Sen's approach to poverty measures and recent developments, paper presented at the Sixth International Meeting of the Society for Social Choice and Welfare (2002). Printed in Chinese in *China Economic Quarterly*, 1, 151–170.

Zhen G, B. (1993). Poverty measurement, statistical inference, and an application to the United States. *Ph.D Dissertation*, West Virginia University, Virginia.

Zheng, B. (1997). Aggregate poverty measures. *Journal of Economic Surveys, 11*, 123–163.

Chapter 7
A Multiple Correspondence Analysis Approach to the Measurement of Multidimensional Poverty in Morocco 2001–2007

Abdeljaouad Ezzrari and Paolo Verme

Abstract The measurement of multidimensional poverty has been advocated by most welfare scholars and is experiencing a growth in interest partly explained by controversial debates emerged across academics and practitioners. This chapter follows one of the least explored approaches—multiple correspondence analysis (MCA)—to assess multidimensional poverty in Morocco between 2001 and 2007. MCA provides two major advantages for the measurement of multidimensional poverty: it generates a matrix of "weights" based on the variance-covariance matrix of all welfare dimensions selected and provides a natural approach for constructing a composite welfare indicator that satisfies essential poverty orderings axioms. The application shows that poverty in Morocco has declined according to both monetary and multidimensional indicators and that these findings are robust to stochastic dominance tests. This chapter concludes that the sustained positive growth that Morocco experienced during the last decade has translated in improvements in living conditions well beyond monetary returns.

A. Ezzrari
Observatory for Living Standards Conditions,
High Commission for the Plan, Rabat, Morocco
e-mail: ezzrari@yahoo.fr

P. Verme (✉)
The World Bank and Department of Economics,
University of Turin, Turin, Italy
e-mail: pverme@worldbank.org

V. Bérenger and F. Bresson (eds.), *Poverty and Social Exclusion Around the Mediterranean Sea*, Economic Studies in Inequality, Social Exclusion and Well-Being 9,
DOI: 10.1007/978-1-4614-5263-8_7,

7.1 Introduction

With various degrees of emphasis and depth, most welfare economists of our time have recognized the importance of considering poverty as a state of deprivation which is multidimensional in nature (Sen 1976; Kolm 1977; Atkinson and Bourguignon 1982; Maasoumi 1986; Duclos et al. 2001). Some important critiques to this approach have also emerged suggesting that several indexes of poverty measured on different dimensions may be more effective than a single multidimensional index (Ravallion 2011). However, the debate around the measurement of multidimensional poverty is not so much about the "if" but about the "how". The Alkire and Foster (2011) index of multidimensional poverty, for example, has recently generated great attention and controversy witnessing both the importance and the complexity of the topic.

The complexity of measurement of multidimensional poverty arises principally from two main issues. The first issue relates to the direction of aggregation in the construction of the index, whether "horizontally" where dimensions are aggregated across individuals first or "vertically" where individuals are aggregated across dimensions first. These two approaches imply different methodologies and can often lead to different results. The second problem is the weighting of the different dimensions. Some scholars prefer to attribute equal and unitary weights to all dimensions while other scholars use various methods to identify dimension specific weights. The weight choice evidently implies a combination of normative and positive criteria, which makes it difficult to argue that one choice is better than another (partly explaining the controversy around multidimensional indexes).

In this chapter we follow the "horizontal" approach by constructing first a composite indicator of welfare and then use this indicator to rank households and measure poverty. In doing so, we follow one of the least beaten tracks by making use of multiple correspondence analysis (MCA) for the construction of the composite indicator. This technique has been recently rediscovered for the study of multidimensional poverty (Asselin 2009; Njong and Dschang 2008) as it provides a method for assigning weights to different dimensions while it responds to some essential welfare axioms.

Our application is on Morocco using 2001–2007 consumption data. During the last decade, Morocco has experienced consistent growth generated by economic reforms and investments in infrastructures. These developments have resulted in improved living conditions as shown by official statistics and reports (Douidich et al. 2009). However, it is typically heard from local and foreign commentators that these gains have been unequally shared and/or that living conditions have been improved in certain areas but not in others. Thus, a multidimensional approach to the study of poverty can provide a better sense of whether the gains in Morocco have been robust and comprehensive.

This chapter is organized as follows: Sect. 7.2 briefly reviews the different approaches to the study of multidimensional poverty. Section 7.3 illustrates the MCA approach and its properties; Sect. 7.4 describes Morocco's economic framework and data; Sect. 7.5 describes the results of the MCA conducted in two steps; Sect. 7.6 provides a multidimensional poverty analysis of Morocco in 2001 and 2007 and Sect. 7.7 concludes.

7.2 Alternative Approaches to the Study of Multidimensional Poverty

A poverty index requires, by definition, a welfare measure that we can use to rank households or individuals and a poverty line that we can use to separate the poor from the non poor. In the case of a unidimensional index of poverty, this exercise is quite straightforward and scholars have reached a certain consensus on the desirable properties that these indexes should have while practitioners have endorsed these properties by using popular measures of poverty that satisfy these properties such the Foster-Greer-Thorbecke (FGT) poverty indexes.

In the case of a multidimensional poverty analysis, the situation is much complicated by the fact that rather than working with one dimension of welfare we work with several dimensions. We pass from a unidimensional space to a multidimensional space. If we think of individuals as represented in rows (r) and welfare dimensions as represented in columns (c), then the new multidimensional distribution can be represented by a rXc matrix. A multidimensional analysis complicates the study of poverty in at least two important respects. First, it is what we could describe as the "direction" of aggregation. The question of how one should proceed in the aggregation of individuals and dimensions, whether by rows first (individuals) or by columns first (dimensions). And second what is the weight that should be attributed to each individual and dimension during the aggregation process.

The direction of aggregation reflects different methodological approaches. We can first define deprivation for a society in each of the deprivation dimensions we wish to use and then aggregate deprivation across dimensions to construct a multidimensional deprivation index. This is the approach followed, for example, by the UNDP Human Development Index where deprivation levels are first measured within the different dimensions at the societal level and then aggregated into one indicator. Alternatively, we can first define a multidimensional indicator of deprivation at the individual level and then aggregate this indicator across individuals at the societal level. Under certain conditions, the end results at the societal level will be the same but this is not necessarily always the case. For example, different approaches to weighting the different dimensions or individuals can result in very different societal indicators.

In this chapter, we follow this second approach. We define first what we call a composite welfare indicator (CWI)[1] at the individual level and then aggregate this

[1] Unlike previous studies, we refer to the composite indicator as 'welfare' rather than 'poverty' indicator. That is because the composite indicator applies to poor and non-poor and is therefore a 'welfare' indicator.

individual welfare indicator across individuals as we would do with standard unidimensional poverty measures. Therefore, the multidimensional question is treated at the individual level with the construction of the CWI while aggregation at the societal level can follow standard procedures used for unidimensional indexes.

The second question of weight is complex and largely unresolved. When we aggregate different welfare dimensions the question is whether each dimension should have the same weight or different weights. Should, say, illiteracy be given the same weight as malnutrition? Some authors have preferred to attribute equal weights to different welfare dimensions. In this case, the problem of weights does not arise as all dimensions are attributed an equal weight (usually one). Other authors have made an effort to attribute different weights to the different dimensions according to various criteria. For example, some authors have imposed arbitrary weights on different dimensions based on economic criteria or simply normative considerations. Some use hybrid approaches where weights are attributed to differentiate income and non-income components but then equal weights are used for non-income dimensions. Other authors have tried instead to use surveys that ask questions to respondents about the importance of different dimensions and used answers to these questions to attribute weights to dimensions. Another stream of the literature tried to seek weighting schemes based on expenditure patterns or other behavioral patterns measured in observed variables. Decancq and Lugo (2010) have reviewed these different approaches to weighting schemes in multidimensional indices of poverty and classified these approaches in no less than eight different categories.

Another approach that has been recently explored with the measurement of multidimensional poverty is a weighting scheme based on factorial or principle component analysis (PCA). The idea is simple. We can consider welfare as a complex and unobserved multidimensional phenomenon that we want to estimate based on a set of observed proxies of welfare. We are searching therefore for a latent variable and our objective is to use and aggregate what we consider as possible dimensions of welfare in a way that they can represent satisfactorily the unobserved complex welfare indicator.

The natural choice for exploring latent variables is a PCA type of analysis. In addition to considering multidimensional poverty as a latent variable, this technique is particularly appealing, because it provides weights for each dimension based on the covariance matrix of the *rXc* matrix. Therefore, the weights are implicitly determined by the procedure and based on the correlation system of all variables vis-à-vis the latent multidimensional variable. PCA type of analyses is particularly indicated for continuous and quantitative variables and this is a drawback if one wants to apply the methodology to the study of multidimensional poverty where non-income variables are typically categorical. A better approach for categorical variables is MCA. This is a technique that is still relatively little explored. Decancq and Lugo (2010) review of weighting methods, for example, does not make any mention of MCA methods, which address part of the concerns raised by the authors in relation to PCAs.

7.3 A Multiple Correspondence Analysis Approach

In this chapter we opt to use MCA developed first by Benzécri (1973) (see also Greenacre 1984). The construction of our CWI will rely exclusively on categorical variables and it is well known in factorial analysis that when dealing with categorical variables a suitable choice is MCA as it makes fewer assumptions than PCA on the underlying distributions of variables. In MCA analysis technique, each modality of each categorical variable is typically broken down into a 0/1 binary indicator.

As PCA, MCA is a data exploration technique that is used to uncover correlation patterns across sets of variables described by single components named principal components. Principal components can be considered as latent unobserved variables that account for the maximum variance of a set of other variables. The first principal component represents the unobserved latent variable that captures the highest variance of all observed variables used in the analysis and that is therefore the best candidate to represent all the variables considered. The first principal component can be visualized as the interpolation line that passes through a cloud of points set in an imaginary n-dimensional space of variables (axes) and that minimizes the square of the distances from all these variables (axes) and points. The main difference of this approach from standard econometric approaches is that the dependent variable is unobserved and cannot be used directly to estimate correlation coefficients. We assume therefore that welfare is a multidimensional latent (unobserved) variable.

There are at least two computational advantages in using MCA for poverty analysis in addition to its suitability for categorical data (Asselin 2009). The first is that MCA gives more weight to indicators with a smaller number of hits. If we have few deprived individuals within any dimension, these individuals are given a higher weight. In the case of deprivation indicators where the state of deprivation is indicated by one, this could be interpreted as giving more importance to minority population groups such as the relatively more deprived. Njong and Dschang (2008) compare PCA, MCA, and the fuzzy approach in a study of multidimensional poverty in Cameroon and find that PCA estimates show unambiguously lower poverty than those obtained from the other two methods. They conclude that MCA and fuzzy methods are more sensitive to deprivation and may be preferred to PCAs approaches for this reason.

The second property is reciprocal bi-additivity. This states that (a) the composite deprivation score of an individual is the simple average of the factorial weights of the deprivation categories and (b) the weight of a given dimension of deprivation is the simple average of the composite deprivation scores of the population units that belong to the given dimension. This second property is particularly important, because it implies that the population ordering in each of the dimensions used is preserved in the composite indicator as discussed below in the context of the basic axioms.

With these properties in mind we can turn to the formulation of the CWI. Let k be the number of dimensions (variables) with $k = (1, 2, \ldots, K)$, j the number of modalities of each dimension with $j = (1,2,\ldots J_k)$ and I the binary (0/1) indicator of each modality. W is the weight determined with MCA (the factor score on the first axe normalized by the eigenvalue λ with $s =$ factor score), and i is the index number indicating households. Then the functional form of the CWI can be described as:

$$CWI_i = \frac{1}{K}\sum_{k=1}^{K}\sum_{j_k=1}^{J_k} W_{j_k}^k I_{j_k i}^k$$

with

$$W_{j_k}^k = \frac{s^k}{\sqrt{\lambda_1}}$$

In essence, the composite indicator is the simple average across dimensions (variables) of the weighted sum of each binary modality of each dimension.

As described in Asselin (2009) a two-steps procedure simplifies somehow the axiomatic requirements for the first component to be a consistent composite index and boils down to two major axioms: (1) *Monotonicity Axiom* stating that the composite indicator must be monotonically increasing in each of the primary indicators. If any individual improves his/her situation in relation to one of the primary indicators, then the composite index should improve (everything else being equal). This axiom implies two requirements: (a) first axis ordering consistency (FAOC-I) for an indicator I, which states that there must be an ordinal consistency between the ordering of categories and the ordering of weights across categories, either increasing or decreasing order; (b) global first axis ordering consistency (FAOC-G), which states that, for all indicators, FAOC-I is met with the same orientation, either decreasing or increasing; (2) *composite poverty ordering consistency (CPOC)* stating that the population ordering of an indicator I is preserved with the composite indicator.

FAOC-I is always met with binary indicators. Consistency is always met if one of the two categories of the binary indicator is empty (such as "have" and "not have" some asset). With multinomial indicators and if FAOC-I is not met, the categories can be reorganized so as to meet the property, for example, by reducing the multinomial indicator into a binomial one. FAOC-G requires instead that all indicators have the same orientation with respect to the first axis. This is more difficult to achieve given that some indicators may represent a subset of deprivation showing different signs versus the first axis. In this case, we might want to re-elaborate the variable or to eliminate the variable altogether. It is also possible to explore further axes as discussed in Asselin (2009) but this is an option that we won't need to follow with our data.

By construction, the weights that result from the MCA procedure average to zero and can have positive and negative values. This is inconvenient for measuring poverty in the second step and it is useful to positively rescale the weight by subtracting the lowest weight from each weight as follows:

$$W_{j_k}^k = \frac{s^k - s_{\min}^k}{\sqrt{\lambda_1}}$$

so that the most deprived category of an indicator is always equal to zero and the composite index is always greater or equal to zero. In this chapter, rather than rescaling the weights, we will rescale the CWI so that it will become easier to compare this measure with standard monetary measures of welfare such as expenditure per capita.

We will also consider multidimensional poverty over time. This adds the complication of the choice of weights given that we need to have consistent weights over time. One possibility is to merge the two datasets and calculate the weight using both surveys. A second possibility is to calculate the weights in 1 year and apply the same weights to the other year. We initially tested both approaches and found that the final results were almost identical with a correlation coefficient calculated between the two CWIs of 0.995 and a very consistent ranking of households. We therefore opted for the second approach and used the weights calculated from the second of the two surveys used in this chapter. This is the same approach followed by other authors (see for example Asselin 2009).

7.4 Economic Context and Data

The study of poverty in Morocco has seen major developments during the last two decades thanks to the increased availability of microdata and thanks to the increased capacity on the part of the government to analyze poverty. The High Commission for the Plan (HCP), for example, has published several studies during the last decade covering a range of topics such as pro-poor growth, geographical targeting, poverty maps, and poverty dynamics. All these studies relied on expenditure as a measure of welfare, took a unidimensional approach to the study of poverty and found consistent results in that poverty has declined in both urban and rural areas between 2001 and 2007.

This positive result is consistent with the general macroeconomic framework and key reforms implemented by the government in recent years. Starting from the 1990s, Morocco has embarked in a process of liberalization of the economy that led to increased foreign investments and trade. The country has also implemented a large program of public investments throughout the decade and has been able to improve significantly on other fronts such as health and education. Average GDP growth has increased to 4–5 % in per year with peaks of 7 %. During the second half of the last decade, the government has also launched a large human development program [initiative for national human development (INDH)] that covers all poor areas of the country and finances local development activities from private entrepreneurship to public recreational centers. The INDH program has now completed its first phase and is currently launching a second phase. The evaluation of the first phase is extremely complex given the nature of the program but stakeholders agree that the program had a visible impact on local communities.

A multidimensional approach to the study of poverty can complement unidimensional studies and also offers a better sense of whether recent reforms and improvements in welfare have concerned most dimensions of welfare or only a few dimensions. We therefore rely on the same data used by other studies including the National Survey on Consumption and Expenditure carried out between 2000 and 2001 and including 14,243 households (henceforth NSCE 2000/2001) and the National Survey on Living Standards carried out between 2006 and 2007 and including 7,062 households (henceforth NSLS 2006/2007). Both surveys are representative at the national, urban–rural and regional levels and they are fully comparable across all variables used in this chapter.

Table 7.1 lists all variables used in the first step of the MCA exercise where we use a large set of variables to test how these variables polarize around the different axes. These variables are grouped in eight broad areas including employment, education, health, housing, utilities, durables, nutrition, and expenditure. The choice of these initial variables is largely normative but based on decades of research on poverty in Morocco where all these variables have been shown to be closely associated with welfare indicators of various types.[2] Most of these variables are also typically used in the literature on the construction of multidimensional poverty indexes (see Decancq and Lugo 2010 for a review).

The MCA procedure is carried out separately for urban and rural areas. Morocco is a country of great divides between urban and rural areas and the sets of variables relevant for constructing the CWI are expected to be different so as to reflect the respective peculiarities of rural and urban areas. Following the first-step procedure, the number of variables is reduced to meet the consistency properties of the principal component. We will see that this variable reduction process leads to different results in urban and rural areas confirming the initial hypothesis that urban and rural areas are better represented by two different models rather than by one model.

7.5 Multiple Correspondence Analysis

The MCA analysis is carried out in two steps. In the first step, we use all variables listed in Table 7.1 to provide a first indication of the association patterns with the principal component. In the second step we remove or adjust variables so as to meet the basic conditions of the MCA discussed in previous sections.

---

[2] See http://www.omdh.hcp.ma/ for a number of poverty studies on Morocco conducted by the High Commission for the Plan.

Table 7.1 List of variables and modalities

Variable	Modality	Variable	Modality
Employment		*Utilities*	
Head of household unemployed	yes	Fresh water	network
	no		fountain
Head of household employed	yes		source
	no		other
Youth unemployment	active	Electricity	individual meter
	unemployed		collective meter
	inactive		no meter
	no youth		no electricity
Household share of employed	none	Used water	network
	less than 1/3		septic tank
	between 1/3 and 2/3		latrine
	more than 2/3		other
Household share of unemployed	none	*Durables*	
	less than half	Satellite dish	yes
	more than half		no
Education		Portable phone	yes
Head of household alphabetized	yes		no
	no	Gas stove	yes
Household share with no education	less than 1/3		no
	between 1/3 and 1/2	Gas hoven	yes
	between 1/2 and 3/4		no
	more than 3/4	Refrigerator	yes
Household share of alphabetized	less than 1/3		no
	between 1/3 and 1/2	Freezer	yes
	between 1/2 and 2/3		no
	more than 2/3	*Nutrition*	
Health		Child stunting	yes
Household access to health	no health problems		no
	total access		no children
	partial access	Child malnutrition	yes
	no access		no
Household medical coverage	yes		no children
	no	*Income*	
Household share of medical coverage	total coverage	Food poverty	yes
	at least half household		no
	less than half household	Income quintiles	Q1
	no coverage		Q2

(continued)

Table 7.1 (continued)

Variable	Modality	Variable	Modality
Housing			Q3
Type of accommodation	villa		Q4
	apartment		Q5
	moroccan house		
	basic dwelling		
	rural house		
	other		
Cohabitation	one household		
	two households		
	three households		
	four households and more		
Bath tub/shower	yes		
	no		
Kitchen	yes		
	no		
Toilet	yes		
	no		
Number of persons per room	one person or less		
	between one and two		
	between two and three		
	more than three		

Step 1. Tables 7.2 and 7.3 illustrate the results of the preliminary MCA for urban and rural areas. The table of eigenvalues highlights the distinction between the first factorial axis and the following axes (labeled as dimensions "Dim" in the tables). In urban areas (Table 7.2), this axis alone explains more than 52 % of the total inertia of the cloud of variables while the explanatory power drops to 9 % for the second axis and to 6 % for the third axis. Most of the variance of the set of variables used for urban areas is captured by one axis. In rural areas (Table 7.3), the first axis is less powerful capturing 46.6 % of the total variance. This is also followed by one axis that captures over 11.5 % of the total variance and a few other non negligible axes.

In order to improve the model in a way that satisfies that basic properties discussed in previous sections, we now need to look at the inertia of each variable and modality on the first axis, which is our latent multidimensional measure of welfare. Table 7.4 provides details of the various poverty proxies for the first axis in urban and rural areas. The state of poverty is described by the indicators negatively correlated to the first axis, while well-being is captured by the indicators that are positively correlated. A review of the most negatively correlated variables to the first axis allows identifying poor households as those households with very low access to education, health, fresh water, sanitation, housing, nutrition, and assets regardless of the place of residence.

Table 7.2 Multiple correspondence analysis (urban areas)

Dimension	MCA (preliminary)			MCA (final)		
	Principal inertia	Percentage	Cumulated percentage	Principal inertia	Percentage	Cumulated percentage
Dim 1	0.0238	52.03	52.03	0.02569	69.22	69.22
Dim 2	0.00425	9.28	61.31	0.00204	5.51	74.72
Dim 3	0.00275	6.01	67.32	0.00127	3.41	78.14
Dim 4	0.00208	4.55	71.87	0.00118	3.17	81.31
Dim 5	0.00139	3.04	74.91	0.00048	1.3	82.62
Dim 6	0.00073	1.6	76.51	0.00037	0.99	83.61
Dim 7	0.00045	0.98	77.49	0.00029	0.78	84.39
Dim 8	0.00034	0.74	78.22	0.00013	0.37	84.76
Dim 9	0.00028	0.61	78.84	0.00006	0.17	84.93
Dim 10	0.00016	0.35	79.18	0.00003	0.09	85.02
Dim 11	0.00013	0.29	79.48	0.00001	0.03	85.05
Dim 12	0.00011	0.25	79.73	0	0.02	85.06
Dim 13	0.00009	0.19	79.92	0	0.01	85.07
Dim 14	0.00006	0.14	80.05	0	0	85.07
Dim 15	0.00005	0.12	80.17	0	0	85.07
Total	0.04575	100	–	0.03711	100	–

Table 7.3 Multiple correspondence analysis (rural areas)

Dimension	MCA (preliminary)			MCA (final)		
	Principal inertia	Percentage	Cumulated percentage	Principal inertia	Percentage	Cumulated percentage
Dim 1	0.01745	46.66	46.66	0.0263	70.34	70.34
Dim 2	0.0043	11.51	58.17	0.0016	4.33	73.67
Dim 3	0.00229	6.12	64.32	0.0011	2.82	77.49
Dim 4	0.00164	4.39	68.68	0.0009	2.33	79.82
Dim 5	0.00109	2.91	71.59	0.0005	1.36	81.18
Dim 6	0.00067	1.79	73.38	0.0003	0.76	81.94
Dim 7	0.0004	1.08	74.46	0.0002	0.57	82.51
Dim 8	0.00029	0.78	75.25	0.0001	0.28	82.79
Dim 9	0.0002	0.52	75.77	0.00006	0.17	82.96
Dim 10	0.00016	0.42	76.19	0.00004	0.1	83.06
Dim 11	0.00011	0.29	76.47	0.00001	0.04	83.1
Dim 12	0.00009	0.23	76.71	0	0.03	83.13
Dim 13	0.00007	0.2	76.91	0	0.01	83.14
Dim 14	0.00004	0.1	77.01	0	0	83.14
Total	0.0289	100	–	0.03738	100	–

More specifically, a low level of welfare is characterized by a low literacy rate of the household head, a small proportion of members who cannot read and write in the household and a high proportion of members with no diploma in the household. In the field of health, sanitation, and drinking water, the MCA shows

Table 7.4 Multiple correspondence analysis (preliminary weights, urban and rural areas)

Area	Variable	Modality	Urban Dim1	Rural Dim1
Employment	Head of household unemployed	yes	−0.4	1.21
		no	0.01	−0.02
	Head of household employed	yes	−0.19	−0.15
		no	0.35	0.73
	Youth unemployment	active	−0.69	−0.29
		unemployed	0.15	1.37
		inactive	0.63	1.08
		no youth	−0.19	−0.35
	Household share of employed	none	0.35	0.9
		less than 1/3	−0.01	0.66
		between 1/3 and 2/3	−0.02	−0.15
		more than 2/3	−0.53	−0.59
	Household share of unemployed	none	−0.12	−0.16
		less than half	0.41	1.54
		more than half	0.19	1.73
Education	Head of household alphabetized	yes	0.79	1.25
		no	−1.12	−0.62
	Household share with no education	less than 1/3	0.97	2.18
		between 1/3 and 1/2	−0.53	0.71
		between 1/2 and 3/4	−1.67	−0.54
		more than 3/4	−2.22	−1.44
	Household share of alphabetized	less than 1/3	−2.06	−1.11
		between 1/3 and 1/2	−0.47	0.51
		between 1/2 and 2/3	0.27	1.67
		more than 2/3	1.32	2.89
Health	Household access to health	no health problems	−0.08	−0.28
		total access	0.16	0.5
		partial access	−0.17	0.31
		no access	−0.34	−0.29
	Household medical coverage	yes	2	4.31
		no	−0.97	−0.32
	Household share of medical coverage	total coverage	2.1	4.48
		at least half household	2.06	4.88
		less than half household	1.65	3.54
		no coverage	−0.97	−0.32
Housing	Type of accommodation	villa	2.73	4.59
		apartment	1.82	−0.97
		moroccan house	−0.12	2.77
		basic dwelling	−3.93	−0.92
		rural house	–	0.93
		Other	−1.67	−0.9
	Cohabitation	one household	0.17	0.03

(continued)

Table 7.4 (continued)

Area	Variable	Modality	Urban Dim1	Rural Dim1
		two households	−1.47	−1.08
		three households	−2.03	−0.35
		four households and more	−2.65	−0.05
	Bath tub/shower	yes	1.68	4.14
		no	−1.36	−0.32
	Kitchen	yes	0.35	0.39
		no	−3.07	−2
	Toilet	yes	0.16	0.97
		no	−5.44	−2.25
	Number of persons per room	one person or less	0.83	0.28
		between one and two	0.4	0.36
		between two and three	−0.9	−0.39
		more than three	−2.2	−1.51
Utilities	Fresh water	network	0.59	2.57
		fountain	−3.34	0.09
		source	−6.09	−1.33
		other	−3.25	−0.2
	Electricity	individual meter	0.7	1.04
		collective meter	−1	0.24
		no meter	−2.79	−0.91
		no electricity	−4.53	−1.91
	Used water	network	0.49	2.75
		septic tank	−1.76	1.4
		latrine	−2.07	1.04
		other	−5.04	−1.84
Durables	Satellite dish	yes	1.2	2.35
		no	−1.79	−0.86
	Portable phone	yes	0.33	0.86
		no	−1.67	−1.53
	Gas stove	yes	0.62	1.02
		no	−0.97	−0.76
	Gas hoven	yes	0.31	1.01
		no	−0.36	−1.09
	Refrigerator	yes	0.67	2.27
		no	−3.06	−1.2
	Freezer	yes	2.92	4.88
		no	−0.17	−0.03
Nutrition	Child stunting	yes	−0.95	−0.86
		no	−0.76	−0.2
		no children	0.31	0.34
	Child malnutrition	yes	−1.21	−1.73
		no	−0.81	−0.38

(continued)

Table 7.4 (continued)

Area	Variable	Modality	Urban Dim1	Rural Dim1
		no children	0.31	0.34
Expenditure	Food poverty	yes	−2.63	−1.47
		no	0.23	0.26
	Expenditure quintiles	Q1	−3.17	−1.59
		Q2	−1.6	−0.4
		Q3	−0.48	0.3
		Q4	0.24	0.91
		Q5	1.56	2.34

that the poor usually do not have access to health care for illness or injury and have no medical coverage. They do not have toilets at home and are not connected to the potable water supply or even to a public fountain nearby (the case of rural households). Poor households also fail to meet their basic nutrition needs to the extent that their expenditure on food is below the food poverty line. These households have also children under age of five suffering from malnutrition and stunted growth.

In the field of housing, poor households live in homes characterized by the absence of basic hygienic conditions and comforts. They have no electricity in their homes, use candles and gas lamps or kerosene for lighting. In terms of communication, poor households do not have televisions, satellite dishes, and cell phones. The percentage of households that do not have these means of communication is greater in rural than in urban areas. The group of poor households also lacks basic durables such as gas stove, gas oven, refrigerator, or freezer. In terms of employment, poor households are those headed by jobless people. There is also a category of non-poor households who may be well off in terms of employment but they are still very poor compared to other dimensions such as education, health, water, and communications. Therefore, the phenomenon of the employed poor is also clearly visible in these data.

In contrast to poor households, wealthy households are generally those who have access to education, health, drinking water, sanitation, decent housing, food, and basic assets. Households that are well connected to the welfare axis are households whose head is literate, who have a high proportion of literate people, and where low levels of education are absent. These same households have also access to basic health services in case of illness or injury and are covered by the health insurance system. They have children under 5 years of age who do not suffer from stunted growth or malnutrition. These households also have homes connected to potable water and sewage systems and have hygienic toilets. They live in decent housing and have electricity and various means of comfort (kitchen and appliances). Affluent households are households that have access to communication (televisions, mobile phones, and satellite dishes), whose head is an employed person and where young children aged 15–24 do not exercise a professional activity.

Table 7.5 Multiple correspondence analysis (final weights, urban and rural areas)

			Urban Dim1	Rural Dim1
Employment	Head of household unemployed	yes	−0.48	–
		no	0.01	–
	Youth unemployment	active	−0.63	−0.08
		unemployed	0.11	0.04
Education	Head of household alphabetized	yes	0.8	1.13
		no	−0.13	−0.56
	Household share with no education	less than 1/3	0.91	1.85
		between 1/3 and 1/2	−0.44	0.71
		between 1/2 and 3/4	−1.49	−0.36
		more than 3/4	−2.32	−1.4
	Household share of alphabetized	less than 1/3	−2	−0.99
		between 1/3 and 1/2	−0.35	0.5
		between 1/2 and 2/3	0.32	1.53
		more than 2/3	1.18	2.38
Health	Household medical coverage	yes	0.15	0.44
		no	−0.13	−0.26
	Household share of medical coverage	at least half household	1.75	3.16
		less than half household	1.2	2.22
		no coverage	−0.79	−0.21
Housing	Type of accommodation	villa	2.47	–
		apartment	1.81	–
		moroccan house	−0.1	–
		basic dwelling	−3.73	–
	Type of accommodation	villa, apartment, Moroccan house	–	2.36
		rural house	–	0.81
		basic dwelling	–	−0.8
	Cohabitation	one household	0.18	–
		two households	−1.49	–
		three households	−2.11	–
		four households and more	−2.56	–
	Cohabitation	one household	–	0.03
		two households and more	–	−0.76
	Bath tub/shower	yes	1.62	3.5
		no	−1.31	−0.27
	Kitchen	yes	0.35	0.38
		no	−3.15	−1.98
	Toilet	yes	0.15	0.93
		no	−5.4	−2.16
	Number of persons per room	two persons or less	0.53	0.29
		between two and three	−0.82	−0.31
		more than three	−2.12	−1.36
Utilities	Sewage	yes	0.5	–
		no	−2.62	–

(continued)

Table 7.5 (continued)

			Urban Dim1	Rural Dim1
	Used water	network	–	2.32
		septic tank	–	1.24
		latrine	–	1
		other	–	−1.71
	Water access	yes	0.6	–
		no	−3.37	–
	Fresh water	network	–	2.22
		fountain	–	0.17
		other	–	−0.47
	Electricity	individual meter	0.7	1.04
		collective meter	−0.92	0.34
		no meter	−2.83	−0.69
		no electricity	−4.77	−1.98
Durables	Television	yes	0.2	0.73
		no	−4.11	−2.38
	Satellite dish	yes	1.2	2.2
		no	−1.78	−0.8
	Portable phone	yes	0.34	0.85
		no	−1.76	−1.51
	Gas stove	yes	0.68	1.18
		no	−1.06	−0.88
	Gas hoven	yes	0.49	1.1
		no	−0.43	−0.74
	Refrigerator	yes	0.69	2.09
		no	−3.15	−1.1
	Freezer	yes	2.73	–
		no	−0.15	–
Nutrition	Child stunting	yes	−0.75	−0.62
		no	0.09	0.15
	Child malnutrition	yes	−1.1	−1.5
		no	0.02	0.07
Expenditure	Food poverty	yes	−2.63	–
		no	0.23	–
	Expenditure quintiles	Q1	−3.17	−1.34
		Q2	−1.49	−0.34
		Q3	−0.39	0.24
		Q4	0.27	0.78
		Q5	1.45	1.96

We can also trace some differences between urban and rural areas. For example, the employment of the head of the household takes different signs in urban and rural areas and so do most modalities for the type of accommodation and for used water. There are also differences in signs for at least one modality of other variables such as education and health variables.

Overall, we can safely argue that the first axis describes a situation of welfare and that this welfare is multidimensional in that many different aspects of welfare such as health and education exhibit very similar behaviors toward this axis. However, much more can be done in further refining the model by removing or adjusting variables so as to better meet the fundamental properties of the MCA approach.

Step 2. Next, we review the behavior of each variable to test whether the model is robust to the underlying properties discussed. The MCA provided us with the basic elements to select the variables that will be used in the construction of the CWI. The main criterion used to reduce the number of variables without losing substantial consistency is the FAOC-I. The variables that have the FAOC-I property are those obeying the rule that welfare deteriorates from a position of wealth to a position of poverty throughout the first axis.[3] We therefore proceeded to eliminate some variables based on FAOC-I. These variables include the employment of the head of the household in both urban and rural areas. This variable had a counter-intuitive sign and was clearly a substitute for other employment variables. Other categorical variables did not meet the FAOC-I property but were kept and restructured so as to meet the property. These variables are access to health care, state of health and medical coverage, type of housing, children's nutritional status, and source of drinking water. Categorical variables representing the proportions of employed in the household and those of the unemployed were eliminated altogether failing to meet FAOC-I. Possession of a freezer in rural areas responds to FAOC-I but the variable was eliminated because of the low frequency of one of the modalities (less than 1 %).

Tables 7.2 and 7.3 also report results for the final MCA exercise. For urban areas (Table 7.2), the MCA performed on the remaining variables led to an increase in the explanatory power of the first factor, which rose from 52 to 69 %. In the final analysis, the poor are situated on the left and the rich on the right so that well-being increases from left to right. In this new plan and as shown in Table 7.5, all variables have the FAOC-I property of ordinal consistency on the first axis. We can also see that all variables show the same direction vis-à-vis the first axis satisfying FAOC-G. There is, therefore, a clear separation of rich and poor households who are opposed to the first factorial axis. We can conclude that the household score generated by the MCA on the first axis is a consistent indicator that can be used for ordering purposes to measure poverty. The same conclusion can be reached for rural areas (Table 7.3) and the gain of the second step procedure is much greater. The explanatory capacity of the first axis passes from 46.7 to 70.3 % with the second axis falling to 4.3 %. The direction of the relation of the

[3] Note that other criteria also used to select relevant variables such as the spread on the first axis, the high frequency of non-responses or low frequencies of responses on certain modalities.

variables with the first axis is also consistent across the variables (Table 7.5) which is what we need to use our new latent variable as a welfare indicator.

It is also instructive to look at the size of the weights in Table 7.5 to see the importance that the analysis attributes to each domain. As a rule of thumb, we take values larger than two as an indication of key welfare factors. In this category, we find households with the share of education above 2/3 for urban areas and the share of alphabetized above 2/3 for rural areas, owing a villa in both urban and rural areas, owing a shower, a television or satellite dish in rural areas, and owing a television and a freezer in urban areas. Also, being connected to the water and sewage systems are major indicators for rural areas while not being connected to these same network is very relevant in urban areas.

7.6 Multidimensional Poverty

7.6.1 The Composite Welfare Index

Following the MCA procedure, we can finally calculate the CWI for each household. As defined in previous sections, the value of the CWI for a household is the average of its weight categories corresponding to the average of standardized scores on the first factorial axis. In other words, it is the factorial coordinates of the household on the first axis that classifies households according to their state of wellbeing. The extreme values of the CWI thus calculated are -1.77 (the poorest households) and 0.75 (the richer households) in urban areas and -1.0 and 1.35, respectively in rural areas. It is worth providing a poverty profile of these extreme cases to have a sense of the actual living conditions of these households.

A household that has the smallest CWI value (-1.77) in urban areas has the following characteristics: the household head is illiterate; more than three-fourth of the members of this household have no grade and less than a third can read and write; members of the household do not have access to health care services and have no medical coverage; children within the household suffer from stunting; housing is precarious and it is not connected to the networks of drinking water, electricity, and sewage; the household does not own basic assets such as shower, kitchen, toilet, television, satellite dish, portable gas stove, refrigerator, and others in terms of expenditure, household food expenditure is well below the food poverty line and total expenditure is not even half the poverty line. Instead, the household with the highest CWI in urban areas (0.75) has the following characteristics: the household head is literate; all household members received education and can read and write; all members of this household have medical coverage; the dwelling of the household is a villa and is connected to all utilities; the household also has all the elements of comfort with all listed durable goods; in terms of expenditure, food expenditure of households is four times the poverty line.

In rural areas, the household with the lowest CWI value (−1.0) lives in very difficult conditions. All members of this household are either illiterate or have no education, they do not have access to health care services and they are not covered by medical insurance. With regard to housing conditions and comforts, this household has no basic equipments or durables. Food expenditure and total expenditure of the household are respectively below the food poverty line and the full poverty line. The rural household with the highest CWI score (1.35) lives in better conditions in terms of access to basic amenities (education, health, housing, and others) and, in terms of expenditure, this household belongs to the wealthiest 20 % of rural households.

7.6.2 Multidimensional Poverty in 2007

The CWI is a numeric variable that measures the level of welfare of households. Like all other variables measuring welfare (expenditure, income, or consumption), this indicator can be used to measure poverty. However, and unlike other monetary measures of welfare, the CWI can take positive and negative values. This question can be simply solved by adding to the score of each household the absolute value of the lowest score in the distribution. Then all we need to do is to determine a threshold that divide the poor from the non-poor and then measure poverty. Therefore, multidimensional poverty in our chapter is better described as a unidimensional poverty exercise based on a multidimensional (composite) measure of welfare. As already discussed, this is different from multidimensional measures of poverty that first aggregate individuals across each dimension at the societal level and then aggregate dimensions with various weighting mechanisms.

As a poverty line, we will use a relative poverty line defined as 60 % of the median CWI ($Z_i = 0.6 * CWI_{i,\ median}$, with $i = 1$ for urban areas and 2 for rural areas). Any study on poverty requires defining a poverty line dividing the poor and non-poor households. For monetary approaches, there are countries that use absolute poverty lines which represent the minimum amount necessary for the individual to meet basic needs. This is also the case, for example, of poverty analysis carried out in Morocco in the past. Other countries use relative poverty lines, often defined as a percentage of median income, which represent the minimum welfare level generally accepted in society. This is the case, for example, of European Union's countries. In the case of the CWI, we opted for a relative poverty line since the CWI indicator is not a monetary indicator where we can find the monetary poverty line. In principle, we could have estimated the CWI for the household closest to the monetary poverty line and used this value as poverty line for the CWI distribution. However, the monetary poverty line is meaningful in that it separates the poor and non-poor on the monetary scale but cannot equally apply to a scale where the household ranking may be different such as the CWI scale. For example, a household that is on the monetary poverty line but owns all possible assets would result in a very high poverty line on the CWI scale. Vice versa, a household that is on the monetary poverty line but owns

Table 7.6 CWI statistics and poverty line

	CWI mean	CWI median	Poverty line
Urban	1,775	1,8546	1,1128
Rural	0,996	0,9902	0,5941

Source NSLS 2006–2007

no assets may result in a very low poverty line on the CWI scale. Hence the choice of a relative poverty line.

To give a flavor of the CWI poverty line selected we looked at the characteristics of the household that is closer to the line in urban and rural areas. In urban areas, we found this household to have the head unemployed, a good share of educated people, no coverage of health insurance, to be well connected to water and electricity, to own many but not all essential items and to be in the first expenditure quintile. In rural areas, the household, the closest to the poverty line suffers from youth unemployment, has low levels of education (but alphabetized), has access to health services but no insurance, lives in basic accommodation and is not connected to electricity, water or sewage networks, and lacks basic items such as a gas stove and a refrigerator. However, this household does not suffer from malnutrition and belongs to the fourth expenditure quintiles among rural households. Although these two households may not be representative of households located around the poverty line on all dimensions, this illustration shows that the profile of an urban and rural poor household may be very different.

Based on the relative poverty line selected, Table 7.6 reports the CWI values of the main moments of the distribution as well as the values of the poverty line for urban and rural areas. It is evident that the mean and the median of the distributions are quite close indicating that the distributions are not very skewed. This is particularly the case in rural areas where the mean and median values are almost identical.

We can now measure poverty using the CWI. In the construction of our CWI, we have considered two variables that were constructed using expenditure per capita, the food poverty variable and expenditure quintiles. Here, we also want to compare poverty measured with the multidimensional indicator with poverty measured with the more traditional expenditure indicator. We can then assess the relation between poverty measured with the CWI and with expenditure per capita.

A simple correlation coefficient between the CWI and expenditure per capita shows that this relation is positive and significant with a coefficient of 0.46. This would suggest that there is a direct linear relationship between the two variables but that this relationship is by no means unitary. High household expenditure per capita is likely to be associated with a high CWI but this is not always the case.

Table 7.7 shows poverty levels according to both CWI and expenditure measures for urban and rural areas. All measures show invariably that poverty is higher in rural areas. This is true for all FGT poverty measures and for both CWI and expenditure measures. The difference in poverty between CWI and expenditure measures is more marked in rural as opposed to urban areas, although this may simply be a scale effect given that poverty is higher in rural areas.

Table 7.7 CWI and Expenditure Poverty

Area	CWI Poverty			Expenditure poverty			Difference		
	FGT0	FGT1	FGT2	FGT0	FGT1	FGT2	FGT0	FGT1	FGT2
Urbain	7.4	1.9	0.8	4.8	0.8	0.2	2.6	1.1	0.6
Rural	18.3	6.2	3.2	14.4	3.3	1.2	3.9	2.9	2
Total	12.1	3.8	1.9	8.9	1.9	0.6	3.2	1.9	1.3

Source NSLS 2006/07. Totals are simple weighted averages of the urban and rural statistics

Table 7.8 Multidimensional poverty versus income poverty

Quintiles of income per capita	CWI poverty		
	FGT0	FGT1	FGT2
1	28.3	10.0	5.4
2	17.1	4.8	2.1
3	9.0	2.8	1.3
4	4.3	1.0	0.4
5	2.0	0.3	0.1
Total	12.1	3.8	1.9

Source Calculations based on data from the NSLS 2006–2007

Table 7.9 CWI versus expenditure poverty

		CWI poverty		
		Poor	Non poor	Total
Expenditure poverty	Urban			
	Poor	2.2	2.6	4.8
	Non poor	5.2	90	95.2
	Total	7.4	92.6	100
	Rural			
	Poor	6	8.4	14.4
	Non poor	12.3	73.3	85.6
	Total	18.3	81.7	100

Source NSLS 2006/07. Totals are simple weighted averages of the urban and rural statistics

The relation between CWI poverty and expenditure poverty has also a mechanical element built into it given that expenditure is one of the dimensions considered in the construction of the CWI. Given that for 2007 we also have income, in Table 7.8 we plotted CWI poverty rates by income per capita quintiles instead of expenditure. We find that to the two higher income per capita quintiles correspond lower levels of poverty as one would expect. CWI poverty is monotonically decreasing in income quintiles with no exceptions for the three FGT indicators. CWI poverty also declines very significantly between the first and second quintile. However, it is also evident that the two types of poverty do not overlap completely. In monetary terms, 60 % of the median value would be situated between the first and second income quintile,

Table 7.10 Double poverty according to various household characteristics

Characteristics of household head	Multidimensional and monetary poverty	Poor multidimensional and non-poor monetary	Multidimensional non-poor and poor monetary	Multidimensional and monetary non-poor	Total
Sex of hh head					
Male	3.9	8.4	5.3	82.4	100
Female	3.5	7.5	4	85	100
Household Size					
One person	0.9	25.3	0	73.8	100
2–3 pers.	1.1	11.7	0.4	86.8	100
4–6 people.	3	8.4	2.6	86	100
7-9 pers.	5	6.6	7.7	80.7	100
10 people and more	7.3	7.8	14.4	70.5	100
Education of hh head					
No level	5.5	11.7	6.6	76.2	100
Fundamental	1.5	3.9	3.4	91.2	100
Secondary	0.7	1	1.6	96.7	100
Higher	0.4	0	0	99.6	100
Employment status of hh head					
Employer	0	2.5	0.6	96.9	100
Salaried	4	8.3	5.7	82	100
Independent	4.9	11.4	5.2	78.5	100
Other	5.1	8.2	3.1	83.6	100
Total	3.8	8.3	5.1	82.8	100

whereas CWI poverty still exists in the third, fourth, and fifth expenditure quintile. Therefore, many non-poor households clearly fall under the CWI poverty line.

It is useful at this stage to assess the extent of double poverty, or the poverty groups that are found to be poor by both the multidimensional and the unidimensional approach. This is shown in Table 7.9. In urban areas, about 2 % of the population is found to be double poor. About 10 % of the population is at least poor in one of the two dimensions while the remaining 90 % is neither CWI nor expenditure poor. In rural areas, the rates are higher as we know by now but the distribution in the various cells of the table is also different. About 6 % of the population is double poor while over 26.7 % is poor according to at least one criteria with only 73.3 % of the population being outside both types of poverty.

Some insights about the main drivers of double poverty can be derived from Table 7.10 where we look at double poverty according to various household characteristics in 2007. Poverty is higher in households headed by men as compared to households headed by women, but this difference is not very large and the number of households headed by women is relatively small. Depending on the size of households, double poverty increases with household size. It displays its maximum when the household is composed of ten or more people with a percentage of 7.3 % against 0.9 % for those living in households with less than four

Table 7.11 Poverty over time

	Multidimensional poverty		Monetary poverty	
	2001	2007	2001	2007
Urban	9.4	7.4	7.6	4.8
Rural	42.3	18.3	25.1	14.4
Total	23.9	12.1	15.3	8.9

Source NSLS 2006/07. Totals are simple weighted averages of the urban and rural statistics

people. Interestingly, households that are poor according to non-monetary criteria but non-poor according to monetary criteria seem to experience higher poverty the smaller is the household size. This might be explained by the fact that smaller families have fewer assets than larger families, income being equal. By grade level of the household head, 23.8 % of people living in households headed by someone with no education are at risk of at least one form of poverty. Double poverty affects 5.5 % of those households headed by people who never attended school. Also, the higher the educational level of the household head the lower is the risk of poverty as one would expect. Overall and considering urban and rural areas together, in 2007 3.8 % of people were below the poverty line according to both CWI and expenditure criteria. This can be considered as the core of poverty in Morocco while 82.8 % of the population is above all forms of poverty.

7.6.3 Multidimensional Poverty Over Time

As already discussed and in order to conduct the MCA analysis over time, we used the weights from the NSLS 2006/2007 for both time periods considered. We also followed exactly the same methodology we used for the 2006/2007 MCA for the 2000/2001 data. The multidimensional poverty line for 2000/2001 is equal to 60 % of the CWI median value and amounts to 1.1 in rural areas and 0.6 in urban areas.

The analysis of CWI and expenditure poverty over time shows that the two forms of poverty declined significantly between 2001 and 2007 (Table 7.11). This decrease is more pronounced with the multidimensional approach as compared to the monetary approach at the national level. The rate of multidimensional poverty decreased from 23.9 % in 2001 to 12.1 % in 2007, an average annual decline of about 12 %. Income poverty has decreased annually by an average of about 9.5 %, from 15.3 % in 2001 to 8.9 % in 2007. Urban and rural areas show different patterns. Poverty in rural areas is much higher and the decline over the 2001–2007 period has been steeper. Multidimensional poverty in particular shows that rural poverty declined by more than half, from 42.35 to 18.3 %.

Table 7.12 Robustness analysis

Threshold z		25	50	75	100	125	150
CWI poverty							
2001	Urban	0.3	1.2	3.9	9.4	19.3	34.4
	Rural	3.8	13.4	27.9	42.3	55.3	68.1
	Total	1.4	6.6	14.5	23.9	35.1	49.2
2007	Urban	0.2	1.1	3.4	7.4	16	31. 0
	Rural	1.4	5.1	10	18.3	27.1	39.3
	Total	0.7	2.8	6.3	12.1	20.8	34.6
Difference	Urban	0.1	0.1	0.5	2	3.3	3.4
	Rural	2.4	8.3	17.9	30.2	28.2	28.8
	Total	0.7	3.8	8.2	11.8	14.3	14.6
Expenditure Poverty							
2001	Urban	–	0.3	2.5	7.6	16.2	24.3
	Rural	0.2	2.3	10.1	25.1	40.9	55.6
	Total	0.09	1.2	5.8	15.3	27.1	38.1
2007	Urban	–	0.07	1.2	4.8	9.9	17.5
	Rural	0.04	1.2	5.7	14.4	25.9	38
	Total	0.02	0.6	3.2	8.9	16.8	26.4
Difference	Urban	–	0.2	1.3	2.8	6.3	6.8
	Rural	0.18	1.1	6.9	10.7	15	29.2
	Total	0.07	0.6	2.6	6.4	10.3	11.7

Source NSLS 2006/07. Totals are simple weighted averages of the urban and rural statistics

7.6.4 Robustness Analysis

The choice we made of poverty lines is an arbitrary choice and changing the poverty line affects results to an extent that poverty trends over time could also be reversed. It is sensible therefore to test the sensitivity of our results to changes in the poverty line. We use stochastic dominance analysis, which ensures that all possible poverty lines are taken into account. This technique can attest that temporal comparisons are valid for a given range of poverty lines and for a given class of poverty measures.

To simplify the presentation of results, we normalized the CWI and expenditure distributions so that the poverty line under the two approaches is equal to 100. To check the conditions of the first order stochastic dominance, we calculated the differences of the effects of poverty distributions calculated for 2001 and 2007, for different values of poverty lines (Table 7.12). It is clear from reading the table that the distribution of 2007 dominates the 2001 distribution nationally according to the two approaches to poverty. Indeed, whatever the poverty line considered, the incidence of poverty is more important in 2001 than in 2007. Expenditure poverty has decreased substantially between 2001 and 2007 regardless of the place of residence, regardless of the poverty line, and regardless of the measure used.

We can now assess stochastic dominance of first-order visually by comparing the cumulative distributions functions (CDFs). Figure 7.1 presents the results.

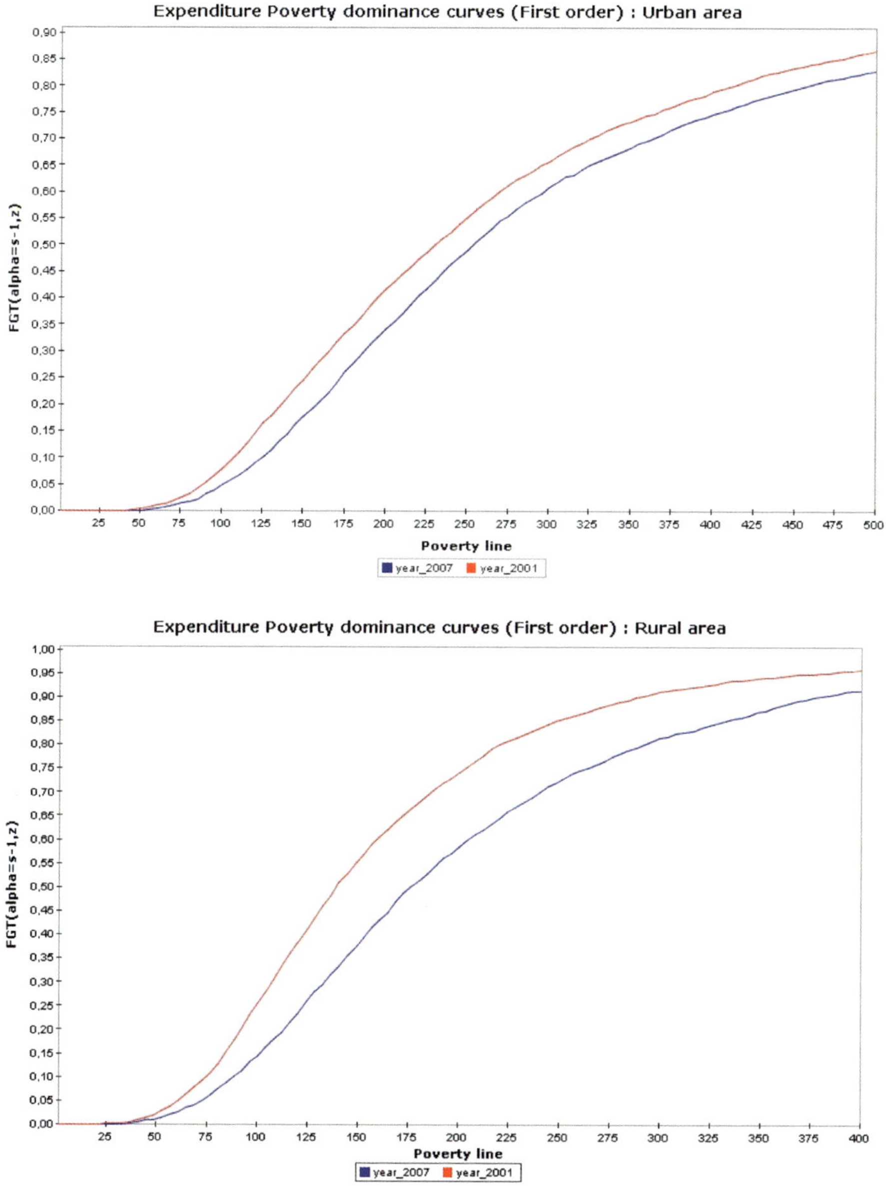

Fig. 7.1 Stochastic dominance (first-order)

Plots are presented for urban and rural areas and for expenditure and the CWI separately. For the expenditure measure, there is a visible gap between the 2001 and 2007 curves in both urban and rural areas. The 2007 curves dominate the 2001 curves in that poverty is lower in 2007 for any poverty line considered. For the

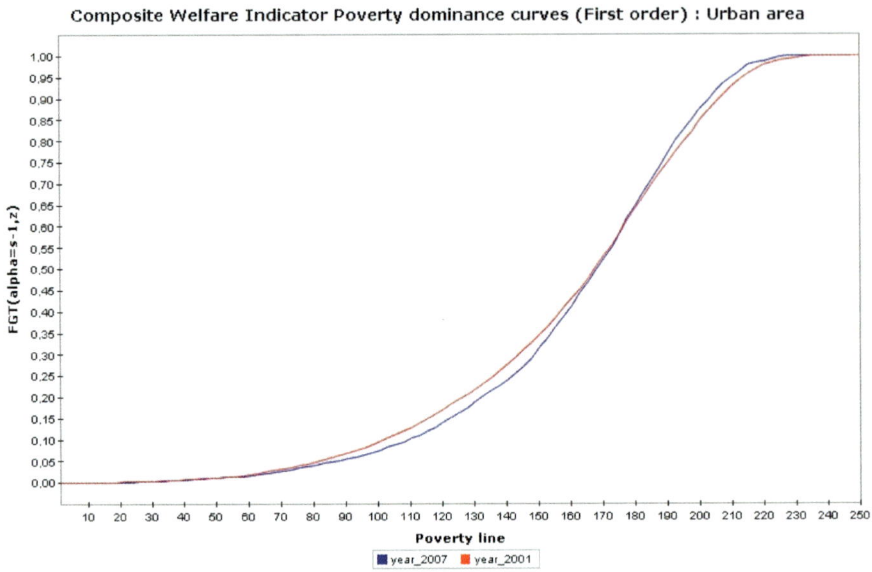

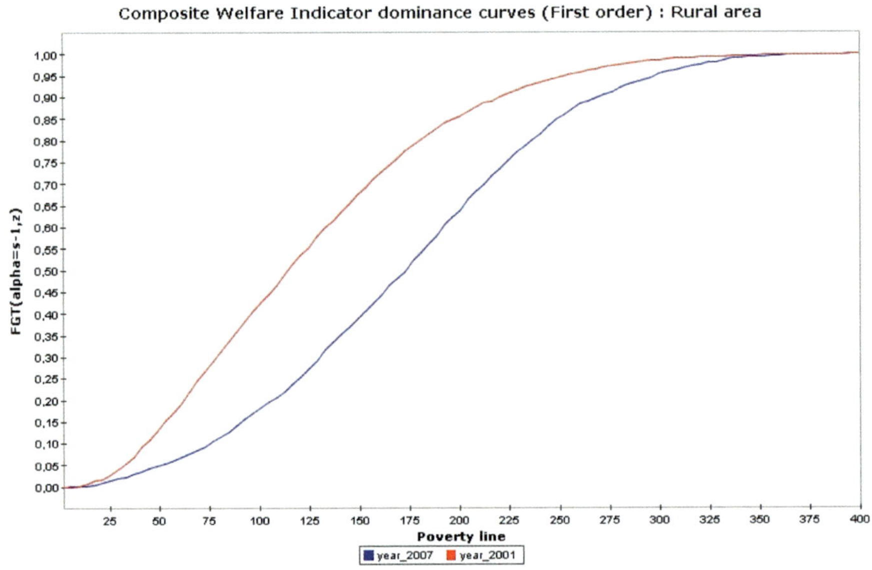

Fig. 7.1 continued

CWI measure, this same phenomenon is even more visible in rural areas where we can clearly see that welfare has improved and poverty has declined for any poverty line considered. However, this is not true for urban areas where the curves intersect around the third quartile suggesting that dominance is reversed along the curve. With very high poverty lines, the poverty headcount was higher in 2007 than in

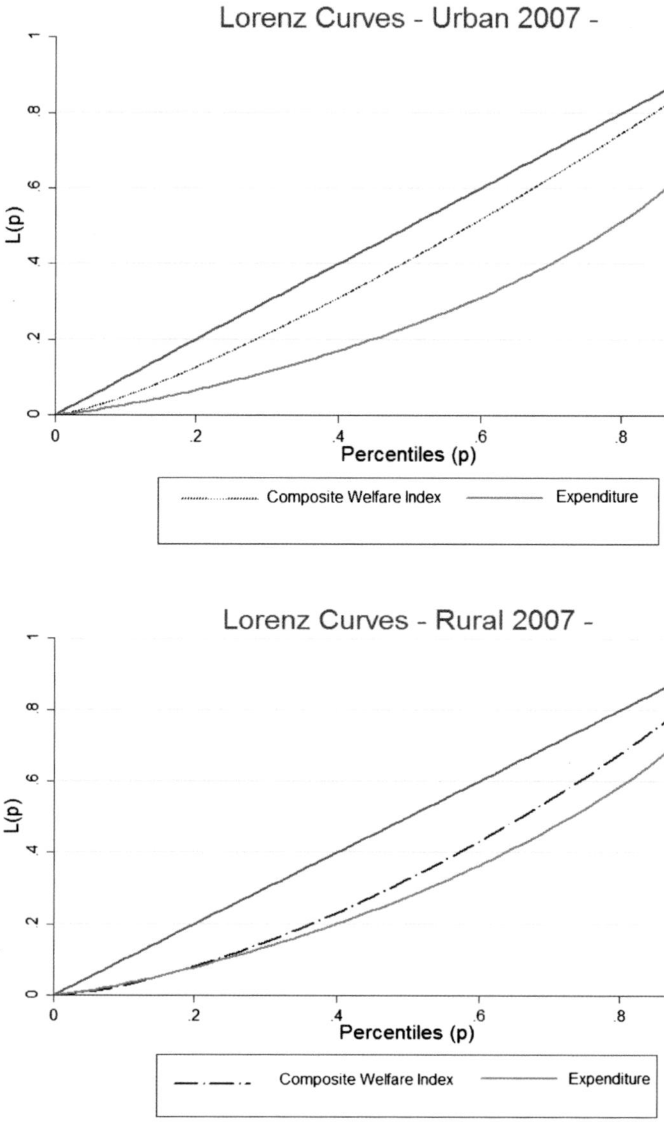

Fig. 7.2 CWI and expenditure Lorenz

2001, although we cannot be sure that this would apply to all poverty indexes. This clearly shows the utility of a stochastic dominance approach. In all statistics presented thus far, we found no indications that poverty might have grown between 2001 and 2007. With first-order dominance, we uncover a small but visible increase in multidimensional poverty for the upper part of the distribution if a very high poverty line is used.

As a final exercise, we focus again on 2007 and compare the expenditure and CWI Lorenz curves, also for urban and rural areas (Fig. 7.2). This is suitable for measures that are intrinsically different such as expenditure and CWI given that both axes are normalized to one. In reality, we cannot exclude some data generation pattern in the MCA approach that would make the CWI scores exhibit consistently lower or higher inequality than a typical expenditure measure. Intuitively, one would expect a measure built on several indicators to show higher inequality than a measure built on one indicator but this is not the case. The CWI measure shows invariably less inequality than the expenditure measure and this is true for both urban and rural areas. This finding is also robust and visible all along the two distributions. We can also see that rural areas exhibit higher inequality than urban areas for both the CWI and expenditure measures. This is particularly true for the CWI measure suggesting that diversity in non-monetary measures is much more pronounced in rural areas than in urban areas.

7.7 Conclusions

This chapter has used MCA to construct a consistent CWI and used this indicator to assess poverty in Morocco between 2001 and 2007. This is one of the least explored paths in the measurement of multidimensional poverty and complements existing studies on poverty in Morocco. It is also an approach that has two important computational advantages. It gives more weight to indicators with a small number of hits, hence giving more importance to deprived minorities, and complies with the welfare axiom of reciprocal bi-additivity (see Asselin 2009).

The MCA procedure was carried out separately for urban and rural areas on the assumption that the MCA models would be different in the two areas. Results of the two-steps procedure showed indeed that the resulting models are different and confirm the known fact about Morocco of a deeply divided society along urban and rural lines. The multidimensional indicators that characterize these two areas for both the poorest households and the least poor are often different, which is an important finding for policy purposes.

Overall, we found that the multidimensional approach largely confirms previous results on monetary poverty. Monetary and non-monetary indicators are positively and significantly correlated but this correlation is by no means unitary suggesting that multidimensional and monetary poverty do not describe the same phenomenon. However, the multidimensional approach corroborates monetary results in various respects. In 2007, poverty is found to be higher in rural areas as compared to urban areas and according to all FGT poverty indicators. And the decline in poverty observed between 2001 and 2007 has not been limited to improvements in monetary indicators but has concerned multiple aspects of living conditions including health, education, and labor market conditions. In fact, the decline in multidimensional poverty appeared to have been more pronounced than the decline for monetary poverty. The multidimensional approach also confirms that the decline in poverty has been stronger in rural as opposed to urban areas, although the starting level was much higher in rural areas.

Robustness tests including stochastic dominance tests validate both the monetary and the multidimensional findings. Interestingly, multidimensional inequality shows invariably less inequality than the monetary counterpart, in both urban and rural areas, which may be explained by the fact that larger families tend to be more monetary poor in per capita terms but also tend to own a larger number of assets on a household basis. This chapter has also confirmed that the urban–rural divide remains very deep in Morocco and should continue to rank high in terms of policy priorities.

Acknowledgments The authors are grateful to Florent Bresson and Valérie Bérenger for very useful comments.

References

Alkire, S. & Foster, J. E. (2011). Counting and multidimensional poverty measurement. *Journal of Public Economics, 95*(7–8), 476–487.

Asselin, L. M. (2009). *Analysis of multidimensional poverty: Theory and case-studies. Economic studies in inequality, social exclusion and well-being* (Vol. 7). Berlin: Springer.

Atkinson, A., & Bourguignon, F. (1982). The comparison of multidimensioned distributions of economic status. *Review of Economic Studies, 49*, 183–201.

Benzécri, J-P. (1973). *Analyse des Donnees* (Vols. 1, 2). Paris: Dunod.

Decancq, K. & Lugo, M. A. (2010). Weights in multidimensional indices of well-being: An overview, Open Access publications from Katholieke Universiteit Leuven, Katholieke Universiteit Leuven.

Douidich M., Ezzrari A. & Soudi K. (2009). Dynamique de la pauvreté au Maroc: 1985–2007, Cahiers du plan N° 26, Haut Commissariat au Plan—Rabat.

Duclos, J.-Y., Sahn, D., & Younger, S. (2001). Robust multi-dimensional poverty comparisons. *Economic Journal, 116*(514), 943–968.

Greenacre, M. J. (1984). *Theory and applications of correspondence analysis*. London: Academic Press.

Kolm, S. C. (1977). Multidimensional egalitarianisms. *Quarterly Journal of Economics, 91*(1977), 1–13.

Maasoumi, E. (1986). The measurement and decomposition of multidimensional inequality. *Econometrica, 54*, 771–779.

Njong, A. M. & Dschang, P. N. (2008). *Characterising weights in the measurement of multidimensional poverty: An application of data-driven approaches to Cameroonian data.* OPHI Working Paper No. 21, wired at www.ophi.org.uk 1.

Ravallion, M. (2011). On multidimensional indices of poverty. *Journal of Economic Inequality, 9*(2), 235–248.

Sen, A. K. (1976). Poverty: An ordinal approach to measurement. *Econometrica, 44*, 219–231.

Part III
Focus on Specific Dimensions of Poverty and Social Exclusion

Chapter 8
Poverty, Education, and Employment Among the Arab-Bedouin in Israel

Suleiman Abu-Bader and Daniel Gottlieb

Abstract The socio-economic situation of the Arab-Bedouin population in the Negev is examined in light of the general Israeli Arab population. Based on the Galilee Society's social survey for 2004 Israeli Arab poverty incidence was found to be 52 % with nearly two-thirds in persistent poverty. Among Bedouins living in villages unrecognized by the Israeli government poverty incidence was nearly 80 %, and poverty severity, as measured by the Sen-Poverty index, was as high as seven times than that among the mainstream population in Israel, i.e., the Jewish population (excluding the—predominantly poor—ultra orthodox Jews). Similarly to international evidence, we found that education, age, family size, employment, and occupation of the household head and the number of income earners in the family are important determinants of the probability of being poor. We show that the rapid increase of Arab women's student enrollment rates not only reduced the education gap compared to Arab men but importantly adds to poverty reduction through various channels. Bedouin households, especially in the unrecognized villages, were found to have much less access to infrastructure compared to other Arabs, thus implying that the lack in infrastructure forms a significant barrier to women's participation in the labor force. This was also found to have an adverse indirect effect on the completion of schooling, and indirectly on mothers' fertility, keeping it relatively high by reducing education's potentially diminishing effect on poverty. A considerable mismatch between skills and employment was found

S. Abu-Bader
Economics Department, Ben-Gurion University of the Negev,
Beer-Sheva, Israel
e-mail: abubader@exchange.bgu.ac.il

D. Gottlieb (✉)
Research and Planning Administration,
National Insurance Institute and Economics Department,
Ben-Gurion University of the Negev, Beer-Sheva, Israel
e-mail: danielgt@nioi.gov.il

V. Bérenger and F. Bresson (eds.), *Poverty and Social Exclusion Around the Mediterranean Sea*, Economic Studies in Inequality, Social Exclusion and Well-Being 9,
DOI: 10.1007/978-1-4614-5263-8_8,

among Arab academics, thus pointing at the possibility of discrimination and segregation in the relevant labor market. Considering various transmission mechanisms, we show that government intervention in infrastructure has a high expected social return, both through direct and indirect channels and could, thus, help in interrupting the vicious circle of poverty.

8.1 Introduction

The purpose of this paper is to describe the socio-economic situation of the Arab-Bedouin society of the Negev on the background of the general Arab society in Israel and to analyze its causes in order to improve the design of poverty reduction.

We use a unique database of the Galilee Society (henceforth GS) that for the first time provides an integrated view of the characteristics of Bedouin villages. This is of particular importance, given the neglect of official data collection by the Israeli Central Bureau of Statistics (henceforth CBS) on the economically weak population of Southern Bedouin community living in localities not recognized by the government.

The way of life in the Arab-Bedouin society has been undergoing a rapid change from a traditional seminomadic agrarian society into a modern sedentary society. This change partly expresses the desire of youths in a traditional society to adapt to Western surroundings, and in part reflects a policy induced by the government. Following the uprooting of Bedouin villages after the war of 1948 from the North-Western region of the Negev to its North-Eastern region, a traumatic and complicated land dispute arose between the Bedouin and the State, which has led the State to turn its back on the most basic infrastructure required for the Bedouin such as water, electricity, education, sewage, garbage disposal, road systems, transportation, and employment infrastructure. Even in the few cases in which the government conceded to recognize a 'unrecognized' village within its existing borders (e.g. Um-Batin in 2004) and develop infrastructure therein, the process became entangled, among other things, in the thicket of conflicting interests within the Arab-Bedouin community. Hence, the inferior economic situation, which manifests in extremely high poverty incidence, is a combined result dictated by the dispute with the State, and intensified by tensions within the Arab-Bedouin society. This state of affairs impedes social development in that region and prevents any significant private investments in it.

Past efforts to resolve the problems between the Bedouin and government have failed. The confidence of the Bedouin in the government's ability to provide adequate public services is at a low. The government is frustrated with efforts to enforce building laws over the land in dispute. This perpetual deadlock has led to recurring episodes of makeshift houses being built by Bedouin and to be demolished rapidly thereafter by the State, while insufficient incentive, meaningful enough, is being created for Bedouin of the unrecognized villages to relocate into permanent communities recognized by the government. Among the Bedouin, this situation deepened the social deterioration and caused increasing alienation toward the state. This failure spells loss for both parties to the dispute: it enhances the social and economic distress of the Bedouin population, while government sovereignty over territories in dispute is weakened. This predicament is deplorable for the Bedouin and for Israeli society in general.

This paper describes the socio-economic situation of the Arab-Bedouin society on the background of the socio-economic situation of general Arab society in Israel, while focusing on poverty under various definitions, especially in the unrecognized villages, the conditions of which have been neglected in official poverty accounts. In sect. 8.2, we review the historical background of the land dispute between the Bedouin and the State of Israel. In sect. 8.4, we describe poverty among Bedouin by using two poverty definitions—the official half-median approach and the Food Energy Intake and Share (FES) approach (Ravallion 1994). Section. 8.4 also describes two other important dimensions of well being—education and the importance of improved access to public infrastructure. Section. 8.5 discusses the empirical results of our analysis of the major determinants of poverty, emphasizing the interdependent nature of the three issues treated in this paper—poverty, education, and access to public infrastructure. We end our discussion with concluding remarks.

8.2 Historical Background

Before the establishment of the State of Israel, the Bedouin population of the Negev counted, according to various estimates, between 65,000 and 100,000 inhabitants. The absolute majority settled in the fertile North-Western region of the Negev (Falah 1989). Most Bedouin residents in this region fled or were expulsed during the war of 1948. Some became refugees in Gaza, the West Bank, or in neighboring Arab countries. According to various estimates, 11,000–17,000 individuals (Porat 1998) remained in Israel. Most of the remaining Bedouin were uprooted from their living area in the Negev's North-West and relocated to its North-East region, referred to as 'the border area', also mentioned as 'Siyagh' or 'Sayej'. These residents joined the tribes that used to live in this region before the establishment of the State and formed the basis of the unrecognized villages. According to Swirski and Hasson (2006), this area spreads over 1.2–1.5 million dunam,[1] of which about 400,000 dunam were arable at the time, about one-fifth of the arable land that the Bedouin cultivated before 1948. Porat (2000) estimates the area at 2 million dunam, while Ghazi Falah (quoted in Porat 1998) estimated the area at 2.7 million dunam. Thereafter, the Western region was expropriated under the Absentee Property Law of 1950, and the region was declared a closed military area from which the Bedouin were restricted. Bedouin residents were consequently

[1] Dunam = 0.247 Acre or 0.1 Hectare.

prevented from living and making use of the land outside the 'Border Area'. Until the termination of military rule in 1966, Bedouin, like the rest of the Israeli-Arab population, needed a special permit to move outside the area designated for them for work, study, or any other purpose.[2]

In 1953, the "Land Acquisition Law—Confirmation of Actions and Compensations" came into effect. Under that law compensation is due to anyone whose land is expropriated by the State. Expropriated lands were transferred to a government development authority and from it to the possession of the State's Settling Authority. Further along, parts of their lands were expropriated in the 'Border Area' for the purpose of establishing seven Bedouin settlements and for general development purposes (including Jewish communities and defense projects).

The Supreme Court (Civil Appeal 218/74) ruled in 1984 that the Negev is State land, and thus on the basis of the Lands Ordinance from the year 1858, installed by the Ottoman regime, and the following British Ordinance concerning the same topic.[3] For land ownership registration, the Bedouin relied on the rules of tradition and did not care to obtain title deeds from the Ottoman or the British authorities, inter alia, due to apprehension of having to pay taxes. In 1962, a government committee recommended the establishment of permanent communities for Bedouin over an area of 7,600 dunam in the Border Area. In 1965, the government approved three sites on which to establish the communities—close to Beer Sheva, Shoval, and Kuseife. The first community, Tel Sheva, was established in 1969, Rahat in 1971, Segev Shalom in 1979. Ar'ara and Kuseife were established in 1982, Lakiya and Hura in the second half of the 1980s. These communities were designed without consulting the Bedouin residents or accounting for their special needs. Al-Huzayel (2004) argues that the purpose of the mentioned government decisions was to take control of Bedouin lands while restricting the area for Bedouin to the smallest possible area of 'Border Area' lands. A survey made in 2001 by Abu-Saad and Lithwick in communities recognized by the State, indicated discontent with urban living conditions there, and distrust by residents toward the government and its conduct in relation to the land dispute.

Alongside the effort to settle the Bedouin in urban communities, the government tried to exercise its formal title by evicting the Bedouin from the lands and relocating them to the aforesaid communities in return for compensation to the families. For that purpose, the government created a mechanism for land settlement. The Bedouin who asserted title in lands were requested to file a claim and then sign a waiver in respect of the land being claimed in return for compensation by the State. Those who did not assert title and conceded to relocation to one of the

---

[2] See Porat (2000).

[3] The Ottoman Ordinance regarded uncultivated lands as "dead" lands ("Mawwat" in Arabic). The Ordinance prescribed that one who cultivates lands can register them in one's name, while the State retains the principal title. In 1921, the British Mandate government published a Lands Ordinance based on the Ottoman Lands Ordinance. The British Ordinance allowed people who cultivate "Mawwat Lands" to register the lands in their name. One who did not register the lands lost his right to the title. See citation of Granovski in Swirski and Hasson, (2005, in Hebrew).

communities, the State offered a plot of land. In 1969, the Ministry of Justice had set a mechanism for land settlement under which anyone who asserts title in land may file a claim with the land settlement officer.[4] 3,000 title claims were filed, asserting title over 991,000 dunam, most during the 1970s. The government had since reached settlements in respect of 140,000 dunam. Hence, the process remains incomplete for an estimated 850,000 dunam, due to disagreement over the rate of compensation and living conditions in the recognized communities, which some of the Bedouin regard as an unsuitable alternative to their living conditions in the unrecognized villages, and due to the denial of a traditional way of life, including traditional sources of income, and the lack of any employment infrastructure in the recognized communities.

The refusal of many Bedouin land claimants to accept stipulations made in the government's offer has led to the enactment of a law which was intended to force a settlement upon those who, thus far, refused the offers.[5] In Tel Malhata (Tel almalh), the "Peace Law" was later effected in slightly improved conditions.[6] Due to continued uncertainty from the government's point-of-view, it froze most areas of public policy and infrastructure investment. This led to a lack of clarity toward the future and as a consequence also prevented private initiative and investment in these places. Even in cases where approved urban planning schemes existed, they were not implemented in many cases, sometimes also due to conflicts within the Arab-Bedouin society concerning title and use of lands by other Bedouin residents.

In 1986, The Administration for the Promotion of Bedouin ("The Bedouin Administration") was established as part of the Israel Land Administration in purpose of negotiating with Bedouin residents who filed title claims in respect of lands. Notwithstanding, the administration defines its functions in a much broader sense: planning and development in existing Bedouin communities, and planning of new communities pursuant to government decisions, allotment of land for agriculture and building of public establishments, settling of agreements for relocating Bedouin from unrecognized communities to recognized ones, administration of the Committee for the Allocation of Drinking Water, etc. The multitude of functions in the hands of the Administration has created a situation, according to which various government offices parted with their responsibility to attend to residents in the ministries' designated areas of responsibility, given the

[4] The State Comptroller (2002, p. 115).

[5] Government resolution No. 179 [BD/1] dated November 18, 1979.

[6] Its formal name is the "Acquisition of Land in The Negev Law—The Peace Treaty with Egypt, 5740—1980". Under this law, people who claimed lands in an extent of up to 100 dunam were offered to vacate and relocate to a developed plot in Kuseife or Ar'ara, or alternatively be awarded full financial compensation.

Table 8.1 The Arab population in Israel by groups, in 2004[a]

Population group	Persons	In % of total Arab population
Total Arab population	1,030,100	100.0
Muslims (in total)	847,967	82.3
Bedouin (in total)	180,146	17.5
Bedouin of the South	131,819	12.8
In recognized communities	78,404	7.6
In unrecognized communities	53,415	5.2
Other Muslims	667,821	64.8
Christians	91,403	8.9
Dauze and others	90,729	8.8

[a] Excludes East Jerusalem and the Golan Heights.
Source of data The Galilee Society, 2004.

Bedouin Administration's extensive assumption of responsibility over the fate of Bedouin residents, in all relevant aspects of public policy.[7] Yet, the Administration finds it difficult to handle the many functions needed in the Bedouin communities, particularly in the light of the ambiguity concerning its powers. The Administration is subject to three bodies at any given time—the Israel Land Administration, the Ministerial Committee for the Coordination of Policy and Operation in the Bedouin Sector, and the Ministry of National Infrastructure. The upshot is that in many areas of public policy the residents do not receive adequate services and the Administration neither has the vocational skills nor the resources required in order to provide adequate services.

In 2004, the Arab population in Israel counted about 1,030,100—82 % Muslims, 9 % Christian, and the remaining 9 % included Druze and other groups (Table 8.1). One-fifth of the Muslim Arabs were Bedouin. Most of them (70.3 %, including the Bedouin of the North) lived in recognized communities and in the South 59 %.

[7] See document by the Director of the Southern District of the Ministry of the Interior, Dudu Cohen, in 2002. "The reality of the Administration for the Promotion of the Bedouin [...], the Authority for Bedouin Education, the Welfare Department for the Bedouin in the Dispersion [...] create a sentiment among the Bedouin of being an aberrant population [...] which is not entitled to receive direct and professional service from the various government offices, as applied to the Jewish population [...]. The Bedouin Administration could (and perhaps should) handle but a single area specific to the Bedouin sector, which is the area of title claims and land settlements. [...] the question is posed whether its organizational emplacement should be with the Israel Land Administration, the Ministry of Justice or the Ministry of Finance..." [Quoted from Swirski and Hasson, (in Hebrew) 2005, p. 16].

8.3 The Data, Sources, and Survey Description

The data used in this paper were drawn from the *Palestinians in Israel Socio-Economic Survey* which was carried out by the Galilee Society (GS), the Arab National Society for Health Research and Services over the year 2004. The survey was the first to cover all the Arab citizens of Israel including the Bedouin in the unrecognized villages, which include about half of them residing in the Negev, where official Israeli data were never collected. To ensure the quality of the survey, the Galilee Society formed a scientific consultative committee with various areas of academic expertise to discuss and identify the topics to be covered and to formulate the questionnaire. A team of experts from the Palestinian Central Bureau of Statistics helped to formulate the questionnaire and accompanied the process of data collection and compilation. The survey covered the following topics: demography, housing, labor, standards of living, education, media and culture, environment, agricultural land holdings, and public order.

The sample design of the survey is a stratified two-stage random sample. In the first stage, the areas are enumerated and selected in one stratification level and in the second stage, 30 responsive households in each of the chosen enumeration areas were selected. The sample was divided into four levels of stratification: Region (North, Center, and South), Community (Urban communities with a population of more than 15,000, urban with population between 5,000 and 15,000, rural communities with population of <5,000, and Bedouin regardless of population size), Locality type (recognized, unrecognized), and locality characteristics (only Arabs, mixed). The sample included 3,270 households (of which 590 were Bedouin from the Negev) with a total of 16,383 persons.

Since the CBS surveys include data on Arabs not from the unrecognized villages we compared income distributions from both surveys, the Galilee Society Survey excluding Bedouin from unrecognized villages and the CBS survey which does not include the Bedouin from the unrecognized villages. The results showed a high similarity of distributions based on variance and mean, and the poverty measures based on the two surveys were similar.

8.4 Poverty, Education, and Access to Infrastructure Among Arab Bedouins

We utilize in this paper, the GS socio-economic survey to study the socio-economic situation of the Bedouin of the Negev. The data reveal severe socio-economic hardship in this community: the rate of unemployment is high, the employment ratio, as a share of the population at working age, is low, and dropout rates of children and youngsters from the education system is substantial.[8] This and other factors described below cause severe and persistent poverty, particularly in unrecognized villages. Poverty severity, as measured by Sen Poverty-Index, deepened over the years.

This section presents stylized data on each of the three related issues of education, infrastructure, and poverty as they emerge from the GS database.

In 2004, their poverty incidence was 79.2 % (for persons), and for sake of comparison, the Sen Poverty-Index was 6.7 times that of the Israeli mainstream (Jewish non-orthodox) population.[9]

8.4.1 Two Approaches to the Measurement of Poverty

There is considerable professional consensus concerning the methods of aggregation of the poor.[10] There is no consensus about the identification of the poor. Poverty measures may differ with respect to their focus on the poverty line and the resource definition or even on the usefulness of a poverty line, as demonstrated in the multidimensional approach. They may also differ with respect to the underlying concept of well being as reflected by use of an absolute or relative approach and by the choice of income or consumption expenditure as a basis for the identification of the poor. The most common poverty line for advanced countries is that of half-median income, as reported by the OECD and also by Israel's National Insurance Institute (henceforth NII).[11] Another increasingly popular poverty line is the one based on the FES approach that was developed in Ravallion 1993, 1994 (particularly appendix 1—A method for setting poverty lines) and 1998. We leave out other important relative indices of poverty, social deprivation or social exclusion for future research, despite their attractiveness for poverty research,

---

[8] Following the official Israeli definition, persons aged 0–18 are referred to as children.

[9] We have selected the non-orthodox Jewish society as a reference group since poverty in the orthodox population is a singular phenomenon that requires separate reference. See also Gottlieb (2007).

[10] The issue of aggregation relates to the calculation of an index on the basis of specific information regarding poverty in households or individuals. The simplest aggregation method is to count the poor. This index, though widely used, ignores differences in the severity of poverty among families. In the professional literature various aggregation methods account for poverty severity such as the Sen-Index, its extension by Shorrocks (1995), the quadratic income-gap measure by Foster et al. (1984) or the to name but a few. The issue of identification deals with questions of "what is poverty" and "who is poor". The discussion about poverty lines relates to the first question while the discussion of income sources relates to the second. See also Sen and Foster (1997), or Ravallion (1994).

[11] See www.btl.gov.il.

much of which has evolved from Sen's seminal capability approach (see Sen 1985) and has become the basis for calculating measures of multidimensional poverty.[12]

In this paper, poverty indices are calculated on the basis of the two above-mentioned measures of the relative approach: (1) by the half-median equivalent[13] net financial income, practiced inter alia, by the NII[14] and (2) the FES.

The FES poverty line is calculated as the sum of an adequate nutritionally oriented food basket and a non-food basket based on the revealed preference of a household at two focal points of the household's decision to allocate expenditure among food and non-food: the first point is at the income level just sufficient to purchase the adequate food basket. Households will typically forego some food consumption in order to acquire essential non-food goods and services. This point being obviously very low is averaged with the second focal point, which is at the income level, at which the household spends the amount needed to consume the nutritionally adequate food basket. The average of the two can be used as the non-food poverty line.

Problems of Measuring Poverty Among the Bedouin

The CBS survey samples do not represent the entire Arab population in Israel, since the Bedouin of the unrecognized villages do not appear in the CBS sampling procedure. This missing population in the CBS sample accounts for some 40 % of the Bedouin population in the South.[15] This is one of the advantages of the Galilee Society's particular focus (oversampling) on the socio-economic situation of residents in unrecognized villages.[16] An important conclusion toward improving

[12] See for example Townsend (1962), Desai and Shah (1988), Saunders (2003) and Deutsch and Silber (2006). Other important approaches, such as the NAS approach, published in Citro and Michael (1995) or the MBM approach described in Human Resources Development Canada, (2003) have been calculated in Gottlieb and Manor (2005) for Israeli data.

[13] We use the official Israeli equivalence scale which is based on the traditional food-share scale of the Engel type: The values of the scale are 1.25, 2, 2.65, 3.2, 3.75, 4.25, 4.75, 5.2 for one, two persons, etc. respectively until it reaches 6 for 10 persons, continuing with an addition of 0.4 for each additional person. Israeli families are in general much larger than those found in Western countries. This is mainly due to Jewish and Muslim religiosity. Therefore, the scale is not truncated for particularly large families. The numbers for the lower sized families are quite similar to those of the OECD scale that prevailed before the OECD switched to using the square root of family size.

[14] Equivalent income refers to the economies of scale ("savings") in expenditures per individual with the increase in family size. Such economies of scale are mainly due to the indivisibility of capital that can be enjoyed by more than one person without infringing the use by other household members, such as the use of the kitchen by all family members, etc.

[15] A few of those living adjacently to recognized communities might be included in the sample.

[16] While the number of Arab households in the income and expenditure surveys of the CBS in 2004 was 1,858 and 646, respectively, the corresponding number in the Galilee Society's survey was 2,680 (exclusive of Bedouin in unrecognized villages). The number of households of unrecognized villages in their sample was 570, reflecting oversampling.

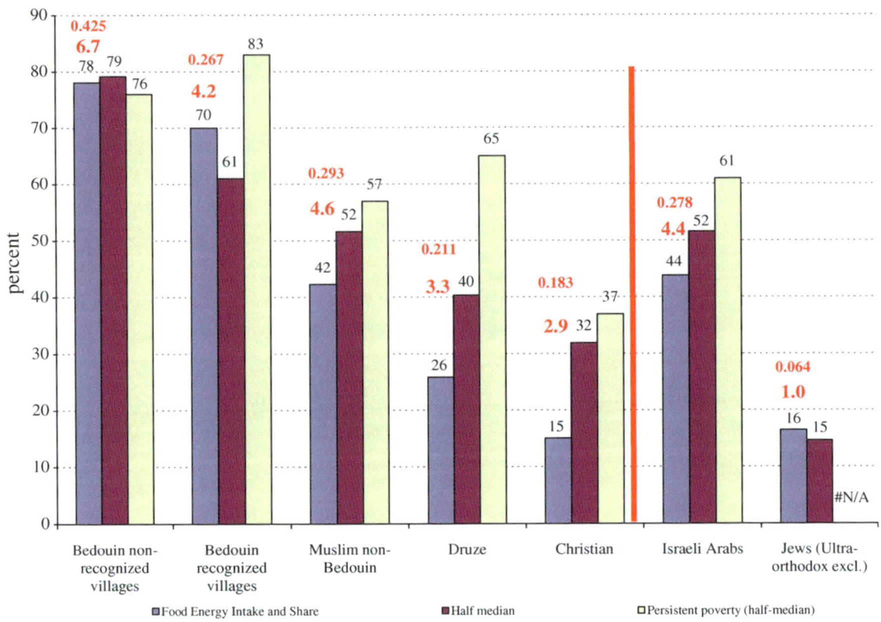

Chart 8.1 The dimensions of poverty in the Arab society in Israel—2004 (The Half Median and Basic Consumption expenditure—FES—Approaches, Including Arabs in Unrecognized Villages). The top numbers (e.g. 0.425 for the unrecognized Bedouin Villages) are values of the Sen-poverty index, the number just below (e.g. 6.7 for the unrecognized Bedouin) are the ratios of the Sen-index relative to that of non-orthodox Jews. The numbers above the column (e.g. 78) indicate the poverty incidence according to the specific poverty definition, except for the number above the column of "persistent poverty (half-median): that number represents the percentage of the of income-poor households who suffer also, from their consumption expenditure being below that same poverty line. Figures to the left of the red bar were calculated based on the Galilee Society Socio-Economic Survey and those to the right of the bar were calculated based on the CBS Income Survey

the infrastructure of information pertaining to the social situation in Israel is to demand the inclusion of unrecognized Bedouin villages in the surveys of the CBS. This is particularly important considering the size and extent of poverty of this ignored population group. At present, 40 % the Bedouin population, many of whom are among the poorest of the poor in Israel, are not represented in official CBS social statistics. This ignorance made it easy for the consecutive governments in Israel to ignore the needs of this population.

8.4.2 Trends in Poverty Among Arabs

Prior to engaging in measuring poverty on the basis of GS data, it is important to clarify a methodological problem concerning the calculation of the poverty estimate in this survey according to the NII poverty line: Given the NII's reliance on

CBS data only, its poverty line is calculated without reference to the Bedouin of unrecognized villages. Hence, the official poverty line has a certain upward bias, since the addition of families, mainly in the bottom sphere of the distribution of incomes, raises the position of the previous median family as compared to the "new" median family after including the unrecognized Bedouin. The poverty line will, thus, be lower than the official line. Poverty incidence is affected by two opposing effects: (1) some of the newly added families who entered the bottom sphere of the incomes' distribution will raise the headcount, and (2) the reduction of the poverty line will exclude some families that were previously considered poor. It seems that the net influence in the half-median approach is very small. In the FES approach, poverty increases at a rate of 1 percentage point. Chart 8.1 below does not account for this effect.

As Chart 8.1 reveals, poverty among the Arab population is very high, especially in unrecognized villages: In 2004, the incidence of poverty in the unrecognized villages was 79.2 % (individuals) and the Sen-Index for the poverty severity was 70 % higher than that of the general Arab population, and 6.7 times higher than that of the "mainstream" Jewish non-orthodox population.

Poverty in the Arab society is severe, and according to the relative approach in 2004 more than half of the Arab population lived in poverty. Chart 8.1 illustrates poverty calculations for the Arab population according to the half-median definition and the FES approach.[17] The Sen-Index, which illustrates the severity of poverty, was 4.4 times that of non-orthodox Jews.

The poorest group according to the poverty incidence is that of the Bedouin in unrecognized villages. The incidence of poverty in recognized villages is the second largest, even though the severity of poverty is lower than that of the remaining Muslim population. The lowest incidence of poverty among Arabs is that of Christians. Notwithstanding, its level is still more than double that of the non-orthodox Jewish reference group. The severity of poverty among the Druze and Christians is more moderate than that of the Muslim population, and it is about three times higher than that of the Jewish reference group.

In order to compare Arab poverty over time to that of the Jewish mainstream population (i.e. excluding the idiosyncratic issue of poverty among Jewish ultra-orthodox people) we need to refer to CBS data, since the GS includes only the Arab population in 2003–2004. However, CBS data do not cover systematically data on Bedouin non-recognized localities, implying that we cannot analyze the poverty situation of this group.

---

[17] As mentioned above, the FES approach can be interpreted as reflecting persistent poverty, which by definition will typically be lower than total poverty, since part of the poor may be in temporary poverty. Calculations of true persistent poverty require panel data on incomes and expenditures over time. Such data is not available in Israel. Here, half-median persistent poverty is approximated by the headcount of individuals, for whom both income and consumption are below the official poverty line (see also the Bank of Israel's Annual Report, 2006, chapter 8 and the NII's poverty report for 2007).

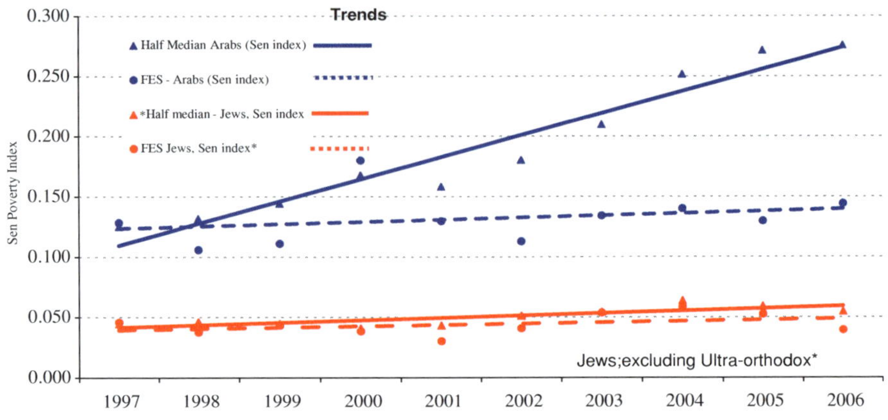

Chart 8.2 Poverty severity (Sen) among Arabs by various poverty measures (compared to the Jewish mainstream society)

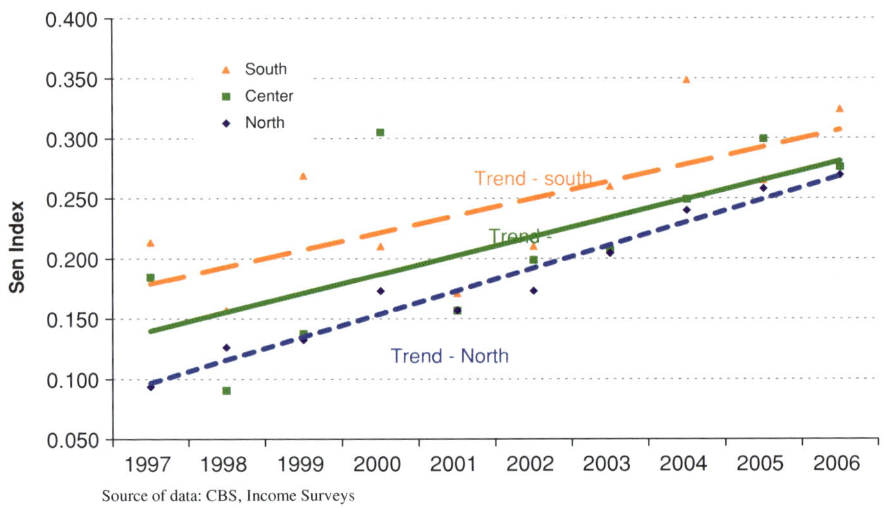

Chart 8.3 Poverty severity (Sen) by regions; Half-median poverty measure. Source of data: CBS, Income Surveys

With these caveats in mind we conclude that poverty severity in the Arab society is significantly higher (about 3–4 times) than that of the non-orthodox Jewish population. Over time, there has been an upward trend in the severity of poverty among Arabs.[18] This conclusion applies to both the half-median and the FES approaches, though the divergence of poverty severity between Jews and

[18] Poverty data over time are based on CBS Data, thus excluding non-recognized villages.

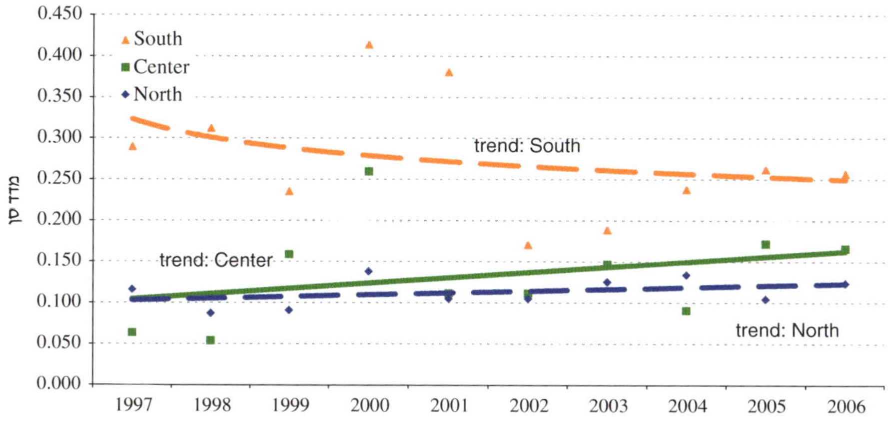

Source of data: CBS Income and consumption surveys for the years 1997 to 2006.

Chart 8.4 Poverty severity (Sen) by regions; FES poverty measure. Source of data: CBS Income and consumption surveys for the years 1997–2006

Arabs is much more severe with the half-median poverty measure than with the FES measure. This result reflects a persistent process of a worsening of the standard of living during the observation period (Chart 8.2).

During the initial years, the severity of poverty fluctuated significantly according to the FES approach, stabilizing thereafter around the level of 0.13.

Through the period 1997–2006 poverty severity increased over time as illustrated in Chart 8.3, among the general Arab population in Israel. According to the FES approach, there seems to be a downward trend of persistent poverty in the South, though since it is accompanied by sizable fluctuations this result seems to be unstable. The severity of poverty according to this approach slightly increased, as illustrated in Chart 8.3, in both the Center and the North. Persistent poverty might include also newly poor people.[19] An upward trend in the severity of poverty is evident in all regions according to the half-median definition (Chart 8.3), and according to the FES the trend of poverty severity has been increasing moderately over time in the Center and the North and showed a downward trend from a high level in the South (Chart 8.4). The substantial impact that developments in the North (due to its relative size) have on the overall result, dictates the overall trend. We conclude that the gaps between the regions suggest a need for affirmative economic policy in the South.

The trends of child poverty are similar in both the half-median approach and the basic needs approach. Among Arab children of the North the deterioration was dramatic since 1997: the incidence of poverty among them increased from 36 to 60 %

[19] This is possible when the young generation of persistently poor families, who find it difficult to escape poverty, create new families. The likelihood of such a phenomenon probably increases inversely with low levels of education and high fertility rates.

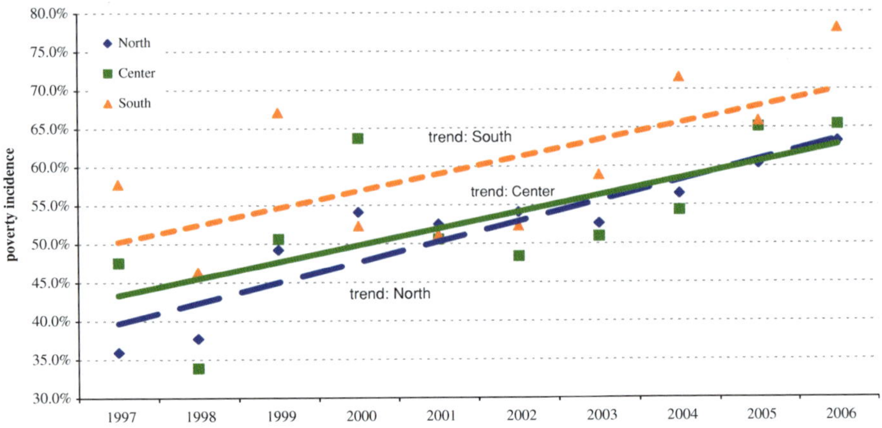

Chart 8.5 Poverty incidence among Arab children by region—half median (Individuals). The source for the calculations: income surveys, the CBS; the data (concerning the South) do not include Bedouin of unrecognized villages

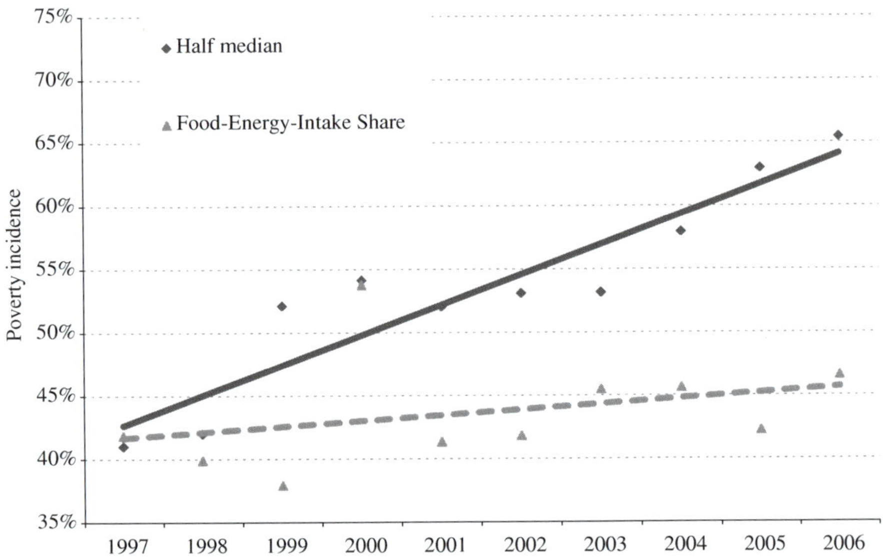

Chart 8.6 The incidence of poverty (individuals) among Arab children

in 2006 (Charts 8.5 and 8.6). Among the children of the South, the deterioration occurred mainly in the years 2002–2004, concomitant with the sharp cutbacks in child allowances and the recession during the first part of the same period. According to this approach, it appears that subsequent growth did not manage to remedy the

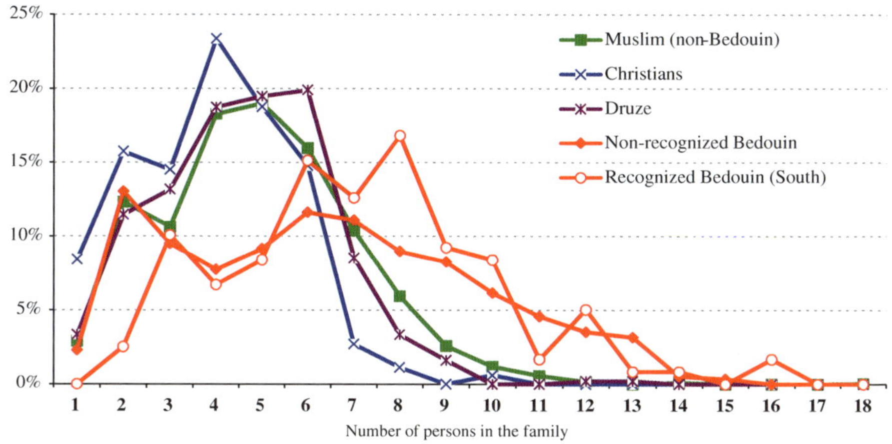

Chart 8.7 Distribution of families by size—different population groups

adverse impact of the harsh social policy of 2002–2003.[20] Among Arab families in the Center of Israel a particularly acute deterioration occurred in 2006. The state of poverty among Arab children toward the end of the observation period (2006) is particularly bad; two-thirds live in poverty (according to the half-median approach). The phenomenon of a converging poverty incidence in the three regions is salient, and instead of convergence taking a course of improvement toward the relatively low levels of the late 1990s, the situation in the North and in the Center has been converging toward the high poverty severity of the South, which most of the time has been the poorest region. High poverty severity among children is particularly harmful in the long run, since it perpetuates persistent poverty in the generation of the children, threatening to affect educational achievements of the young generation's capability to escape persistent poverty (see analysis below). The alternative approaches toward measuring poverty emphasize the interdependence between the choice of the poverty definition and the impact of growth on poverty. It appears that growth had a moderating influence on poverty among children according to the FES approach, while the half-median approach, at most, indicates an effect of moderation on poverty dimensions.

8.4.3 Family Size and Fertility

The interest in the relationship between fertility and family size is well established (see Becker 1965), in which such decisions are analyzed in a dynamic model. Dasgupta (1993) relates the decision concerning childbirth and family size to an

[20] One of the criticisms made against the relative approach pertains to its diminished sensitivity to growth, since the poverty line adapts each year to developments in the median income.

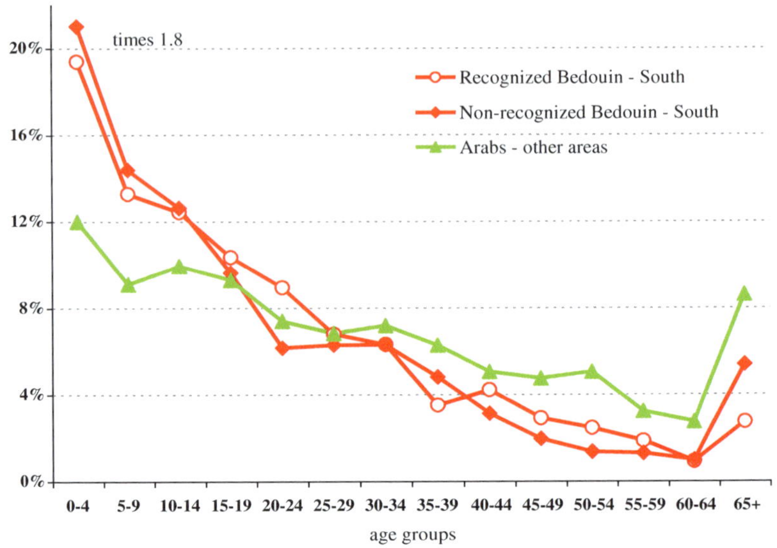

Chart 8.8 The distribution of Arab population by age and groups

additional important dimension not mentioned in the neoclassical model—namely the cultural consideration latent in fertility patterns practiced in traditional societies. In recent years, the demand for education of women has been recognized to play an important role in the fertility decision in traditional societies which are in a process of gradual change toward the woman's enhanced independence.[21]

The age distribution among population groups of the Arab society in Israel reflects the differences in fertility between the various groups (Chart 8.7).

The share of families with more than seven persons is significantly higher among the Bedouin of the South, than among the three other groups (Christians, Druze, and non-Bedouin Muslims). Thus for example, nearly all families with more than nine persons are Bedouin. The smallest families are among Christian Arabs and the distribution of family size among the Druze and the non-Bedouin Muslims is quite similar. These differences are manifest in the age distribution in Chart 8.8: thus for example, the share of children under the age of 5 years among the Bedouin in the South is about 1.8 times that of the remainder of the non-Bedouin Arab society. Given the gaps in education, the differences in the rate of population growth emphasize persistent poverty. One way to counter this influence is by affirmative action in favor of the Arab-Bedouin population.

[21] See for example Basu (2002).

Table 8.2 School enrollment rates—by population groups, gender, and age groups

Years of schooling	Unrecognized villages	Recognized villages-Southern Israel	Bedouin	Muslims (non-Bedouin)	Druze	Christians
Men and Women—All ages						
Up to 6	75.4	84.6	83.1	91.8	92.7	93.7
Up to 8	64.8	75.6	74.8	84.6	85.8	87.7
Up to 10	44.9	59.6	57.1	66.5	69.6	74.1
Up to 12	32.2	44.9	42.9	55.5	58.8	63.9
Men and Women—20 years old and younger						
Up to 6	96.9	98.5	98.3	99.4	98.4	99.0
Up to 8	91.0	94.7	94.6	98.3	97.9	98.2
Up to 10	72.3	89.9	86.2	93.1	95.8	95.6
Up to 12	47.5	69.6	67.8	80.8	89.4	89.1
Men and Women—44 years old and older						
Up to 6	11.5	30.4	28.6	69.3	74.4	82.4
Up to 8	8.3	14.5	19.3	54.7	62.2	69.6
Up to 10	4.2	8.3	9.2	28.8	31.9	45.3
Up to 12	3.5	7.0	6.6	23.0	22.8	34.2
Men—20 years old and younger						
Up to 6	97.5	100.0	98.9	99.3	98.5	100.0
Up to 8	94.8	96.9	96.2	97.9	97.9	99.1
Up to 10	80.6	95.4	89.5	90.3	97.1	94.8
Up to 12	54.1	69.1	68.6	71.7	93.3	88.9
Men—44 years old and older						
Up to 6	23.8	46.9	45.0	82.2	89.6	89.1
Up to 8	17.4	24.1	30.7	65.8	81.9	74.8
Up to 10	9.7	16.1	15.9	36.2	51.0	47.7
Up to 12	8.1	13.4	12.7	29.7	38.6	38.0
Women—20 years old and younger						
Up to 6	96.0	97.1	97.6	99.5	98.2	98.0
Up to 8	85.8	93.1	92.8	98.7	97.8	97.4
Up to 10	60.9	85.5	82.7	95.9	94.4	96.4
Up to 12	38.7	69.9	66.9	90.3	85.9	89.4
Women—44 years old and older						
Up to 6	2.0	12.5	13.9	55.6	59.2	76.8
Up to 8	1.4	4.2	9.0	43.0	42.4	65.2
Up to 10	0.0	0.0	3.2	21.0	12.6	43.4
Up to 12	0.0	0.0	1.1	15.8	7.0	31.1

Source of data for calculations The GS, 2004. The rates are defined as the ratio of school-enrollment of persons who successfully completed studying the indicated number of years relative to the potential rate for the relevant age group. The potential group for ages 20 years and younger in the category of up to 12 years of study is limited from below by the age of 17 years

Table 8.3 The change in school enrollment rates between generations by gender

Pop. group Schooling years	Unrecognized villages	Recognized villages-Southern Israel	Bedouin	Muslims (non-Bedouin)	Druze	Christians
among men (increase (+) in percentage points), age 20—compared to age 44+						
Up to 6	73.7	53.1	53.9	17.1	8.9	10.9
Up to 8	**77.4**	72.7	65.5	32.1	16.1	24.3
Up to 10	71.0	**79.3**	**73.7**	**54.1**	46.1	47.1
Up to 12	46.0	55.7	55.9	42.0	**54.7**	**50.9**
among women (increase (+) in percentage points), age 20—compared to 44+						
Up to 6	**94.1**	84.6	83.7	43.9	39.0	21.2
Up to 8	**84.5**	**88.9**	**83.9**	55.8	55.4	32.2
Up to 10	60.9	85.5	79.5	**74.9**	81.8	52.9
Up to 12	38.7	69.9	65.8	74.5	78.9	**58.3**
Gap in study-rates between men and women—between generations (percentage points)						
The 20-year old and younger age group						
Up to 6	1.5	2.9	1.3	−0.3	0.3	2.0
Up to 8	8.9	3.8	3.4	−0.8	0.2	1.7
Up to 10	**19.8**	9.9	6.9	−5.6	2.7	−1.6
Up to 12	15.5	−0.7	1.6	−18.5	7.5	−0.6
The 44 year old and older age group						
Up to 6	**21.8**	34.4	31.1	26.5	30.4	12.3
Up to 8	16.0	19.9	21.7	22.8	39.5	9.6
Up to 10	9.7	16.1	12.7	15.2	38.3	4.2
Up to 12	8.1	13.4	11.5	13.9	31.6	6.9

8.4.4 Education

Primary and junior high school education among the Arab population witnessed substantial changes in recent decades. Literacy, which was the privilege of a few three decades ago, quickly became available to almost all. The greatest leap in this respect is among the Bedouin. Its main achievement was the increase in the rate of school attendance in elementary school and junior high school, whereas at higher levels rates of school attendance are still lower, especially among the Bedouin of unrecognized villages (see Table 8.2 below).

The School-Dropout Problem

One of the salient problems in the area of education is the share of dropouts (a mirror image of the school-enrollment rate) from the education system at various education levels. Among young Christians up to the age of 20 years, the dropout rate from 12 school years is 11 %. Among the Druze the dropout problem is almost as that among Christians and among the Muslims the dropout rates are significantly higher—20 % among non-Bedouin Muslims, and about 32 % among the Bedouin. The problem is severe among Bedouin of the unrecognized villages (over 50 % dropout among 20-year olds and younger). The comparison between dropout rates of people aged 44 years and older and the dropout rates of 20-year olds and younger indicates a substantial improvement over the years. The biggest improvement (94 percentage points) was achieved in the reduction of dropout rates at the level of elementary school among Bedouin women of the unrecognized villages. This success is primarily due to the establishment of elementary schools in a number of these villages. On the other hand, the least progress was achieved with respect to dropout ratios of Bedouin women for up to 12 years of schooling—only 36 % (Table 8.3).

A major success in recognized villages is the sharp increase in school enrollment rates of up to 10 study years' education. The increase in school enrollment rates of up to 12 study years' education is also impressive among women—85.5 and 69.9 %, respectively (Table 8.2). This advantage in favor of the recognized communities could be a meaningful incentive to relocate from an unrecognized to a recognized community, particularly so, if, as reported in Abu Saad et al. (2007), Bedouin mothers tend to attach a high value to their children's education.

The data suggest that the reduction in dropout rates in unrecognized villages may be attributed to the government's investment in elementary and junior high schools in unrecognized villages. Whether the government to take similar action with respect to high schools, we might eventually witness an improvement at that level too.

8.4.5 Access to Infrastructure

One of the main functions the State is expected to provide is adequate infrastructure for its citizens—basic infrastructure such as water, electricity, sewage system, garbage disposal, public transportation, as well as a higher level of infrastructure in a civilized society, related to health, education, and culture. The importance of public investment in infrastructure to the household can be ranked by the household's ability to substitute for its inadequacy. The degree of substitutability may be reflected by the ability and the cost incurred by the household when accessing a given service independently, when the public infrastructure is lacking. For example, while there is a very high cost involved in the private substitution of an inexistent road, since building the road is complicated and expensive, requiring also official permits, etc. Hence, the private and social loss involved in the absence of roads or their inadequacy is particularly high. On the other hand, the lack of a public transportation system is smaller for families with higher income, since they can substitute more easily for public transportation. For the poor or the disabled its lack will be much more costly, for example lowering their chances of finding employment. Investment in basic infrastructure for the Arab sector has been neglected for many years. The reasons for this are partly rooted in discrimination practiced by the State and public institutions in areas such as water, electricity, and transportation, against the Arab population, and in particular, against the Bedouin population in unrecognized villages. The problem is further complicated by conflict and mismanagement within the local authorities. Furthermore, the lack and inadequacy of basic infrastructure also reduces private incentives for building infrastructure.

Difficulties of access to basic infrastructure are typical for very poor countries, or remote regions in very big countries. One might, thus, expect that such problems should not occur in a small and advanced country such as Israel. The existence of such problems in Bedouin communities thus indicates that discrimination might be part of the explanation.[22]

Government statistics concerning the distribution of investment in infrastructure by regions and communities are not available. However, the GS provides a possibility to calculate indirect indicators for access to infrastructure through information gathered on the households' access to different types of infrastructure.

Part of the infrastructure, such as roads and schools are usually provided directly by the government, while services in the fields of education, welfare, religious matters, and health are provided by local authorities and health insurance companies respectively, with the government participating financially. With regard to year 2004, Shahor (2007) found that the balance grants[23] of the government should have been more generous for the economically weak local authorities. He examined the collection of incomes derived from municipal taxes in Jewish, Arab, and Druze communities while accounting for the socio-economic situation of the authorities. He concluded that the shortage in incomes derived from municipal taxes in Arab communities was to some extent due both to the residents' unwillingness to pay taxes, and to the fact that the government fulfilled its financing function insufficiently, refusing in part to support economically weak Arab authorities. Convenient access to public transportation is crucial for the residents' successful integration in employment, as well as for the pursuit of other economic and social activity. Transportation infrastructure has three aspects: (1) roads, railroads, streets, and inner pathways; (2) connection of the inner road system to that of interurban roads; (3) an infrastructure of a network of transportation systems within the community, and between the community and the rest of the country.

The GS survey does not afford an opportunity to study all three aspects since the questions presented in the survey only inquire about the household's access to transportation services. The possible answers in the survey are: the nearest station is (1) at least 1 km away; (2) 1–5 km away; (3) more than 5 km away.

[22] Alesina and Glaeser (2004), find heterogeneity in the population to be an important determinant of poverty. They argue that heterogeneity leads to discrimination that arises from the fact that minorities, little appreciated by the "mainstream" groups, will be underrepresented in government and so is their share of federal and local government budgets.

[23] Economically weak local authorities have been entitled to grants in order to provide services to the public.

Table 8.4 Households' access to infrastructure by type of infrastructure

Type of infrastructure	The population with accessibility problems						
	Total Arab population						
	The rate out of the total relevant population (in percent)						The number of individuals
	South			Center	North	Total	
	Recognized	Unrecognized	Total				
The distance from the **clinic** closest to the place of residence is more than 5 km	0.0	65.7	26.5	0.7	6.4	8.2	84,913
The distance of the **public clinic** closest to the place of residence is more than 1 km	85.0	89.0	86.7	20.8	40.5	43.8	451,265
The distance of the **school** from the place of residence is more than 1 km	28.7	76.9	48.0	11.3	30.4	30.1	310,070
The distance of the **public transportation stop** is more than 1 km	47.5	77.6	59.3	28.4	33.1	35.9	369,825
The distance of the **public transportation stop** from the place of residence is more than 5 km	1.9	51.8	22.7	0.0.	5.5	7.0	72,208
A source other than the Electric Corp. for the supply of **electricity** (generator)	0.0	90.6	36.5	0.0	0.9	5.4	55,191
Not connected to the **sewage system**	1.5	99.5	41.0	1.9	15.7	17.1	176,328
An independent source of **water supply** (other than through Mekorot)	0.0	40.3	16.3	0.0	0.9	2.8	28,598
Garbage disposal other than by the Authority	9.6	89.3	41.6	1.2	19.5	19.9	204,903
No **telephone line** connection	78.2	99.3	86.8	22.3	23.4	31.4	324,151

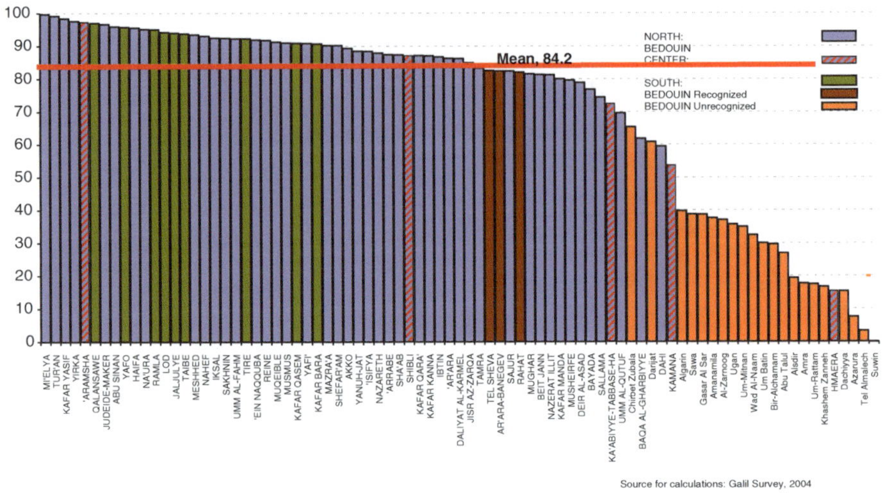

Chart 8.9 Ranking of the Arab communities by access to basic infrastructure (100 = full access, 0 = no access)

Table 8.4 emphasizes the disadvantage of Bedouin localities, both recognized and unrecognized, from other Arab localities, no matter which type of infrastructure we consider.

Table 8.4 indicates that the biggest problem is in the area of health: the distance to the nearest public clinic for about 44 % of the Arab population is more than 1 km away from home, and for more than 8 % of them the nearest clinic is more than 5 km away. With regard to about 30 % of the population the distance to the nearest stop of public transport is more than 1 km and for 7 % the distance is >5 km. This diminished access is even sharper when contemplating the low rate of private vehicle owners: The table in Annex 1 reveals that in recognized Bedouin communities and among the other Muslims only 64 % own a car. In the unrecognized villages, only 45 % own a private vehicle. Ownership of a commercial vehicle is at a rate lower than 11 % through all observed population groups. 30 % of individuals are more than 1 km away from elementary school and about 17 % of the observed population is not connected to the sewage system. Nearly 20 % of the Arab public has to dispose garbage privately. About 32 % of the population is not connected to Bezeq's landline telephone system. Hence, the biggest problems of accessibility are in the areas of health, transportation, education, and communication. Among the basic infrastructure, the most difficult situation is in the area of sewage and electricity.

Infrastructure Level Index

As was outlined earlier, access to basic infrastructures is considered a major determinant of poverty, as a lack of such infrastructure will hinder the establishment of businesses and services and make it harder for those seeking to find a job in a nearby Jewish locality. In the absence of a variable that can describe the level of household's access to basic services, we applied the principal components analysis to nine variables describing the accessibility of basic services, to construct such a measure. The aim is to account for most of the variability in the original variable using a small number of components. The nine infrastructure variables capture the access of a household to running water, telephone, electricity, the sewage system, and the distance between the household and the nearest public transportation stop, clinic, hospital, elementary school, high school, and shopping area. All variables are binary, each of the first four variables takes the value one if there is an access to the specific infrastructure and zero otherwise. Three variables take the value one if distance from the house to the closest transportation stop, clinic, or elementary school is <1 km and 0 otherwise; one variable takes the value one if the distance from the house to high school or the shopping area is between 1 and 5 km and zero otherwise; and the last variable takes the value one if the distance between the house to the nearest hospital is between 5 and 15 Km and zero otherwise. We chose the principal component as the raw infrastructure measure. This variable accounted for more than 85 % of the sum of the variances of the original variables. Then we applied a linear transformation for the constructed raw infrastructure measure such that the values of the new variable fall in the range 0–100, where the values 0 and 100 of the index demonstrate the lowest, and the highest infrastructure level, respectively, available for the specific household. The infrastructure level for a locality was calculated as a weighted average of the infrastructure level of households in that locality with weights being the relative households' size. Chart 8.9 illustrates the distribution of the infrastructure level index in the Arab localities in the GS survey.

The calculations show that the lack of infrastructure is extremely severe in unrecognized villages in the South where the highest score reached by the index is 65 points and the lowest is virtually zero (chart 8.9). In three of the unrecognized villages in the sample the index reaches <10, and in 8 of them the value is below 20. In the recognized Bedouin communities, which are part of the sample, the situation is better—about 80. On the other hand, the situation of some of the Bedouin communities of the North is similar to that of unrecognized villages of the South. Chart 8.9 also clarifies that the situation cannot be blamed solely on the land dispute between the Bedouin and the State, since the chart shows that inadequate access is a feature common to almost all of them, though by varying degree. Notwithstanding, discrimination is more severe in the South. Furthermore, one should keep in mind that the underinvestment in roads and junctions both in the unrecognized villages and between them and main roads in their vicinity has not been included in the GS survey.

8.5 Empirical Results: Major Determinants of Poverty

In recent years, a number of researches have examined the probability of a household to live in poverty. Obviously, such research is sensitive to the poverty definition that is chosen. When using the GS database we cannot employ the FES

Table 8.5 The probability of a household to live in poverty [£]—a logistic model

(A Logit model, below the poverty line = 1, above the poverty line = 0)

The name of the explanatory variable	Model 1	Model 2
Constant	−0.10	0.79[b]
Age of the household's head	−0.02[c]	−0.02[c]
Family size	0.36[c]	0.36[c]
A household headed by a woman (1 = yes, 0 = no)	0.37[b]	0.37[b]
Secondary education of the household head (1 = yes, 0 = no)	−0.41[c]	−0.40[c]
Post-High-school education of the household head (non-academic) (1 = yes, 0 = no)	−0.90[c]	−0.87[c]
Academic education of the household head (1 = yes, 0 = no)	−1.39[c]	−1.35[c]
One provider in addition to the household head (1 = yes, 0 = no)	−1.12[c]	−1.10[c]
Two income earners in addition to the household head (1 = yes, 0 = no)	−1.90[c]	−1.88[c]
A mixed community (1 = yes, 0 = no)	0.31[b]	0.38[b]
Household is employed within the community (1 = yes, 0 = no)	0.34[c]	0.36[c]
Household head is chronically ill (1 = yes, 0 = no)	1.01[c]	1.02[c]
The community is located in the Center of Israel (1 = yes, 0 = no)	0.04	0.10
A unrecognized Bedouin community of the South (1 = yes, 0 = no)	0.39[c]	−0.20
A recognized Bedouin community of the South (1 = yes, 0 = no)	−0.18	−0.20
An academic occupation (1 = yes, 0 = no)	−0.72[c]	−0.72[c]
A freelance occupation (1 = yes, 0 = no)	−1.10[c]	−1.08[c]
A professional occupation in industry, construction and other branches (1 = yes, 0 = no)	−1.04[c]	−1.02[c]
A non-professional occupation in industry, construction and other branches (1 = yes, 0 = no)	−1.02[c]	−1.00[c]
Access to infrastructure (10 types of infrastructure, index 0–100)	–	−0.01[c]
McFadden R-squared	0.2612	0.2638
Number of observations with the dependent variable = 0	1,644	1,642
Number of observation with the dependent variable = 1	1,518	1,581
Percent of accurate observations (dependent variable = 0)	76.46	77.77
A percent of accurate forecast (dependent variable = 1)	70.75	71.15
A percent of accurate forecast (total)	73.72	74.59

The source of data for calculations The Galilee Society survey, 2004

[£] according to the relative approach of half the median net disposable income

[a], [b], [c] indicates significance at levels of 10, 5, and 1 %, respectively

definition discussed above, so we restrict ourselves to the half-median approach. We apply a logistic functional form to represent the probability of a household living in poverty. Such an approach is used for example in Kyereme and Thorbecke (1991) who analyze poverty by use of cross-sectional data for households in Ghana. They use the gap of a household's calories consumption from a nutritional-normative level of calories as an indicator of poverty and try to explain it by economic, demographic, and geographic variables. Rodriguez and Smith (1994) also use a logistic regression for poverty estimation in Costa Rica and

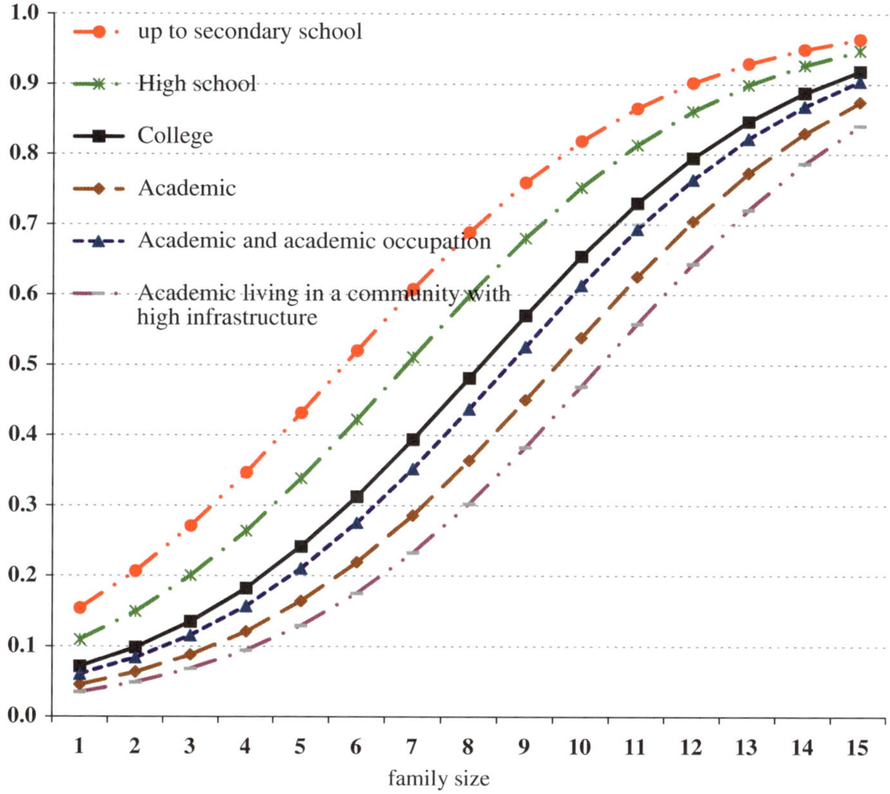

Chart 8.10 The probability of a household to live in poverty—by household size and household head education level

Note Background data: the head of the household is a 46-year-old man (the average age) who is employed in a professional occupation (exclusive of the third graph from the bottom), a single provider, free of illnesses, lives in a community which is not mixed, employed outside the community he lives in, and lives in a community with a medial level of infrastructure (10 types)

Coulombe and McKay (1996) adopt a similar approach for Mauritania based on household surveys made in year 1990.

The main findings in this literature are (1) a significant negative relation between poverty and the level of education of the household head, (2) a positive relation between poverty and the dependency ratio or family size, and (3) a positive relation between poverty and living in the periphery.[24] (4) The state of employment of the household's head, (5) the number of other income earners in the home and (6) the occupation of the household's head and (7) gender (household head being a woman) were also found to be of relevance.[25] A number of papers explain poverty *inter alia*

---

[24] See for example Geda et al. (2005).

[25] See for example Benhabib et al. (2006).

by (8) the household's access to basic infrastructure. This was typically examined with regard to very poor third world countries.[26]

For the Israeli economy, empirical research on poverty has been carried out by Shaayo and Vaknin (2000) who analyzed persistent poverty and by Flug and Kassir (2001), who analyzed in particular the role of labor force participation. The situation of single mothers was discussed in Flug and Kassir (2001). Silber and Sorin (2006), and Deutsch and Silber (2006), analyzed the issue of dimensions of poverty in a multidimensional view. Gottlieb (2007) analyzed poverty and labor market behavior of ultraorthodox ("Haredi") Jews in Israel.

The purpose of our empirical analysis is, among other things, to examine the importance of the above mentioned variables, typically used for explaining poverty. More specifically, we include age and age-squared, education level, gender of the head of the household, geographic area where the household is located, the number of income earners, and for the first time, at least in the Israeli context, we add the household's access to infrastructure as an explanatory variable.

Empirical results are presented in Table 8.5. The second model expands the first (which resembles the standard models found in the literature) by adding a variable of households' access to infrastructure.[27]

Our findings show that the chances of a household to be poor fall with age. The coefficient of age squared, though negative, as expected, is not statistically significant at the level of 10 %.[28]

Unsurprisingly, family size significantly raises the probability of living in poverty (Chart 8.10). Family size affects the probability of being poor both directly and indirectly: the direct effect is due to the fact that a given income is divided among a higher number of people the larger the household size. The indirect effect is due to the interaction between family size and the supply of working hours of the household as a whole, either due to the need to take care of the small children or elderly parents living in the household.

The head of the household being female was also found to be significant. This result reflects in particular the economic pressure on single mothers, being the single potential or actual adult provider. Possibly, this might also reflect discrimination in the labor market against women, either through a negative premium on the wage or through a higher probability of involuntary part-time work.[29] However, we did check whether this result interacted with the type of occupation and the level of education of the head of the household. We found that it was not affected by them.

The level of education (years of schooling) was found to be a highly significant explanatory variable of the probability of being poor. Notwithstanding, illiteracy

---

[26] Respective examples are the papers of Wodon (1998); Datt et al. (2000), as well as Datt and Jolliffe (2005). Many of these researches use the FES definition of poverty (see Sect. C' above).

[27] Variables with a significance level above 10 %, except for the Area variable, were omitted from the table.

[28] Such a result, which is consistent with the life-cycle hypothesis, was significant in part of the regressions, but not in the ones presented in the table.

[29] These issues have not been studied in this work.

(0 years of study) in contrast to a small number of school years (1–8 years) was not found to add to the explanatory power of the regression. This finding fits the fact that those employed with up to 9–10 years of schooling typically work in low-paid jobs for the unskilled, and due to other reasons described above they may even face workers with higher education competing for jobs, suitable for unskilled workers.

Secondary education significantly helps to reduce the chances of poverty. Post-high-school education helps more than secondary education, and academic education importantly reduces the chances of being poor.

The triangular relationship between the level of education, the occupation of the head of the household and the chances of being employed is salient. Academic education yields a very high chance of being employed. Among women, the probability exceeds 90 %. Furthermore, academic education provides higher earnings potential even when the employment is in a job requiring lower skills, probably reflecting the lack of opportunities for above average skills.

The number of income earners in the household is found to be a significant variable. The existence of additional income earners sharply reduces poverty. In Arab households, there are fewer income earners per capita than in Jewish households, mainly reflecting the low employment rate among women. Arab families' dependency ratio is also higher. As mentioned above, gaps in education and professional skills, but also lack in infrastructure and possibly discrimination in the labor market for high-skilled workers create barriers to an improved integration of Arabs in the labor market in Jewish areas.

The occupation of the head of the household constitutes a central factor in determining the probability of a household to live in poverty. As expected, the chances of a family with the head of the household being in a managerial occupation are very low. The probability of poverty drops also if the head of the household not only holds an academic degree but also works in an academic job. The probability of living in poverty decreases with work of the head of the household no matter whether he is employed in a professional or non-professional occupation in industry or construction.

The location of employment is found to influence the probability of living in poverty: Employment within the community raises the probability, probably due to the lower level of wages within the community than compared to that outside.

As mentioned above in the logistic model the coefficients do not allow for a direct interpretation in terms of the marginal contribution of a specific variable to the probability in question. An example of such a marginal influence on the probability of living in poverty is depicted in Chart 8.10 for the variable of family size, based on the regression in Table 8.5. The other variables are held constant at the average of each of the other variables in the regression. For binary ("dummy") variables the level is either 0 or 1.

The chart shows that at any given family size the probability of the household to live in poverty falls with the level of education. The largest reduction in the probability of being poor due to a rise in the level of education occurs in families sized 6–8 persons. The two bottom lines in the chart show the importance of the access to infrastructure to poverty reduction: if the access to infrastructure is improved from an intermediate level (75 points out of 100) to the top level, the probability of a household to be poor drops by some 8 percentage points. This implies an elasticity of about one quarter. However, this elasticity reflects only a partial effect of the full infrastructure effect on poverty, since we showed above that infrastructure will reduce the children's probability of dropping out of the education system. As a consequence, the improved educational achievement among girls will in the medium run translate into a lower rate of fertility among women (see discussion above).

8.6 Concluding Remarks

The various governments over the last years have not only been neglecting infrastructure in Arab communities, but there is also a lack of general transparency concerning statistics about infrastructure. An improved transparency in infrastructure statistics for the whole economy is a first and necessary step in order to ameliorate inequality in infrastructure and access to it. Furthermore, the government does not seem to have any strategy for closing the gap in the Arab sector. The "Agenda" document of the Prime Minister's office can testify to that by its disregard of the issue of infrastructure.[30]

In this paper, we examine the socio-economic situation of the Arab-Bedouin population of the Negev and compare it to the situation of the rest of the Arab population in Israel. The Arab society in Israel is living in economic and social conditions that are seriously inferior to those of the Jewish society. The majority of Arab communities are rated at the bottom clusters of the socio-economic index published by the Central Bureau of Statistics. This situation is strongly reflected in poverty data calculated here, using the unique data set of the Galilee Society for the year 2004 and using CBS data for analyzing trends. Over the period of 9 years, the severity of poverty in the Arab society, measured by the Sen-Index, registered an upward trend and was on average about three times higher than the comparable figure in the Jewish (non-ultraorthodox) sector. In year 2004, poverty incidence among persons was 52 % and poverty severity according to the Sen-Index was more than four times that prevailing in the Jewish population. The situation is quite similar when using an alternative definition of poverty (FES) which may be interpreted as stressing persistent poverty, since it reflects a basic needs approach resting solely on data of consumption expenditure.

The database enables for the first time a comprehensive analysis of the socio-economic situation of the entire Arab-Bedouin population, i.e., including the

[30] It only mentions the "lack of convenient and regular transportation to work places". Moreover, the State has not bothered thus far to study the topic through an official survey of the CBS.

residents in villages not recognized by the state. This population's situation is particularly harsh for both poverty measures, with poverty incidence by the relative (half-median) measure reaching nearly 80 % in 2004. The severity of poverty among the unrecognized Bedouin was about seven times higher than that of the Jewish (non-ultraorthodox) population. An empirical analysis of poverty indicates that similar to findings from international poverty research, the probability of living in poverty in the Arab society is well explained by the standard variables such as gaps in education, the age of the head of the household, family size and the low number of income earners. A major innovation of this paper is in the quantification of the relevance of access to public infrastructure in Arab communities. This variable examines the household's access to a wide variety of services which are vital for daily functioning. Such basic infrastructure barely exists in the unrecognized villages but is also below full access in most Arab communities.

The lack of access to infrastructure affects poverty both directly and indirectly: It is shown to affect dropout rates from the education system negatively and also worsens the chances for employment. The negative effect on educational achievements on girls has an additional expansionary effect on fertility in the medium run and thus on persistent poverty. These effects raise poverty and perpetuate the dependency of the residents on the welfare system. The sharp worsening of the social security net in Israel's social policy in the years 2002–2004 has sharply raised poverty incidence and poverty severity in the Arab society. The adverse effect is particularly strong since the high rate of child poverty combined with the problems in education infrastructure implies that their poverty will eventually become a persistent phenomenon due to the effect on the children's future earnings capacity.

The shortage in public transportation infrastructure is particularly regressive, since it affects poorer households more strongly than others, given their low rate of ownership of private cars. The lack of the government's resolve to improve infrastructure in Arab communities thus create a particular disadvantage among the poor, pushing them deeper into poverty and lowering their participation in the labor market. The adverse effect of a reduced access to public infrastructure is even greater, since such government policy sends a negative signal to domestic and foreign private and corporate investors. These results present a challenge to governments which declare enhanced labor force participation as their top priority.

A positive aspect found in this study is that the rates of school enrollment (the rate of students in proportion to the population at the relevant age) have been improving consistently over the generations, especially among women. As a result, the educational gap has been reduced both by gender and by age groups.

One might have expected that the urbanization initiated by the government several years ago among the Bedouin of the South should have created a wider array of employment opportunities, particularly in other than the traditional sectors. However, the insufficient access to basic public infrastructure has created a significant barrier to the labor force participation of women and had thus worsened poverty and its outlook. The weak state of infrastructure was found to have had negative influences on dropout rates from the schools. Also found was an indirect influence of the accessibility to infrastructure on fertility, since the influence of dropping out from school reduces the well-known influence of female premarital education on fertility, thus reducing the leverage of the positive effect that female education could have exerted on poverty.

Acknowledgment The Hebrew version of the paper was written for the program "Economics and Society" headed by Professor Arie Arnon at the Van Leer Institute and financed by the National Insurance Institute Research Fund. Special thanks to Alex Fruman for excellent research assistance and Sarah Gargi for preparing the English version for publication. We thank Ismael Abu-S'aad, Aamer El-Huzayel, and Aamer Abu-Qarn, Valérie Bérenger, Florent Bresson, Ramsis Gera, Jacques Silber, and an anonymous referee for useful comments. Opinions expressed are of the authors and should not be attributed to any of the mentioned institutions.

Annexes

Tables Annex

Table A.1 Annex 1: Main indicators by population groups

Accommodation (households)

	Non-Bedouin Muslims	Bedouin	Christians	Druze	Recognized Bedouin	Unrecognized Bedouin	Bedouin of the South	Recognized Bedouin of the South
The average number of individuals per room	1.21	1.61	1.03	1.05	1.53	1.80	1.64	1.55
The average number of rooms in a residential unit	4.03	4.05	3.83	4.41	4.31	3.52	4.15	4.61
Household title over a residential unit or tenancy								
Title	91.7	95.9	81.4	98.8	94.4	98.8	95.1	92.4
Tenancy	8.3	4.1	18.6	1.2	5.6	1.2	4.9	7.6
The rate of households that require at least one residential unit in the next 10 years	59.2	71.5	57.3	53.2	72.3	70.0	72.8	74.8
The rate of households which are incapable of supplying at least one residential unit of those required in the next 10 years	26.3	30.5	22.7	23.8	25.5	40.8	27.8	18.5
The type of accommodation								
Villa	7.2	5.6	5.3	10.8	7.9	0.9	3.3	5.0
House	65.9	64.0	41.5	68.7	75.8	39.8	65.4	84.0
An apartment in a building	26.4	5.6	53.2	20.5	8.3	0.2	4.0	6.7
A separate room	0.3	0.2	NA	NA	NA	0.5	0.2	NA

(continued)

Table A.1 (continued)

Accommodation (households)

	Non-Bedouin Muslims	Bedouin	Christians	Druze	Recognized Bedouin	Unrecognized Bedouin	Bedouin of the South	Recognized Bedouin of the South
A tent	NA	0.1	NA	NA	NA	0.4	0.1	NA
A shack	0.1	22.2	NA	NA	7.8	51.8	24.2	4.2
Other	0.0	2.3	NA	NA	0.3	6.5	2.7	NA
Connection to public infrastructure								
Water infrastructure (Mekorot Corp.)	99.9	84.9	99.3	100.0	96.7, 3.3[1]	60.4, 39.1[1]	83.3, 16.4[1]	100.0
Electricity infrastructure (the electric network)	100.0	67.8	99.3	99.5	97.0, 3.0[2]	7.7, 89.4[2]	61.2, 37.6[2]	100.0
Sewage infrastructure (system)	88.4	54.3	99.2	88.4	80.3	0.7	57.3	98.3
Sewage infrastructure (cesspit)	11.2	42.1	0.8	10.9	19.4, 0.3[3]	88.7, 10.6[3]	38.2, 4.4[3]	1.7
Telephone line	74.6	21.3	89.9	88.4	31.3	0.7	11.5	19.3
At least one computer at home	49.9	23.9	56.9	57.4	30.3	10.7	21.6	29.4
Internet service at home	23.8	5.3	45.4	25.2	7.8	0.2	4.5	7.6
Playground or public park in the area	17.3	2.0	30.5	12.8	2.8	0.4	2.1	3.4
Title over a car, library								
A private car	64.2	56.0	63.6	75.4	61.4	44.7	55.8	63.9
A commercial car	9.2	5.7	11.4	8.4	5.5	6.0	6.4	6.7
A library at home	32.6	15.5	44.9	48.2	19.1	8.1	15.6	21.0

Relocation, confiscation, and demolitions	Non-Bedouin Muslims	Bedouin	Christians	Druze	Recognized Bedouin	Unrecognized Bedouin	Bedouin of the South	Recognized Bedouin of the South
The rate of people relocated from their original place of residence in Israel	16.5	17.9	14.5	0.1	20.7	11.6	22.7	30.5
Confiscation of property since 1947	22.6	32.4	18.6	27.2	29.3	38.7	37.2	36.1
Demolition of buildings in the recent 12 month	0.1	1.7	NA	NA	0.6	4.0	2.2	0.8
The number of demolitions	161	604	NA	NA	139	465	604	139
The rate of demolitions out of the total number of residential units' owners in the sector	0.2	1.9	NA	NA	0.7	3.9	2.3	1.0
Kinship between married women and their husbands								
First degree kinship	20.4	34.1	11.4	28.5	31.8	39.3	37.4	35.7
Other degree of kinship from the same clan	15.8	27.4	7.2	13.5	25.3	32.8	26.9	23.2

Comments:

[1]special connections [2]A generator [3]None

Household Families p Details

Source of data The Galilee Society survey

References

Abu-Saad, K., Tamar, H., & Abu-Saad, I. (2007). Weaving tradition and modernity—bedouin women in higher education. The center for bedouin studies and development, Media Gal, Tel-Aviv, Israel.

Alesina, A., & Glaeser, E. L. (2004). *"Fighting poverty in the US and Europe, a world of difference"*. Oxford: Oxford University Press.

Al-huzayel, A. (2004). "The unrecognized villages in the negev, from a prohibiting political discussion to an appreciative planning discussion", A lecture at the Truman Institute, the Hebrew University. (in Hebrew).

Basu, A. M. (2002). Why does education lead to lower fertility? A critical review of some of the possibilities. *World Dev, 30*(10), 1779–1790.

Becker, G. S. (1965). A theory of the allocation of time. *Econ. J., 75*(299), 493–517.

Benhabib, A., Ziani, T., & Maliki, S. B. -E. (2006). Poverty evaluation in algeria: a logit-probit model applied to a multidimensional field survey in the region of Tlemcen. In M. Petmesidou & C. Papatheodorou (Eds.), *Poverty and social deprivation in the mediterranean*. New York: Zed Books.

Citro, C. F., & Robert, T. M. (eds.). (1995). Measuring poverty: A new approach, national research council, National Academy Press, Washington.

Coulombe, H., & McKay, A. (1996). Modeling poverty determinants in Mauritania. *World Dev, 24*(6), 1015–1031.

Datt, G., & Jolliffe, D. (2005). Poverty in Egypt: Modeling and policy simulations. *Econ Dev Cult Chang, 53*(2), 327–346.

Datt, G., Kenneth, S., Sanjukta, M., & Gabriel, D. (2000). Determinants of poverty in mozambique: 1996–97, FCND discussion papers, International Food Policy Research Institute (IFPRI).

Dasgupta, P. (1993). *An inquiry into well-being and destitution*. Oxford: Oxford University Press.

Deutsch, J., & Jacques, S., (2006). The "fuzzy set" approach to multidimensional poverty analysis: using the shapley decomposition to analyze the determinants of poverty in Israel. In A. Lemmi., & G. Betti. (eds.), *Fuzzy set approach to multidimensional poverty measurement*. New york, NY: Springer.

Falah, G. (1989). Israeli State policy toward bedouin sedentarization in the negev. *Journal of Palestine Studies, 18*(2), 71–91.

Flug, K., & Kassir, N. (Kaliner). (2001). "On poverty, work and what's in between", Discussion paper series, Research department, Bank of Israel, 1–34 Aug, (in Hebrew).

Foster, J. E., Greer, J., & Thorbecke, E. (1984). A class of decomposable poverty measures. *Econometrica, 52*(3), 761–766.

Geda, A., de Jong, N., Kimenyi, M.S., & Mwabu, G. (2005). "Determinants of poverty in Kenya: a household level analysis", working paper 2005–44, University of Connecticut, Department of Economics.

Gottlieb, D., & Roy, M. (2005). "On the choice of a policy-oriented poverty measure: the case of Israel 1997–2002", mpra.ub.uni-muenchen.de/3842/. (in Hebrew).

Gottlieb, D. (2007). "Poverty and labor market behavior in the ultra-orthodox population in Israel", economy and society, Van Leer Institute, Jerusalem. http://www.vanleer.org.il/econsoc/pdf/1_research_mdiniut4.pdf, pp. 1–55. (in Hebrew).

Kyereme, S. S., & Thorbecke, E. (1991). Factors affecting poverty in Ghana. *Journal of Development Studies, 28*(1), 39–52.

Porat, H. (1998). The development policy and the question of the bedouins in the negev during the first years of the state 1948–1953. *Studies of the Restoration of Israel, 7*, 389–438. (in Hebrew).

Porat, H. (2000). The actions taken by the State of Israel and the alternative solution of the left to the question of the Bedouins in the negev during the first years of the state 1948–1953. *Studies of the Restoration of Israel, 10*, 420–476. (in Hebrew).

Ravallion, M., & Bidani, B. (1993). *"How robust is a poverty profile", Policy research working paper, 1223* (pp. 1–36). World Bank: November.

Ravallion, M. (1994). *Poverty comparisons*. Chur, Switzerland: Harwood Academic Publishers.

Rodriguez, A., & Smith, S. (1994). A comparison of determinants of Urban, Rural and Farm Poverty in Costa Rica. *World Development, 22*(3), 381–397.

Saunders, P. (2003). "Can social exclusion provide a new framework for measuring poverty?" SPRC discussion paper no. 127, 1–20 Oct 2003.

Sen, A. K. (1985). *Commodities and vapabilities*. Amsterdam: North Holland.

Sen, A. K. (1997). "On Economic Inequality", Expanded edition with a substantial annexe by James E. Foster and Amartya Sen, Clarendon Press, Oxford, UK.

Shaayo, M., Michael, V. (2000). "Persistent poverty in Israel: first results from the consolidated data of the population census, 1983–1996". *The Economic Quarterly, 47*(4), 597–628. (in Hebrew).

Shahor, T. (2007). "Incomes of the local authorities in Israel derived from municipal taxes: a comparison between Jewish and Arab local authorities", forthcoming in the Economic Quarterly. (in Hebrew).

Shorrocks, A. (1995). Revisiting the sen poverty index. *Econometrica, 63*(5), 1225–1230.

Silber, J., & Sorin, M. (2006). Poverty in Israel: Taking a multidimensional approach. In M. Petmesidou & C. Papatheodorou (Eds.), *Poverty and social deprivation in the mediterranean*. New York: Zed Books.

Townsend, P. (1962). The meaning of poverty. *British Journal of Sociology, 13*(3), 210–227.

Wodon, Q. (1998). *"Micro determinants of consumption, poverty, growth and inequality in bangladesh", poverty reduction and economic management sector unit, South Asia Region, World Bank, Washington*. USA: DC.

Chapter 9
On Social Exclusion and Income Poverty in Israel: Findings from the European Social Survey

Naama Haron

Abstract Although there is a growing awareness for the need to take a multi-dimensional approach to the measurement of poverty, most of the research on poverty in rich countries still relies mainly on a traditional income- or expenditures-based approach. This is also the case in Israel where only a few studies have taken a look at other aspects of poverty such as social exclusion. The goal of this chapter is to measure the extent of social exclusion in Israel and compare the results obtained with those derived from a traditional income-based approach to poverty measurement. Also, an analysis of risk factors for social exclusion in Israel is presented. The empirical analysis is based on the first round of the European Social Survey- 2002, a database which provides several indicators of social exclusion. These indicators are aggregated into a single score, which is a weighted average of the various elementary indicators. The measure of social exclusion of the individuals in the survey is then compared with their income data from the survey, and linked to various factors assumed to have an impact on income poverty and social exclusion. The results, as might have been expected, do not show a strong correlation between income poverty and social exclusion. Still, known risk factors for poverty such as a low level of education, young age, poor health, cultural and religious sectors, and female gender have a strong impact on social exclusion. A cluster analysis reveals the existence of three subgroups in Israel which could be labeled as "rich", "middle class", and "poor". While there is a significant income gap between the "rich" and the other two clusters, when looking at social exclusion the main gap is between the "poor" and the other two clusters. These findings seem to confirm that social exclusion in Israel is a distinct phenomenon that deserves to be measured and addressed in addition to traditional poverty.

N. Haron (✉)
School of Social Work, Bar-Ilan University, Ramat Gan, Israel
e-mail: naama.haron@gmail.com

V. Bérenger and F. Bresson (eds.), *Poverty and Social Exclusion Around the Mediterranean Sea*, Economic Studies in Inequality, Social Exclusion and Well-Being 9,
DOI: 10.1007/978-1-4614-5263-8_9,

9.1 Introduction

The growing need for a multi-dimensional approach to the identification and measurement of poverty contributed considerably to the development of the social exclusion concept at both research and policy making level. Still, most of the information available on poverty in developed countries relies on financial data of income and expenditure (Israel Central Bureau of Statistics 2008; Laderchi et al. 2003; Nolan and Whelan 2009). This is also the case in Israel where only a few studies have taken a look at other aspects of poverty such as social exclusion (Katan 2002; Striar 2001). The focus of the studies that have been published was an examination of populations that were considered to be suffering from social exclusion, such as women, youth, Bedouins, and Palestinians (Ronen et al. 2008).

One reason for the small amount of research is the difficulty in implementing the concept of social exclusion, with much of the difficulty originating from the ambiguity of the concept itself. Social exclusion has no precise definition and there is an ongoing debate on its theoretical foundations and how research can be applied to it (Silver 1994). It is, however, widely accepted that social exclusion refers to people who are unable to participate in various aspects of life in their surrounding society (Bossert et al. 2005). This includes people who are excluded from society for reasons such as racism and discrimination and also because of poverty (Burchardt et al. 2002).

When compared to the vagueness of the definition of social exclusion and the disagreement on how to measure it, the monetary approach to poverty measurement comes up with convenient measures, such as the one currently used in Israel. The National Insurance Institute in Israel (NII) publishes poverty figures each year. The poverty line in Israel is defined as being equal to 50 % of the median disposable cash income (disposable income: income plus transfer payments minus direct taxes) of a standard person.[1] This definition was originally adopted in the early 1970s as a temporary method but since then no changes have been introduced (Kehunai 2004). The definition used in Israel is the most popular definition and is also the official measure used by the OECD. The European Union uses a variation of this measure, based on the average cash income.

To the best of our knowledge, no attempt has been made to derive an integrative measure that would allow examination of the extent of social exclusion in the Israeli population. Our study attempts to address this shortage of knowledge on the subject in Israel, as social exclusion suggests a wealth of new and wider information concerning disadvantaged populations, beyond the knowledge acquired by income-based poverty research (Devicienti and Poggi 2007; Nolan and Whelan 2009).

[1] Furthermore, the NII also publishes expenditure based poverty calculations, such as those proposed by the American National Academy of Science and the Canadian Market Basket Measure.

The present study proposes a social exclusion index for the general population and compares it to an income-based poverty measure. We also present some of the major factors that have an impact on poverty, and analyze their influence on social exclusion. Using cluster analysis we provide a combined analysis of income poverty and social exclusion.

In Sect. 9.2, we present a definition of social exclusion used as a base for the operationalization of the concept in this study. In Sect. 9.3, we briefly review some of the major determinants of poverty and existing information on exclusion in Israel. The method for the empirical investigation is presented in Sect. 9.4, and the results are described in Sect 9.5. The paper ends with a discussion of the results in Sect. 9.6.

9.2 Definition of Social Exclusion

Today, there is a vast literature on social exclusion but as yet there is not one agreed upon definition of the concept or its relation to the concept of poverty. We shall not attempt to review this literature here, or exhaust all definitions. An informative review on the origins of social exclusion and some of the questions concerning the concept can be found in (Burchardt et al. 2002; Morgan et al. 2007; Saunders et al. 2007; Silver 1994). Still, it is worth mentioning the ground breaking work of Townsend who defined poverty as follows:

> When Individuals, families and groups in the population… lack the resources to obtain the type of diet, participate in the activities and have the living conditions and amenities which are customary, or are at least widely encouraged or approved, in the societies to which they belong. Their resources are so seriously below those commanded by the average individual or family that they are, in effect, excluded from the ordinary living patterns, customs and activities (Townsend 1979).

Much of the poverty and social exclusion research in the developed world relies on Townsend's basic concepts of poverty and exclusion. Townsend's influence can be recognized in the EU definition of poverty (EEC 1985) and in the EU definition of social exclusion:

> Social exclusion is a process whereby certain individuals are pushed to the edge of society and prevented from participating fully by virtue of their poverty, or lack of basic competencies and lifelong learning opportunities, or as a result of discrimination. This distances them from job, income and education opportunities as well as social and community networks and activities. They have little access to power and decision-making bodies and thus often feel powerless and unable to take control over the decisions that affect their day to day lives (European Commission 2004).

There are some agreed upon elements of social exclusion in the literature. It is multidimensional—as it is considered to encompass multiple aspects of life in society (social, political, cultural, etc.). Also, it is dynamic—social exclusion is a process[2] that influences the chances of people to be included in society in the future as well as in the present. Also, it is a relative concept and it implies an agency—a factor outside the person that impacts the exclusion (Tsakloglou and Papadopoulos 2002).

In order to operationalize the concept of social exclusion in the presented study, we relied on the paradigm of social quality (Berman and Phillips 2000). The concept of social quality is useful to the understanding of the way an individual lives in a given society. This paradigm includes four components: social-economic security, social inclusion (exclusion), social cohesion, and empowerment. Here, we shall focus on the first two components, as it is these components which theoretically encompassing the phenomena we researched.

The paradigm allows a clear distinction between social-economic security (entailing poverty and all its measures) and social exclusion. Social exclusion in this context refers to the extent people have access to social relations and institutions. It is tied to the principles of equality and social justice, and to the reasons inherent in a society for their nonexistence. We chose this relatively unfamiliar paradigm because, being the first study to present an empirical measure of social exclusion in Israel, it was preferable to operationalize the concept in as distinct and different a way as possible from the concept of poverty known in Israel (poverty measured by income).

Berman and Phillips (2000) have created a useful outline of the indicators used to examine social exclusion by distinguishing between two levels of connection and attachment of the individual to a society. The first level, 'Demos', examines the degree of social inclusion at the macro level—the level of the state and nation. This level is universal and formal, and deals with the realization of civil rights. The meaning of social exclusion at this level is the lack of access to rights and services the person is entitled to in his/her society.

The second level, 'Ethnos', refers to the micro level and represents a connection to one's community. The concept of community is not necessarily geographic but may be based on sector, culture, language, and so on. The connection with a community is not based on formal rights, but on social and emotional elements. Individuals choose to identify themselves with and participate in a community. Therefore, indicators of social exclusion at this level will include not only subjective indicators such as a person's feelings about society and relationships with friends, but also objective indicators such as membership of organizations. At this level, social exclusion means a lack of identification with and participation in a community—and stems from "a weakness of social bonds" (Berman and Phillips 2000).

Following the outline set above, the social exclusion indicators in the present study were based on their ability to measure lack of access to civil rights and services or the connection level to one's community, friends, and family.

---

[2] This study, as a lot of the studies on social exclusion, is based on a cross sectional survey. Regretfully, this limits the possibility of measuring a process and so we focus on the state of social exclusion rather than on the process.

The database used here is the European Social Survey, which was also conducted in Israel in 2002. An overall social exclusion index was developed based on indicators available in the survey. The social exclusion score of individuals in the population was then compared with their level of poverty (measured by income). Possible determinants of social exclusion were also examined. These determinants are known risk factors for poverty in Israel and they are presented in the next section.

9.3 Poverty and Social Exclusion in Israel

When looking at the characteristics of the poor population in Israel, it is possible to identify several groups with specific characteristics. Among the poor one finds unemployed people, individuals with a low level of education, large and single parent families, immigrants, Ultra-Orthodox Jews, Muslims, and Bedouins (Abu-Bader and Gottlieb 2009; Alfandari 2005; Dahan et al. 2006; Flug and Kasir-Kaliner 2003; Gara and Cohen 2001; Gottlieb 2007; Gottlieb and Kasir-Kaliner 2004; Heller 2002; Mei-Ami 2004; National Insurance Institute of Israel 2010; Peleg 2005; Sadan 2006; Shaayo and vaknin 2000; Shachak 2004).

According to the National Insurance Institute (2010), the current poverty rate for households in Israel is 20.5 %. In comparison, 32.3 % of single parents' households, 42 % of households headed by low-educated parents (up to eight years of schooling), 53.5 % of Muslim households, 56.9 % of the Ultra-Orthodox Jews, 59.9 % of households with four children or more, and 68.9 % of households with no breadwinner, suffer from poverty.

The numbers above show that the most important determinant of poverty in Israel is the labor market status. The problem of poverty in Israel is particularly serious because the country has low employment rates by comparison to other OECD countries. Several population subgroups do not participate in the labor force, particularly Ultra-Orthodox Jewish men, who in accordance with their spiritual rabbinical leadership, study in theological seminars for many years well into their prime working age. Another population with a very low-employment rate is that of Muslim women (Dahan et al. 2006; Gottlieb and Kasir-Kaliner 2004), possibly not only due to cultural values but also due to low education, lack of basic infrastructure, and thus low employment opportunities, particularly in the south of the country (Abu-Bader and Gottlieb 2009).

One of the most influential factors on labor market status is education. There is a consensus that education is the key in reducing socio-economic gaps (Heller 2002), and works to reduce poverty by increasing employment rates and wages (Flug and Kasir -Kaliner 2003). However, one should also remember that the heterogenous quality of education yields different returns in the labor market. The latter is therefore quite low among Ultra-Orthodox Jews (Flug and Kasir -Kaliner 2003) because their education has little value in economic terms (Dahan 1998).

The relationship between education and the labor market is also influenced by gender. While the labor force participation rate of university graduates does not seem to depend on gender, the data show that as the level of education decreases, fewer women are employed with respect to men and their exposure to poverty grows (Central Bureau of Statistics 2009).

Another factor influencing the labor market in Israel is the fact that the age of entry into the labor force is quite high because of the length of the compulsory military service for both men and women (Central Bureau of Statistics 2009).

The risk of poverty is also influenced by the household type. Households headed by single parents are at high risk from poverty because of the difficulty of raising children alone (Bradshaw et al. 2000). Moreover, the vast majority of single-parent households are headed by women. In 2001, 90.7 % of single-parent families with the youngest child aged under 18 were headed by women (Svirski et al. 2002). This increases the risk of poverty, as in Israel women receive on average lower wages than men (Svirski et al. 2002). Also, households with more than 3 children under the age of 18 are more vulnerable to poverty than the average, and vulnerability rises with the number of children in the family (National Insurance Institute of Israel 2010).

The factors described above are significant in understanding the poverty gaps between different sectors in Israel. The sectors most prone to poverty, as already mentioned, are Muslims and Ultra-Orthodox Jews (Abu-Bader and Gottlieb 2009; Flug and Kasir-Kaliner 2003; Gara and Cohen 2001; Gottlieb 2007; Livyatan 2003). According to the annual report of the Bank of Israel (Bank of Israel 2009) the Ultra-Orthodox sector is the poorest sector (60 % are poor) with the largest income gap (39 %).[3]

Muslim households are over 3.5 times more likely to suffer from poverty than Jewish households (National Insurance Institute of Israel 2010). A problematic phenomenon is an apparent discrimination against this sector. An increase in the level of education among Muslims does not improve their position in the labor market or lower the unemployment rate in this sector as much as may be expected. Gara and Cohen (2001) found that even when controlling for education, family size, and the number of wage-earners in a household, the poverty rate among non Jews is greater. The authors suggest two possible reasons: discrimination in the labor market (this includes barriers to enter defense-oriented high- and medium-tech industries) and the lower than average quality of education provided to this population.

The vulnerability of both these sectors to poverty stems from a combination of the factors described above, of poor integration in the labor market and a high amount of children per household.

If we consider other traditionally "at-risk" populations, in 2009 elderly households did not suffer from a greater poverty rate than the rest of the population. Still, elderly households are considered to be more exposed to the risk of

[3] Actually, the poorest population group in Israel is the Bedouin population in the south. Data on this group are insufficiently collected by the Israeli Central Bureau of Statistics. Abu-Bader and Gottlieb (2009) used a specially collected unique database to come to this conclusion.

poverty than the general population since they generally do not participate in the labor force, and have growing medical needs which raise their costs of living (Gal 2004). In this context, age should be mentioned as a risk factor for poverty as well. Apart from the elderly, the younger population (45 and less) has a higher poverty rate compared to the general poverty rate of the population, while the intermediate population (46 to retirement) has a smaller poverty rate compared to the rest of the population due to their better position in the labor market.

The poverty headcount ratio is not higher among immigrants (17.4 %), but they are considered to be an at-risk population, because their professional background often does not correspond to the needs of the Israeli labor market and because of language barriers and hence integration difficulties (Heller 2002). More than a quarter of the children of immigrants live below the poverty line (25.3 %) compared with Israeli born Jewish children (23.8 %) (The Israel National Council for the Child 2008). Elderly immigrants endure even greater social and economic distress and often confront health problems and loneliness (Mei-Ami 2004). All the risk factors to poverty described above will be tested to see if they are also connected to social exclusion in Israel.

As mentioned previously, not much research has been carried out on social exclusion in Israel, but still, a few studies should be mentioned. The Israel Center for Social Justice published a comparison of social exclusion indicators between Israel and the EU (Hovav 2007). These indicators mostly include measurements of economic poverty and inequality, but also indicators of employment, education, health, and life expectancy. Unlike the study presented here, they do not include indicators of direct social exclusion (such as exclusion from social relations).

Some clues on the relationship between income poverty and social exclusion can be found in a major study that was conducted on Income Support benefit recipients (King and Shavit 2005; Morganstein et al. 2003). These benefit recipients had almost no income apart from their benefits, and their characteristics matched the risk factors to poverty described previously. However, since this study did not compare income support recipients with the overall population, it is difficult to draw strong conclusions from it. It was found that, although this population was entitled to a range of free or discounted services (e.g., tutoring children, transportation to school, dental care, rehabilitation programs), there was a limited take up ratio for these services, mainly due to a lack of awareness of these services and/or how to take advantage of them. The social support level for recipients of income support was also examined, and it was found that most of them were in touch with some family and friends. Still, they were found to be less engaged in leisure activities compared to the general population (general population data were derived from a different survey of Israel's Central Bureau of Statistics), among benefit recipients as well as in the general population—those who are employed and have higher education engage in more leisure activities compared with unemployed people with fewer years of schooling. The study also measured the level of social involvement (measured by variables such as the frequency of listening to the news and engaging in military service or volunteer work), which was found to increase with age. No association was found between employment and the degree of social involvement and almost no association was found between the level of social support and social involvement.

These results point toward a link between economic poverty and exclusion from various civil services. However, there is no clear evidence of a relationship between poverty and exclusion from social relations. Looking at the research conducted in Europe, the findings confirm that economic poverty and social exclusion are related but there are significant differences in the findings between countries and the link appears mainly in dimensions of exclusion that are related to the material aspects of life. Indeed, exclusion from social relationships has only a weak correlation with economic poverty (Bohnke 2008; Bradshaw et al. 2000; Burchardt et al. 1999; Larsen 2004; Perry 2002; Saunders et al. 2007; Tsakloglou and Papadopoulos 2002).

This short review highlights the importance of measuring social exclusion in Israel, as the current measures of poverty in the country do not capture the social aspects of living in poverty. The current study seeks to explore two subjects: first, the relationship between economic poverty (as measured by income level) and social exclusion, and second, the identification of possible risk factors for social exclusion. The review above points toward discrimination of the Muslim population in Israel, but apart from that it is hard to hypothesize about social exclusion in other population groups.

9.4 Measuring Social Exclusion

This section presents the operationalization of the social exclusion concept and the research method of this study. First we shall present the survey which provides the data for this study, and then moves on to the social exclusion measurement.

9.4.1 Data Source

The research is based on data from the ESS: European Social Survey—2002 round 1 (R Jowell and the Central Co-ordinating Team 2003). This survey is an initiative funded by the European Science Foundation and the European Union. Over 20 countries participate in the survey, and Israel participated in the 2003 survey as an observer attached to the European Academy of Science. The field work and the preparations involved were undertaken by the B.I and Lucille Cohen Institute for Public Opinion Research at Tel-Aviv University. The purpose of this survey was to create a database on attitudes, values, behavior, and changes in these over time. The survey questionnaire contains more than 200 questions on issues such as political behavior, social and political trends, social capital and trust in institutions, discrimination, subjective well-being, basic human values, attitudes toward social exclusion, religion, identity (ethnic and civic), and more. The questionnaire also includes detailed demographic information on the individual, his/her spouse and parents. The survey data are published on the European Social Survey website and are accessible to the general public.

The population surveyed: The population surveyed was a sample of the Israeli population aged 15 and above, and over 3,600 households, representing all households in Israel (including the Jewish population living in Judea, Samaria, and the Gaza territories), were sampled. The final sample concluded 2,499 cases (71 %) ("B.I. and Lucille Cohen Institute for Public Opinion Studies—ESS", 2005). In this research, only data on the population aged 18 and over were used, including 2210 respondents (weights, according to the ESS data files, were used for statistical analysis).

9.4.2 Measuring Social Exclusion and Poverty

A social exclusion index was developed to measure the exclusion level for each individual. A total of 30 indicators on the ESS questionnaire were found appropriate for testing for social exclusion and were aggregated into an overall social exclusion index. The list of all indicators is presented in Table 9.1. The index is an additive measure of the normalized scores of these 30 indicators of exclusion. Each indicator has its own weight. Scores range continuously from zero (full inclusion) to one (full exclusion).

Most of the indicators were chosen because they describe the level of connection and the identification one has to the society around oneself. Only a few indicators answer the criteria of realization of civil rights per se. These indicators are voting in the general election, personal safety in the streets and exposure to crime, and being discriminated against. Indicators of political activity can also be accounted as realization of rights (such as the right to freedom of speech) and also as a measure of taking interest in life and connecting with the society one lives in.

A total of 25 (out of 30) indicators were grouped by domain into four subindices. Three of these subindices have a reasonable internal reliability, as tested by Cronbach's Alpha test.[4] The last five indicators cover different topics and were not grouped into a subindex.

There is an inherent meaning to the identification of exclusion for a specific domain, and also to the *overall score of exclusion* (comprised of the sum score of the 30 indicators) which is assumed to be a multi-dimensional expression of the accumulation of exclusion in different aspects of life (which limits the relative importance of the Cronbach's Alpha test results in this study).

---

[4] Cronbach's Alpha reliability test results ranges from 0.00 to 1.00. Alpha greater than 0.6 is considered high enough, taking into account that these are indicators from a survey not designed especially for the study (Berthoud et al. 2004).

Table 9.1 Components of the social exclusion index[a]

Interest in current affairs *α-Cronbach* = 0.66	Frequency of TV viewing of current affairs/political programs (0.55)
	Frequency of listening on current affairs/political Programs on the radio (0.15)
	Taking time to read about current affairs/politics in the newspaper (0.03)
	Frequency of conversation about politics and current affairs (0.44)
	Level of interest in politics (0.63)
Political Activity *α-Cronbach* = 0.63	Wore a sticker (0.12)
	Signed a petition (0.18)
	Boycotted products (0.15)
	Purchased products for political/moral/ environmental reasons (0.18)
	Participated in a demonstration (0.10)
Connection with organizations (contributed/participated/volunteered/ friend) *α-Cronbach* = 0.69	Connection to sports club (0.27)
	Connection to culture/leisure organization (0.24)
	Connection to trade union (0.17)
	Connection to a professional organization (0.09)
	Connection to a consumer club (0.26)
	Connection to a humanitarian organization (0.10)
	Connection to an environmental organization (0.08)
	Connection to a religious organization (0.10)
	Connection to a political party (0.10)
	Connection to scientific/education organization (0.14)
	Connection to a social club (0.18)
	Connection to a voluntary organization (0.13)
Social relations *α-Cronbach* = 0.46	Frequency of social gatherings with friends/family/ co-workers (0.88)
	Frequency of participation in social activities (0.65)
	Someone to talk to about personal matters (0.91)
Other indicators	Frequency of Internet use (0.61)
	Voting in elections[b] (0.79)
	Part of a group that suffers discrimination (0.83)
	Victim of robbery/assault (0.83)
	Sense of personal safety—can walk alone after dark in their residential neighborhood (0.84)

Total index *α-Cronbach* = 0.73

[a] The numbers in brackets indicate the weight for each item

[b] An interesting finding is that the Alpha Cronbach test results show that if the "voting in elections" indicator is added, it reduces the reliability of the political activity index, and apparently has a different meaning to respondents than other political activities

As can be seen in Table 9.1, the number of indicators considered appropriate in each subindex varies. This is one of the limitations of the study, it relies on existing data and so cannot address all possible domains of exclusion or give equal attention to the domains portrayed in the subindices. Still, we believe there is much merit in analyzing all the existing data.

The weight of an indicator was determined by the number of respondents included (not excluded), divided by the overall number of respondents who gave an answer to the question from which the indicator was derived (i.e., the percentage of the respondents who were included on the indicator). This method was suggested in a report of the British Department for Work and Pensions on developing deprivation indices (Willitts 2006) and draws on the work of Desai and Shah (1988). After multiplying each indicator by its weight, all indicators scores were added together and divided by the sum of the weights (normalized).

The idea of weighting the individual indicators scores is based on the theoretical framework put forward by Townsend (1979, 1987) which stated that a person is excluded from society when he/she cannot participate in the activities which are customary in the society to which he/she belongs. As more people in society are included in a particular activity, the more significant the exclusion from it becomes. For example, the more people who vote in elections (i.e., are included), the greater the theoretical weight that should be given to the case of a person who did not vote. The weights also limit the impact of the arbitrary element in the choice of the indicators made by the researcher. Exclusion variables that have a real significance for the population will have a larger weight.

Measuring poverty by income: only a crude measure of income is available in the ESS. The indicator of a household's total net income is divided into 10 levels of income, ranging from none to over 46,000 NS ILS (Israeli new Shekels) per month. The level of household income was then divided by the number of standard persons (an equivalence scale used by the Israeli Central Bureau of Statistics and in official poverty measurements in Israel) in the household to allow comparison between the exclusion score for the individual and his/her income level.

We shall now turn to the empirical section of this chapter where a comparison is made between social exclusion scores and level of income.

9.5 Empirical Results

First, we shall present an analysis of the relationship between social exclusion and economic poverty. To evaluate the relationship between income and social exclusion, a Pearson correlation coefficient was computed and it was clearly negative ($r\,(1420) = -0.282$, $p < 0.001$). The variance accounted for exclusion by income is 7.9 %.

Another way to learn about this relationship was to examine the overlap level of the phenomena in the population. For this exercise, we chose a cutoff point of the bottom quintile scores of income and exclusion in the population. The reason for this is that 25 % is in accordance with the poverty estimation of individuals in Israel. It was found that only 10.5 % of the total survey population suffers from both low income and a high level of exclusion. This means that only 42 % (10.5 % out of 25 %) of those who suffer from one phenomenon also suffer from the other.

In order to address the question of risk factor for social exclusion, a series of one way analysis of variance with Scheffé post hoc tests and independent samples *t*-tests were conducted. Table 9.2 shows the connection between each risk factor and the overall social exclusion index as well as subindices of social exclusion. It can be seen that as a whole, the overall index accurately represents the trends of the subindices.

The results show that factors associated with economic poverty are also connected to social exclusion. Belonging to a sector other than Non-Orthodox Jews increases the risk of social exclusion. Muslims suffer significantly more exclusion compared to Non-Orthodox Jews in all subindices. In contrast, Ultra-Orthodox Jews are more excluded in the "connection to organizations" and "interest in current affairs" subindices but do not differ significantly from Non-Orthodox Jews on the subindices of "political activity" and "social relations".

Household structure also influences the level of social exclusion. Small families are significantly less excluded compared to large families (four or more children) and single parent families. The difference stems from the "interest in current affairs" and "social relation" sub-indices. In the other two subindices no significant difference between households was found.

Immigrants (up to ten years in the country) suffer greater social exclusion compared to veteran citizens in Israel. This is based on a significant difference between the groups in the "connection to organizations" and "political activity" subindices. The unemployed are more excluded compared to employed respondents in all subindices except the "political activity" subindex. Compared to healthy individuals, those who suffer some debilitation also suffer from social exclusion in all subindices except in the "interest in politics" subindex. Women are more excluded in the "political activity" and "interest in current affairs" subindices compared to men.

Education (years of schooling) has a strong link to exclusion ($r(1738) = -0.333$, $p < 0.001$). Thus, the variance accounted for exclusion by education is 11.1 %. When testing for levels of education, the main difference is found between those who have lower education (non- high school graduate, high school graduate, matriculation certificate) and those with higher education (tertiary education and up).

These findings do not however imply any causality, in part because of the overlap that exists between the various factors. For example, many respondents who are Ultra-Orthodox or Muslim have a lower level of education and live in households with many children.

To better understand the impact of various risk factors on social exclusion we estimated a stepwise regression. This type of regression, and not logistic regression, was chosen, since even though the social exclusion index varies continuously from zero (minimal level of social exclusion) to one (maximum level of social

[5] The regression was also performed against the original social exclusion index, where the results were almost identical.

Table 9.2 Averages, standard deviations and mean comparison test scores of social exclusion indices

	Social exclusion level		SE—interest in current affairs		SE—political activity		SE—connection with organizations		SE—social relations	
	M	SD	M	SD	M	SD	M	SD	M	SD
Religion										
Christian	0.457	0.091	0.574	0.205	0.933	0.145	0.845	0.185	0.286	0.173
Non ultra-orthodox jews	0.402	0.097	0.520	0.202	0.834	0.241	0.809	0.192	0.280	0.177
Ultra-orthodox jews	0.462	0.097	0.672	0.211	0.891	0.189	0.877	0.189	0.307	0.195
Muslims	0.484	0.100	0.600	0.212	0.911	0.178	0.864	0.179	0.326	0.196
Other	0.441	0.106	0.592	0.213	0.944	0.136	0.837	0.168	0.277	0.198
	$F(4,1415) = 9.79$		$F(4,2136) = 22.96$		$F(4,2140) = 9.68$		$F(4,2085) = 6.928$		$F(4,2127) = 3.24$	
	$p < 0.001$		$p < 0.001$		$p < 0.001$		$p < 0.001$		$p < 0.05$	
Household										
Single	0.415	0.106	0.527	0.215	0.870	0.213	0.839	0.163	0.292	0.202
Couple without children	0.408	0.101	0.483	0.211	0.860	0.232	0.813	0.185	0.299	0.176
Large family[a]	0.429	0.100	0.565	0.214	0.856	0.217	0.832	0.214	0.333	0.199
Small family[b]	0.394	0.093	0.507	0.188	0.848	0.231	0.794	0.194	0.291	0.164
Single parent	0.451	0.127	0.556	0.220	0.854	0.229	0.807	0.204	0.330	0.201
	$F(4,1778) = 32.14$		$F(4,1647) = 6.81$		$F(4,1649) = 0.39$		$F(4,1591) = 2.78$		$F(4,1626) = 3.14$	
	$p < 0.001$		$p < 0.001$		$p > 0.05$		$p > 0.05$		$p < 0.05$	
Seniors[c]	0.411	0.101	0.482	0.237	0.926	0.234	0.834	0.193	0.327	0.207
Other	0.414	0.101	0.543	0.204	0.843	0.168	0.819	0.168	0.283	0.178
	$t(1787) = 0.26$		$t(2152) = 3.07$		$t(2156) = -5.70$		$t(2099) = -1.09$		$t(2140) = -2.55$	
	$p < 0.005$		$p > 0.05$		$p < 0.001$		$p > 0.05$		$p < 0.05$	
Immigrants	0.444	0.101	0.539	0.206	0.937	0.159	0.876	0.192	0.312	0.179
Veteran	0.411	0.099	0.538	0.220	0.842	0.234	0.815	0.170	0.284	0.181
	$t(1787) = 0.26$		$t(1999) = 0.03$		$t(2156) = -6.89$		$t(2099) = -4.16$		$t(2140) = -1.87$	
	$p < 0.001$		$p > 0.05$		$p < 0.001$		$p < 0.001$		$p < 0.05$	

(continued)

Table 9.2 (continued)

	Social exclusion level		SE—interest in current affairs		SE—political activity		SE—connection with organizations		SE—social relations	
	M	SD	M	SD	M	SD	M	SD	M	SD
Education										
Non high school graduate	0.460	0.099	0.560	0.242	0.925	0.160	0.902	0.146	0.360	0.223
High school graduate	0.436	0.098	0.564	0.209	0.882	0.208	0.871	0.173	0.300	0.188
Matriculation certificate	0.433	0.095	0.591	0.220	0.899	0.188	0.864	0.162	0.256	0.170
Tertiary education	0.403	0.094	0.531	0.188	0.835	0.835	0.789	0.193	0.267	0.159
BA	0.375	0.088	0.504	0.181	0.750	0.750	0.754	0.186	0.212	0.156
MA or higher	0.363	0.101	0.454	0.147	0.778	0.778	0.711	0.231	0.277	0.155
	$F(5,1773) = 42.08$		$F(5,2136) = 17.78$		$F(5,2137) = 33.26$		$F(5,2083) = 52.89$		$F(5,2123) = 18.06$	
	$p < 0.001$		$p < 0.001$		$p < 0.001$		$p < 0.001$		$p < 0.001$	
Work										
Unemployed	0.449	0.095	0.554	0.214	0.848	0.245	0.864	0.174	0.317	0.218
Employed	0.396	0.097	0.522	0.191	0.837	0.233	0.796	0.196	0.277	0.168
	$t(1092) = 6.09$		$t(1278) = 1.98$		$t(1285) = 0.54$		$t(1257) = 4.53$		$t(1264) = 2.24$	
	$p < 0.001$		$p < 0.05$		$p > 0.05$		$p < 0.001$		$p < 0.05$	
Health										
Limited health	0.446	0.103	0.533	0.224	0.900	0.188	0.868	0.186	0.352	0.208
Healthy	0.407	0.099	0.540	0.204	0.838	0.238	0.809	0.191	0.271	0.171
	$t(1778) = 6.13$		$t(1768) = -0.53$		$t(2141) = 5.44$		$t(2083) = 5.42$		$t(2125) = 7.02$	
	$p < 0.001$		$p < 0.05$		$p < 0.001$		$p < 0.001$		$p < 0.001$	

(continued)

Table 9.2 (continued)

	Social exclusion level		SE—interest in current affairs		SE—political activity		SE—connection with organizations		SE—social relations	
	M	SD	M	SD	M	SD	M	SD	M	SD
Gender										
Women	0.425	0.101	0.573	0.214	0.861	0.223	0.825	0.187	0.288	0.183
Men	0.400	0.099	0.498	0.192	0.836	0.239	0.814	0.196	0.283	0.178
	t (1787) = 5.40		t (2152) = 8.58		t (2156) = 2.54		t (2099) = 1.37		t (2140) = 0.659	
	$p < 0.001$		$p < 0.001$		$p < 0.05$		$p > 0.05$		$p > 0.05$	

[a] Four or more children
[b] Three children or less
[c] All members of household are 65 years old and above.

Table 9.3 Regression of exclusion with risk factors

	Step					
Factor	β_1	β_2	β_3	β_4	β_5	β_8
Education	−0.313	−0.318	−0.292	−0.308	−0.283	−0.272
Male		−0.152	−0.169	−0.170	−0.155	−0.141
Jewish			−0.160	−0.146	−0.149	−0.156
Age				−0.125	−0.229	−0.214
Health					0.218	0.208
Orthodox jew						0.110
Small family						−0.098
Immigrant						0.095
R^2	0.10	0.12	0.14	0.16	0.19	0.23
$R^2\Delta$		0.02	0.02	0.03	0.04	0.07

All values are significant at $p < 0.001$

exclusion), it does not indicate the probability of being excluded but the actual exclusion index (*EI*) score. The independent parameter for the regression was[5]:

$$\ln \frac{EI}{1 - EI}$$

Table 9.3 contains six of the eight regression models that were estimated: the first model includes only the predominant risk factor—education, while the last model takes into account all the relevant risk factors.

The variables introduced in the last regression accounted for 23 % of the variance of social exclusion. Education is the most important factor, and after that comes age and health. Being an Ultra-Orthodox Jew raises the probability for social exclusion almost as much as being a non-Ultra-Orthodox Jew lowers it. Interestingly, the "participating in the labor market" variable (though entered into the regression) did not have enough impact on social exclusion to be represented in the final regression model.

In order to further study the connection between income and social exclusion and their manifestation in Israeli society, these variables were used in a cluster analysis. This analysis aims to distinguish between subgroups with homogeneous characteristics in the population. In cluster analysis, the observed cases are distributed into a decreasing number of subgroups so that the cases in one subgroup are similar according to the parameters chosen. The final number of subgroups in the analysis is determined using one of several possible criteria.

Cluster analysis was previously used by Ferro-Weber et al. (Ferro-Luzzi et al. 2006) to distinguish between subgroups in Switzerland suffering from multidimensional poverty. Ferro-Luzzi et al. used hierarchical clustering, where the distance between the cases is a Euclidean distance (the geometrical distance between two points—for example the root of the sum of squares of the difference between parameters). We chose to use two-way clustering. This method was introduced by Zhang et al. (Zhang et al. 1996), and developed in order to enable

Table 9.4 Characteristics of the clusters

Parameter	Cluster									
	1 "Poor"		2 "Middle class"		3 "Rich"		Cluster average			
	M	SD	M	SD	M	SD	M	SD	Significance test	Significance
Income	1401	1080	2077	1026	5918	1253	2438	2011	F(2,1416) = 822.3	$P < 0.001$
Exclusion Index	534	060	366	060	350	075	441	100	F(2,1416) = 1158	$P < 0.001$
Size of household	4.62	2.43	4.05	1.83	3.24	1.44	4.10	2.03	F(2,1416) = 34.53	$P < 0.001$
Years of education	11.62	3.589	13.61	3.39	15.45	3.25	13.29	3.65	F(2,1377) = 92.09	$P < 0.001$
Age[a]	40.89	17.11	44.17	15.70	45.36	15.54	43.36	16.19	F(2,1416) = 7.60	$P < 0.01$
Subjective health[b]	2.20	1.07	1.97	0.93	1.88	0.830	2.02	966	F(2,1414) = 10.97	$P < 0.001$
	N	[c]%	N	%	N	%	N	%	Significance test	Significance
Cases in cluster	426	30	785	55.3	208	17.7				
Employed	166	74.1	440	88.2	148	98	251	86.3	$\chi^2(2) = 47.09$	$P < 0.001$
Immigrants	50	11.7	51	6.5	3	1.4	35	7.3	$\chi^2(2) = 23.49$	$P < 0.001$
Elderly household[d]	26	6.1	59	7.5	11	5.3	32	6.8	$\chi^2(2) = 1.77$	$P > 0.05$
Gender (male/female)	166/261	38.9	362/423	46.1	118/91	56.5	646/775	45.5	$\chi^2(2) = 17.80$	$P < 0.001$

[a] There is no significant difference between age and health of the "middle class" and "rich" clusters
[b] 1- good health, 5- poor health
[c] The parameter "cases in cluster" signifies the percentage of all cases; for other parameters it signifies the percentage within the cluster
[d] All members of the household are 65 years old and above

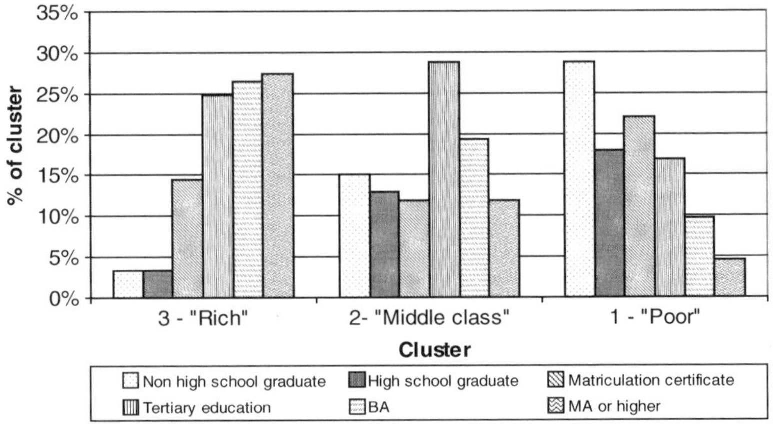

Graph 9.1 Education distribution within clusters

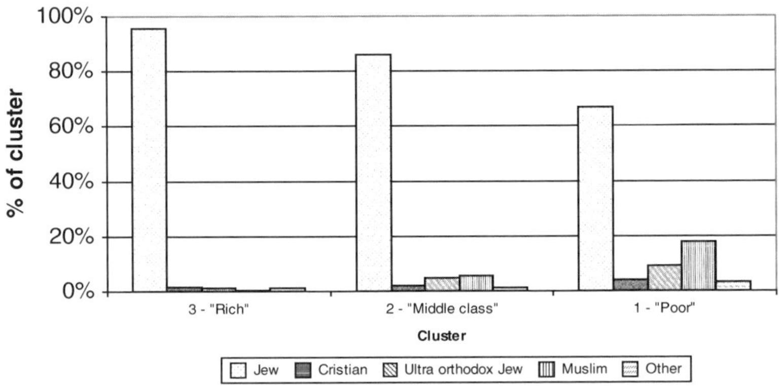

Graph 9.2 Religion distribution within clusters

clustering when there is a very large number of cases (the two-step nomenclature refers to the pre-clustering and clustering steps in the analysis algorithm). In our analysis we chose maximum likelihood as the distance between the cases (the largest probability that the two cases belong to the same sub-group).

The two-step cluster analysis also provides a criterion for determining the number of clusters: the Akaike Information Criterion (AIC) and the Bayesian Information Criterion (BIC). These two criteria determine the compatibility of the model with the data by estimating the amount of information lost in each model compared to its complexity. This means there is a preference for the simpler model with the minimal loss of information. We chose to use the BIC criterion (which also introduces a dependence on the size of the population), even though the results were found to be similar for both criteria.

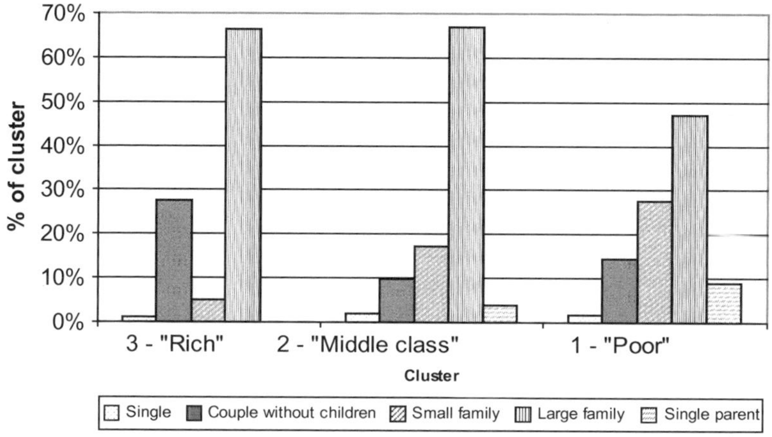

Graph 9.3 Household-type distribution within clusters

The results of the analysis indicate that the population is divided into three distinct sub-groups: Low income and high social exclusion—the "poor" sub-group, intermediate income and low social exclusion—the "middle class" sub-group, and the "rich" sub-group characterized by a high income and very low social exclusion index (see Table 9.4). A statistically significant difference was found between the mean income of the three clusters ($F(2,1416) = 822.31$, $p < 0.001$). A significant difference was also found between the mean social exclusion index score for each cluster ($F(2,1416) = 1158.33$, $p < 0.001$), but a Scheffé significance test showed that while the difference in social exclusion between the "poor" cluster and either the "middle class" or "rich" clusters was high ($p < 0.001$), the difference between the "middle class" and "rich" clusters for Social Exclusion was not significant ($p > 0.05$). The following are the characteristics of the three clusters.

As can be seen from Table 9.4, the "rich" cluster is characterized by smaller households, a higher level of education, almost 100 % employment, almost no immigrants and more men than women. No significant health differences were found between the "rich" and the "middle class" clusters (perhaps due to the relative efficiency of the public health system in Israel that still offers high quality services to most of the public).

The following graphs (Graphs 9.1, 9.2, 9.3) illustrate differences between the clusters with respect to levels of education, religion, and size of households. The better off the cluster—the better the level of education. While the "rich" cluster includes almost exclusively non Ultra-Orthodox Jews, most Muslims and Ultra-Orthodox belong to the "poor" cluster. Looking at household types one may observe that in the "rich" cluster there are more couples without children but almost no single parent or large families, obviously because of the economic difficulties which these types of households have to face.

9.6 Discussion

The goal of the present research was to examine the phenomenon of social exclusion in Israel. Two main questions have been addressed in the study. The first concerns the relationship between social exclusion and income poverty. The findings show that social exclusion exists and the population which is exposed to social exclusion is not necessarily identical to that exposed to income poverty. These findings are consistent with social exclusion research findings from other developed countries (Bradshaw et al. 2000; Burchardt et al. 1999; Saunders et al. 2007; Tsakloglou and Papadopoulos 2002).

The cluster analysis results also point toward a different of income and social exclusion behavior. It seems that high income may act as a shielding factor against social exclusion—as seen in the case of the "rich" cluster, but low income does not necessary entail high social exclusion as well. This is shown in the relative similarity of income level between the "middle class" cluster and the "poor" cluster, and the clear difference in the social exclusion level of these clusters. These findings certainly support and illustrate the need for a multidimensional measurement of poverty and disadvantage.

The second question concerned the risk factors for social exclusion. The analyses performed revealed that a low level of education, young age, poor health, cultural and religious sectors, and being a female have a strong impact on social exclusion. In fact, higher levels of social exclusion were also found in single parent households, in families with a large number of children, in jobless households, and in immigrants' households. It seems that all poverty risk factors checked in this study also act as risk factors for social exclusion. Still, the emphasis is different.

Low education is a known risk factor for poverty, usually because of its impact on the status of a person on the labor market. From the results in this study, it seems that the impact of education on social exclusion is direct, while the employment status does not have a significant impact on exclusion. Individuals who suffer from poor health suffer from social exclusion mainly in aspects concerning their physical ability to participate in society (as their interest in current affairs is not significantly different from the average healthy person). Vice versa, belonging to the female gender excludes mainly from "extra curriculum" activities such as interest in current affairs, which are time consuming, but not from social relations. Addressing the social exclusion of the populations described above should yield somewhat different policy strategies than those dedicated to addressing traditional economic poverty, which usually emphasizes increasing the household or individual income.

Addressing the issue of social exclusion in Israel draws attention to the unique status of two sectors—the Muslims and the Ultra-Orthodox Jews. Being a Muslim entails a high level of social exclusion in all subindices, independent of factors such as education and health. Developing policy to promote this sector should take into account the discrimination this sector suffers from (Abu-Bader and Gottlieb 2009; Dahan et al. 2006) as well as its great economic poverty.

The Ultra-Orthodox Jews present a high level of social exclusion compared to the non-Ultra-Orthodox Jews but only in subindices relating to their connection to the general society (interest in current affairs and connection to organizations) and not in indices relating to their connection to their local community and family and their ability to claim their political rights. Apparently, the exclusion of this sector has basis in the choice the Ultra-Orthodox Jews have made to live as separately as possible from the general population in order to preserve a very demanding religious life. We shall not discuss here the implication of social exclusion by choice but it is obvious that a different policy should be tailored for this sector, one that takes into account more than its economic poverty.

This research opens only a small window onto the issue of social exclusion in Israel and may raise more questions than it answers (such as those on other aspects of exclusion, such as geographical exclusion and access to various civil rights, and on other excluded populations, such as the non-civilian population in Israel), but its findings clearly show the relevance of the concept of social exclusion to the Israeli society and the importance of further research.

Acknowledgments This Study is a part of a Dissertation research project supervised by Prof. J. Silber and Dr. M. Monnickendam, Bar-Ilan University.

References

Abu-Bader, S., & Gottlieb, D. (2009). *Poverty, education and employment in the Arab-Bedouin society: A comparative view*. Retrieved December 10, 2010 from http://www.btl.gov.il/Publications/research/Documents/mechkar_98.pdf.

Alfandari, Y. (2005). *Absolute poverty measure in Israel*. Jerusalem: Central Bureau of Statistics. (Hebrew).

Bank of Israel. (2009). *Welfare policy issues, bank of Israel annual report- 2008*. Retrieved October 5, 2010 from http://www.bankisrael.gov.il/deptdata/mehkar/doch08/eng/pe_8.pdf (Hebrew).

Berman, Y., & Phillips, D. (2000). Indicators of social quality and social exclusion at national and community level. *Social Indicators Research, 50*(3), 329.

Berthoud, R., Bryan, M., & Bardasi, E. (2004). *The dynamics of deprivation: The relationship between income and material deprivation over time*. London: Department for Work and Pensions.

B.I. and Lucille Cohen Institute for Public Opinion Studies- ESS. (2005). Retrieved November 23, 2006 from http://www.bicohen.tau.ac.il/ (Hebrew).

Bohnke, P. (2008). Are the poor socially integrated? The link between poverty and social support in different welfare regimes. *Journal of European Social Policy, 18*(2), 133.

Bossert, W., D'Ambrosio, C., & Peragine, V. (2005). Deprivation and social exclusion. *Economica, 74*(296), 777–803.

Bradshaw, J., Levitas, R., & Finch, N. (2000). *Lone parents, poverty and social exclusion. Poverty and social exclusion survey of Britain*. Retrieved June 1, 2006 from http://www.bris.ac.uk/poverty/pse/99PSE-WP11.pdf.

Bradshaw, J., Williams, J., Levitas, R., Pantazis, C., Patsios, D., Townsend, P., et al. (2000b). *The relationship between poverty and social exclusion in Britain*. Poland: Paper presented at the International Association for research in Income and Wealth Conference Cracow.

Burchardt, T., Le Grand, J., & Piachaud, D. (1999). Social exclusion in Britain 1991–1995. *Social Policy and Administration, 33*, 227–244.

Burchardt, T., Le Grand, J., & Piachaud, D. (2002). Introduction. In J. Hills, J. Le-Grand, & D. Piachaud (Eds.), *Understanding social exclusion*. Oxford: Oxford University Press.

Central Bureau of Statistics. (2009). *Society in Israel- report no. 2* Retrieved October 1, 2010 from http://www.cbs.gov.il/publications09/rep_02/pdf/h_print.pdf (Hebrew).

Dahan, M. (1998). *Haredi population and the local authority. Part 1: Revenue sharing in Jerusalem.* Jerusalem: Jerusalem Institute for Israel Studies Jerusalem. (Hebrew).

Dahan, M., Achdut, L., Blasha, S., Ben-Shalom, Y., Bar Siman Tov, M., De-Shalitt, A., et al. (2006). *Why is poverty widespread in Israel.* Paper presented at the 14th Caesarea Forum, Caesarea, Israel. http://www.idi.org.il/PublicationsCatalog/Documents/BOOK_5002/%D7%9E%D7%93%D7%95%D7%A2%D7%99%D7%A9%D7%99%D7%95%D7%AA%D7%A8%D7%A2%D7%A0%D7%99%D7%99%D7%9D%D7%91%D7%99%D7%A9%D7%A8%D7%90%D7%9C.pdf (Hebrew).

Desai, M., & Shah, A. (1988). An econometric approach to the measurement of poverty. *Oxford Economic Papers, 40*(3), 505–522.

Devicienti, F., & Poggi, A. (2007). Poverty and Social Exclusion: Two Sides of the Same Coin or Dynamically Interrelated Processes?, from http://econpapers.repec.org/paper/ccawplabo/62.htm.

EEC. (1985). *85/8/eec: Council decision of 19 December 1984 on specific community action to combat poverty* Retrieved 23.03.2011, from http://eur-lex.europa.eu/Notice.do?val=115378:cs&lang=en&list=364792:cs,534756:cs,118174:cs,365456:cs,117734:cs,363966:cs,363561:cs,115378:cs,&pos=8&page=3&nbl=28&pgs=10&hwords=&checktexte=checkbox&visu=#texte.

European Commission. (2004). *Joint report on social inclusion.* Retrieved 04 November, 2009 from http://ec.europa.eu/employment_social/spsi/docs/social_inclusion/final_joint_inclusion_report_2003_en.pdf.

Ferro-Luzzi, G., Fluckiger, Y., & Weber, S. (2006). *A cluster analysis of multidimensional poverty in Switzerland.* Retrieved November 30, 2008 from http://ssrn.com/paper=918744.

Flug, K., & Kasir -Kaliner, N. (2003). Poverty and employment, and the gulf between them. *Israel Economic Review, 1*, 55–80.

Gal, J. (2004). *Social security in Israel.* Jerusalem: Magnes. (Hebrew).

Gara, R., & Cohen, R. (2001). Poverty among Arab population and sources to inequality between Arabs and Jews. *Economic Quarterly, 48*(4), 543–571. (Hebrew).

Gottlieb, D. (2007). *Poverty and labor market behavior in the ultra-orthodox population in Israel. Policy paper 4.* Jerusalem: Van Leer Institute.

Gottlieb, D., & Kasir-Kaliner, N. (2004). *Poverty in Israel and proposed strategy for its reduction: Employment expansion and changed in welfare system.* Jerusalem: Bank of Israel—Research Department. (Hebrew).

Heller, E. (2002). *Report and review of the development of social gaps in the last twenty years in Israel.* Jerusalem: Israeli Parliament—Research and information center. (Hebrew).

Hovav, H. (2007). *Poverty and social exclusion in Israel and the EU countries: Comparative look.* Jerusalem: Israeli Center for Social Justice. (Hebrew).

Israel Central Bureau of Statistics. (2008). *Report of the developing additional poverty indices team.* Jerusalem (Hebrew).

Katan, Y. (2002). *Poverty: Components, causes and coping strategies.* Jerusalem: The Henrietta Szold Institute. (Hebrew).

Kehunai, S. (2004). *Measurement of poverty in Israel—the current and alternative methods.* Jerusalem: Israeli Parliament—Research and information center. (Hebrew).

King, J., & Shavit, G. (2005). *Quality of life of recipients of income support.* Jerusalem: National Insurance Institute & JDC—Brookdale Institute. (Hebrew).

Laderchi, C. R., Saith, R., & Stewart, F. (2003). Does it matter that we do not agree on the definition of poverty? A comparison of four approaches. *Oxford Development Studies, 31*(3), 243.

Larsen, J. E. (2004). *Social inclusion and exclusion-conceptual issues and measurement of inclusion and exclusion in Denmark 1976 to 2000.* Paper presented at the RC 19 annual conference: Welfare state restructuring: processes and social outcomes, Paris, France.

Livyatan, O. (2003). *The effect of 'redistribution' of poverty depth (1987–2001).* Jerusalem: Bank of Israel—Research Department. (Hebrew).

Mei-Ami, N. (2004). *The condition of elderly immigrants in Israel*. Jerusalem: Israeli Parliament-Research and information center. (Hebrew).

Morgan, C., Burns, T., Fitzpatrick, R., Pinfold, V., & Priebe, S. (2007). *Social exclusion inclusion and mental health: A conceptual and methodological review* Retrieved November 30, 2010 from http://www.haps.bham.ac.uk/publichealth/methodology/docs/invitations/JH27_Stefan_Preibe_Final_Report.pdf.

Morganstein, B., Schmelzer, M., Haron, T., King, J., Aharoni, T., & Neon, D. (2003). *The employment status of recipients of income support: Resources, barriers, and requirements to facilitate integration into work*. Jerusalem: National Insurance Institute and JDC—Brookdale Institute. (Hebrew).

National Insurance Institute of Israel. (2010). *Poverty and social gaps in 2009, annual report*. Retrieved March 1, 2011 from http://www.btl.gov.il/English%20Homepage/Publications/Poverty_Report/Documents/poverty%20report%202009.pdf.

Nolan, B., & Whelan, C. (2009). *Using non-monetary deprivation indicators to analyse poverty and social exclusion in rich countries: Lessons from Europe?*. Dublin: UCD School of Applied Social Science.

Peleg, D. (2005). *Poverty, the elderly, and pensions*. Tel-Aviv: Adva Center. (Hebrew).

Perry, B. (2002). The mismatch between income measures and direct outcome measures of poverty. *Social policy journal of New-Zealand, 19*, 101–127.

R Jowell and the Central Co-ordinating Team. (2003). *European Social Survey 2002/2003: Technical report*. London: Center for Comparative social Surveys, City University.

Ronen, Y., Doron, I., & Slonim-Nevo, V. (Eds.). (2008). *Human rights and social exclusion in Israel*. Tel Aviv: Ramot—Tel Aviv University. (Hebrew).

Sadan, E. (2006). *The part and contribution of the Arab sector to the economy*. The Abraham Fund Initiatives. Paper presented at the Herzelia Convention, Israel. (Hebrew).

Saunders, P., Naidoo, Y., & Griffiths, M. (2007). *Towards new indicators of disadvantage: Deprivation and social exclusion in Australia*. Sydney: Social Policy Research Centre, University of New South Wales.

Shaayo, O., & Vaknin, M. (2000). Persistent poverty in Israel—initial results from file merger of 1983 and 1995 household censuses. *Economic Quarterly, 47*(4), 597–624. (Hebrew).

Shachak, Y. (2004). *Economic distress of single mothers*. Jerusalem: Israeli Parliament- Research and information center. (Hebrew).

Silver, H. (1994). Social exclusion and social solidarity: Three paradigms. *International Labour Review, 133*(5–6), 531–579.

Striar, R. (2001). From exclusion to social involvement. *Society and Welfare, 21*, 213–232. (Hebrew).

Svirski, S., Connor-Attias, E., Ktaus, V., & Herbst, A. (2002). *Single mothers in Israel*. Tel-Aviv: Adva Center. (Hebrew).

The Israel National Council for the Child. (2008). *Data from "Children in Israel-2008" Annual*. Retrieved October 27, 2009 from http://www.children.org.il/Files/File/leket2008.doc (Hebrew).

Townsend, P. (1979). *Poverty in the United Kingdom. A survey of household resources and standards of living*. Harmondsworth, England: Penguin Books.

Townsend, P. (1987). Deprivation. *Journal of Social Policy, 16*, 125–146.

Tsakloglou, P., & Papadopoulos, F. (2002). Aggregate level and determining factors of social exclusion in twelve European countries. *Journal of European Social Policy, 12*(3), 211–225.

Willitts, M. (2006). *Measuring child poverty using material deprivation: Possible approaches*. Working paper no.28. London: Department for Work and Pensions.

Zhang, T., Ramakrishnan, R., & Livny, M. (1996). Birch: An efficient data clustering method for very large databases. *SIGMOD, 25*, 103–114.

Chapter 10
Empowerment and Poverty in Rural Syria: A Case Study of Twenty-Four Villages

Zaki Mehchy and Nader Kabbani

Abstract Syria experienced healthy economic growth rates between 2000 and 2010. However, the country continues to suffer from deficits in social-political dimensions of human development, as marginalized and disempowered groups, including rural communities, face barriers to participation, and accessing opportunities. This study seeks to identify the determinants of empowerment and its association with money-metric poverty in 24 villages in rural Syria, using data collected in 2008 and 2009 through surveys designed with the participation of local communities. Indices were constructed using operational *elements* of empowerment that include access to information, inclusion and participation, and local organizational capacity. The results find that men are more empowered than women; young people are less empowered than adults; educational attainment is positively associated with level of empowerment, especially among women; and landowners are more empowered than others. We find a positive correlation between empowerment and poverty, measured in terms of spending per capita; however, the correlation losses its statistical significance when controlling for other socio-economic factors. Thus, on its own, increasing empowerment does not necessarily lead to improved economic livelihoods. The findings suggest that policies and programs aimed at alleviating social-political and social-economic dimensions of deprivation may be complimentary and should be introduced in tandem.

Z. Mehchy (✉)
Syrian Center for Policy Research, Damascus, Syria
e-mail: zmehchy@gmail.com

N. Kabbani
Silatech, Doha, Qatar
e-mail: nkabbani@silatech.com

V. Bérenger and F. Bresson (eds.), *Poverty and Social Exclusion Around the Mediterranean Sea*, Economic Studies in Inequality, Social Exclusion and Well-Being 9,
DOI: 10.1007/978-1-4614-5263-8_10,

10.1 Introduction

Between 2000 and 2010, Syria underwent a period of substantial economic reform, accompanied by healthy economic growth rates of 4.5 % per year. During this time, poverty rate estimates at the national level declined from 14.3 % in 1997 to 12.3 % in 2007 (UNDP 2005, 2009a). However, the country continued to suffer from deficits in the social-political dimensions of human development, as marginalized and disempowered groups, including rural communities, faced barriers to economic participation, and accessing opportunities. Indeed, between 1997 and 2007, poverty rates fell far more sharply in urban areas of the country (by 21 %) as compared to rural areas (only 5.5 %), suggesting that the latter had not benefited as much from the economic reforms taking place.

Access to opportunities and participation are both part of the construct of empowerment, which can be defined as “enhancing an individual’s or group’s capacity to make choices and transform those choices into desired actions and outcomes” (Alsop and Heinsohn 2005). This study attempts to introduce an empowerment dimension into an examination of poverty in rural Syria using 24 villages as a case study. The paper is part of a pilot exercise that eventually aimed to introduce human development indicators at the national level, covering dimensions of monetary deprivation, empowerment, equity, and sustainability that can help to assess progress toward national development goals. This would help NGOs and national and local governments to design more effective development programs.

Mahbub ul Haq (2004) identified empowerment as one of four pillars of human development, along with productivity, equity, and sustainability. Productivity focuses on economic growth, equity on distribution and inclusiveness, sustainability on ensuring that growth takes social and environmental factors into account, and empowerment on participation and opportunities. Indicators of economic growth, poverty, inequality, and environmental degradation abound. Among the four pillars, empowerment is arguably the least studied empirically. This is a severe limitation when attempting to assess human development in countries with barriers to participation and access to opportunities that may be limiting the benefits of economic growth from reaching marginalized populations. In the case of Syria, there is anecdotal evidence on the persistence of such barriers, but little empirical evidence regarding changes in levels of individual or community empowerment. The purpose of this paper is to initiate an empirical investigation of empowerment and contribute to filling a knowledge gap in the literature.

In this chapter, empowerment is conceptualized and broken down into elements that can be measured and monitored through indicators. An empowerment index is constructed that can be used to study the correlation between empowerment and other socio-economic indicators, including money-metric poverty. The results are expected to help identify less empowered groups and reasons behind their low levels of empowerment. Moreover, practitioners and policymakers often argue that reducing poverty can lead to greater empowerment and/or vice versa. Empirically examining the association between poverty and empowerment may shed light on this claim; eventually helping to design programs and policies that can improve human development in all its dimensions.

Section 10.2 provides an overview of empowerment in the Syrian context. Section 10.3 presents the concept of empowerment used in this paper. Section 10.4 presents the results of an empowerment survey conducted in 24 villages in rural Syria. Section 10.5 constructs empowerment indices and examines their association with individual and household characteristics. Section 10.6 examines the association between empowerment and money-metric poverty. A final section concludes.

10.2 Empowerment in the Syrian Context

For the past five decades, Syria has followed a public-sector-led development model that combined increases in the social provision of goods and services with restrictions to personal economic, social, and political freedoms. This model reflected, to an extent, the norm in other Arab countries and could be thought of as part of an "authoritarian bargain", by which citizens accept weaker political rights in exchange for economic security (Desai et al. 2009). Thus, while Arab countries experienced significant improvements in social-economic dimensions of human development over the past four decades (UNDP 2011; Rauch and Kostyshak 2009), they continue to suffer from deprivation in social-political dimensions (UNDP 2002, 2009b). However, over the past two decades, as a result of population pressures, rising expectations commensurate with higher standards of living, and weak economic performance due to poor institutions, many Arab governments were not able to keep up with the social-economic needs of their populations, resulting a fall in the quality, if not the quantity, of public services.

The Syrian case falls well within this model. Syria experienced notable increases in human development over the past five decades in terms of improvements in education, health, and income indicators, especially during the 1970s. During the 1980s and early 1990s, a period of economic stagnation in which GDP per capita barely increased from 1,062 USD in 1981 to 1,073 USD in 2000 (World Bank 2008), improvements in human development remained on par with World averages, mainly driven by improvements in health indicators (UNDP 2011).

However, improvements in the provisions of social services masked severe deficits in the social-political dimensions of human development. On the social front, the public-sector-led development model contributed to a culture of disempowerment and dependency. Government erected barriers to community participation and initiative. For example, few new NGOs were licensed between 1970 and 2000. By 2000, there were only 510 licensed NGOs in Syria, less than three NGOs per 100,000 people (Barout 2010) as compared to 30 NGOs per 100,000 people in Egypt and 50 per 100,000 people in Jordan (CIVICUS 2010). During this time, people and communities came to rely on government for services, jobs, information, organization, and more. On the political front, there were virtually no opportunities for political participation outside the constitutionally-sanctioned one-party system. Voice and dissent were, and remain, severely repressed; Syria consistently ranks among the bottom 10 % of the countries in the World with regard to voice and accountability (World Bank 2011).

Beginning in 2001, the Syrian government initiated a series of reforms to move the country toward a "social market" economy. The government introduced elements of a new social contract that relied less on government intervention and more on private sector solutions. The most important reforms were economic. Government removed barriers to private sector entry for many industries, permitted the licensing of private secondary schools, universities, and banks, introduced legislation to reform the country's rigid regulatory framework, and more. However, overregulation of the private sector continued. As a result, the main benefits of economic reform tended to flow to individuals who were politically connected and could circumvent restrictions. The "social" part of the reform effort involved loosing barriers to civil liberties, while maintaining a central regulatory role for the state. For example, between 2000 and mid-2010, the number of registered NGOs doubled to 1,074 (Barout 2010). However, licensing required security clearances and most new licenses were granted to groups that had some measure of political connectedness. Progress on political reforms was far more limited. In the mid-2000s, the government had drafted laws that would have allowed for multi-party elections. However, as of late 2010, these laws had yet to be discussed or enacted.

While the impact of these reforms has yet to be assessed formally, since 2001, economic growth has averaged a healthy 4.5 % per year. In addition, poverty rates, as measured by household spending per capita, declined from 14.3 % in 1997 to 12.3 % in 2007 (UNDP 2005, 2009b), providing encouraging signs for the direction of the country. However, given the restrictions to full economic, social, and political participation, these findings need to be assessed through a lens of access to opportunities and full participation. Indeed, even a simple disaggregation suggests that economic growth was not inclusive. While rural areas of the country suffer from higher poverty rates than urban areas (15 % vs. 9.9 % in 2007), between 1997 and 2007, poverty rates declined far more sharply in urban areas (by 21 %) as compared to rural areas (only 5.5 %), suggesting that rural areas were not benefiting as much from the economic reforms taking place.

The fact that rural communities, and potentially other marginalized groups, have not been able to benefit fully from economic growth, suggests that they may be facing significant barriers to accessing opportunities and participation in the development process of the country. This motivates our conceptual and empirical investigation of empowerment and its associations with other dimensions of human well-being and development.

10.3 Conceptualizing Empowerment

This section defines empowerment, identifies its elements, and operationalizes the concept. Aggregating up from individual to community empowerment is discussed together with ways of measuring individual empowerment and community empowerment in the Syrian context.

10.3.1 Empowerment Definitions and Elements

Empowerment has been defined in many ways, most of which are quite similar. In simple terms, empowerment can be thought of as "enhance[ing] the possibilities for people to control their own lives" (Rappaport 1981). Possibilities and areas of control can cover many dimensions—economic, political, social, and more. Most conceptual definitions focus on the idea of gaining control over decisions and resources that determine quality of life. Many conceptual definitions also take into account structural inequalities that affect entire groups rather than individuals (World Bank 2002). However, most definitions are not very operational (Hamyaran 2006). How does one "enhance possibilities"? How do we measure "control" or progress toward achieving this outcome?

For the purposes of this paper, we adopt an operational definition proposed by the World Bank: Empowerment is "enhancing an individual's or group's capacity to make choices and transform those choices into desired actions and outcomes" (Alsop and Heinsohn 2005). This definition is well-aligned with human development theory, wherein the basic purpose of development is to enlarge people's choices and enhance their capabilities (Sen 1999, 2005) and reflects the evolution in conceptualizing human development as a multidimensional phenomenon in which empowerment is a key element (ul Haq 2004).

To assess whether individuals or communities are becoming more or less empowered over time, we need to identify core elements which are measurable and which relate choices, actions, and outcomes. For example, World Bank (2002) identifies four key elements:

- *Access to information.* Informed citizens are better equipped to take advantage of opportunity, access services, understand rules and regulations, and exercise their rights.
- *Participation and inclusion.* Ensuring opportunities for all members of a community to participate in decision-making is critical to ensuring that the use of public resources builds on local knowledge and priorities and brings about local commitment to change.
- *Accountability.* Individuals should be able to hold local representatives accountable for their policies, actions, performance, and use of funds. People are less likely to participate if it will make no difference because providers are not held accountable.
- *Local organizational capacity.* Community residents must be able to work together, organize themselves, and mobilize resources to solve problems of common interest. Organized communities are more likely to be heard and have their demands met.

One problem in moving from a concept to an operational definition is that some aspects of the original concept may be lost or weakened. In the concept of empowerment, the focus changed from power and "control" over decisions to (1) making choices and (2) transforming choices into desired actions and outcomes.

It is difficult to measure degrees of power and control. However, choices, actions, and outcomes are all observable and potentially measurable. Superficially, the distinctions between the definitions seem minimal. However, differences may emerge at the implementation stage and should be examined through validity tests.

The domain in which people are engaged, and the level at which they are acting should also be considered. Alsop and Heinsohn (2005) identify three *domains*: the State (civic activity), the Market (economic activity), and Society (social activity), and three main *levels*: Local, Intermediary, and Macro (national level). An individual or community may be empowered in one domain and/or level but not in another. Due to data limitations and resource constraints, a specific examination of domains and levels is not included at this stage of our analysis.

10.3.2 Community Empowerment

Many indicators at the community level can be constructed by asking key informants, such as the number of schools, the presence of community councils, linkages with other communities and central authorities, etc. However, for many indicators, it is best to aggregate responses from the individual level to the community level. This is the case with unemployment rates, money-metric poverty rates, and key elements of empowerment, such as access to information. Much of our discussion of empowerment so far applies to both individuals and communities. However, while the concepts of individual and community empowerment are closely related, the exercise is not as simple as estimating the share of "empowered" community members. Community empowerment does primarily involve empowering individuals in those communities. However, the concept of community empowerment is more complex and necessitates clarifying the relationship between individuals and their communities.

A key difference between individual and community empowerment is that the latter is *not only* about "enhancing a group's capacity to make choices and transform those choices into desired actions and outcomes." It is also about the distribution of power among members of a group. Thus, an additional element of community empowerment involves rebalancing the structure of power within a community by addressing control of resources, decisions, information, and actions (INTRAC 2001). Holding elections to community council is one mechanism for dealing with this issue. However, elections may result in selecting the most popular candidates instead of the most qualified, leading to less empowered communities. There is no clear way of dealing with such issues; strategies are often based on local contexts.

A second difference between individual and community empowerment is that individuals are more mobile. Empowered individuals may choose to leave their communities. Far from empowering communities, the drain of valuable human resources may lead to less empowered communities. A third element in community empowerment is the importance of different individuals and sub-groups being able to work together effectively. Even if every individual in a community happened to be empowered, the community itself may not be empowered if internal divisions and/or a lack of "local organizational capacity" prevent individual empowerment from transferring to the group. A final element of community empowerment is vertical and horizontal linkages with other communities and with commercial, political, and social centers. An empowered community is one that has established healthy interdependent relationships with other communities and regional centers of authority.

Thus, individual empowerment is mainly concerned with (i) access to opportunities and information, (ii) participation and inclusion, and (iii) ability to hold others accountable. Community empowerment includes: (i) individual empowerment, (ii) the distribution of empowerment, (iii) local organizational capacity, and (iv) relationship with the outside world. The first is arguably the most important since individual empowerment is a prerequisite for community empowerment. All other elements are either part of, or can be incorporated into, the four basic elements of empowerment discussed above. While the long-term aim of this research effort is to conduct (eventually) an analysis of community empowerment across a large number of villages, in this paper the analysis is undertaken at the individual level.

10.3.3 Economic Development and Community Empowerment

Empowerment is a social-political not a socio-economic construct. Enhancing a person or group's capacity to make choices and transform those choices into desired actions and outcomes says nothing about the actual choices being made or outcomes being pursued. It is possible that a person or community will *choose* not to pursue developmental goals at all. Empowerment is about *capacity* to make effective choices, whatever those choices may be. Thus, empowerment is an important element of the human development framework, but not necessarily an element of the more traditional economic development framework.

Community empowerment can be an intermediate link between inputs in the development process and development outcomes. However, it can also be by-passed altogether; it need not be an integral part of a development framework. Thus, on the one hand, empowerment is one of many sub-processes in the development process and can be understood as a means to an end, such as improved livelihoods. On the other hand, it can be treated as an end in itself. In fact, empowerment can refer to a process and an outcome at the same time (Khwaja 2005; Alsop and Heinsohn 2005). But, in neither case it is a stand-alone development framework.

A broad definition of poverty includes monetary and non-monetary components, and the latter touches on the concept of empowerment. Thus, one can consider empowerment as an element of multidimensional poverty. As an empirical exercise, in the final section of this paper, we examine the association between empowerment and a money-metric indicator of poverty.

10.3.4 Measuring Empowerment

In order to monitor progress toward human development goals, key indicators must be developed. Indicators should correspond to measurable elements of each concept. The details of constructing the empowerment indicators and indices are presented below. We began by introducing indicators corresponding to the four elements of community empowerment proposed by Alsop and Heinsohn (2005): (1) access to information, (2) participation and inclusion (3), accountability, and (4) local organizational capacity. However, in keeping with the values of local community participation and empowerment that motivate the study, it was important to allow elements to be changed by the communities themselves (Laverack 2006). Thus, our approach to measuring community development and empowerment relied on working with local communities to refine the indicators. This came at the expense of strict conformity with the elements above; but, we argue, this served to strengthen the study.

One of the main adjustments made, based on testing and local community feedback, was to remove most of the questions related to "accountability" and some questions related to "participation and inclusion" and "local organizational capacity". This was because many people did not feel comfortable answering questions touching on politics or political participation, such as holding (unelected) local administrators accountable. Indeed, including such questions may have drawn unwanted attention from local government officials, placing interviewers at the risk of being questioned and of banning the survey activity altogether. Likewise, local community feedback encouraged us to include indicators related to "capability" which cover individuals' ability to make or control her/his decisions. It made sense to include this basic element of individual empowerment, even in constructing community-wide indicators. Finally, we merged questions dealing with "inclusion and participation" and "local organizational capacity" together in order to come up with empowerment elements that have roughly the same number of questions (indicators).

10.4 Twenty-Four Villages in Rural Syria

The Research Team of the Syria Trust for Development conducted an empowerment survey in 24 villages of rural Syria. The 24 villages are administratively part of four different governorates (out of 14 governorates in Syria): Lattakia, Idleb, Aleppo, and Rural Damascus. In 2008, the total population of these villages was estimated to be about 51,000 inhabitants living in 6,700 households; the population of each village varied between 500 and 8,500 people. Nearly 60 % of the population in the villages (Fig. 10.1) under study was under 24 years old.[1]

[1] Based on official numbers from each village municipality database.

Fig. 10.1 Location of the studied villages. *Source.*: Universes in Universe (2011), with Author's additions to identify village clusters.

These villages are not necessarily representative of rural Syria. They were included in the study based on their program relevance; they are part of a group of villages in which the Syria Trust, a local NGO, is conducting rural development activities and projects. The Trust bases its village selection criteria on location, community support, and the availability of human resources within reasonable travel distance. That said, we have not identified any systematic observable features to distinguish these villages from others across the country. Even so, this study remains an exploratory exercise that aims to introduce an empowerment dimension into an examination of poverty in rural Syria. Future work will refine the indicators and methods used and ensure their application to other parts of the country.

The questionnaire was designed with the participation of local communities in the selected villages. Moreover, the research team trained local community members to conduct the survey (under close supervision). The questionnaire was divided into two main sections. The first one focused on basic household information, including demographic information on family members, such as: age, gender, marriage, educational attainment, employment, migration, and overall household characteristics, such as: assets, local infrastructure, coping strategies, health indicators, financial status, land, and livestock. The second section included questions that were related to empowerment elements, mainly (1) access to information, (2) capability, (3) participation and inclusion, and (4) local organizational capacity.

The survey targeted individuals aged 15 years and above. The selection of households to include in the sample within each village was random, covering all streets in each village. Whenever possible, the survey included three persons from each household: the head of household, the spouse, and one of the household's children above age 15, taking gender balance into account. The sample covered 3,721 individuals in 1,514 unique households. The sample size was distributed evenly by gender, and about 16 % of the surveyed individuals were between 15 and 24 years old, and 50 % were between 25 and 44.

Based on results from the conducted survey (Table 10.1), the respondents have a relatively high illiteracy rate that reaches around 28 % of the sample. Illiteracy rates among women are much higher than among men. Indeed, about 75 persons of illiterate respondents are women. About half of respondents completed primary school or less. Only 10 % completed preparatory school, 10 % completed secondary school, and 12 % completed post-secondary intermediate institutes or university. Gender differences in educational attainment persist at all levels of schooling.

Only 43 % of respondents were employed. There are substantial gender differences here, with 81 % of male and only 35.5 % of female respondents employed. The unemployment rate among the surveyed respondents was 37 %; the majority of them are female. Around 35 % of employed people work in the agricultural and construction sectors and need little experience or special skills.

Government water networks supply 76 % of households in the studied villages with drinkable water; the other 24 % are supplied through wells, springs, and buying water. Government electricity networks cover 99 % of households, but only 42 % have Government sanitary service, and the other 58 % depend on covered and uncovered pits.

10.5 Constructing the Empowerment Index

As mentioned above, the survey instrument covered four empowerment elements (access to information, capability, inclusion/participation, and local organizational capacity). The primary objective was to obtain both qualitative and quantitative information on empowerment in the region. Due to the lack of sufficient indicators, the study combined the inclusion/participation and local organizational capacity questions into one indicator, generating a total of three empowerment indicators (Mehchy and Kabbani 2007). Each of the three empowerment indicators is analyzed separately. Additionally, an overall empowerment index is constructed to measure the total empowerment level across groups.

The study first grouped individual indicators or "indicator sets" (in cases when more than one question is involved in constructing the indicator) into "thematic indices" covering the areas of "access to information", "capability", and

Table 10.1 Descriptive statistics of the sample

Variable (No. of observations)	Governorates								Total	
	Idleb		Aleppo		Lattakia		Rural Damascus			
Gender (15+)	2153	*100 %*	*780*	*100 %*	*545*	*100 %*	243	*100 %*	3721	*100 %*
Male	1099	51 %	401	51 %	253	46 %	123	51 %	1876	50 %
Female	1054	49 %	379	49 %	292	49 %	120	54 %	1845	50 %
Education	*2144*	*100 %*	779	*100 %*	544	*100 %*	*243*	*100 %*	3710	*100 %*
Illiteracy	560	26 %	349	45 %	94	17 %	33	14 %	1036	28 %
Reading and writing	340	16 %	98	13 %	30	6 %	33	14 %	501	14 %
Primary	805	38 %	262	34 %	199	37 %	102	42 %	1368	37 %
Preparatory	191	9 %	42	5 %	91	17 %	37	15 %	361	10 %
Secondary and above	248	12 %	28	4 %	130	24 %	38	16 %	444	12 %
Age Categories	*2131*	*100 %*	*774*	*100 %*	*541*	*100 %*	*240*	*100 %*	*3686*	*100 %*
15–24	373	18 %	126	16 %	73	13 %	35	15 %	607	16 %
25–44	1141	54 %	328	42 %	254	47 %	119	50 %	1842	50 %
45–64	525	25 %	261	34 %	157	29 %	68	28 %	1011	27 %
64<	92	4 %	59	8 %	57	11 %	18	8 %	226	6 %
Household	*919*	*100 %*	*302*	*100 %*	193	*100%*	*100*	*100%*	*1517*	*100 %*
1–5 persons	309	34 %	123	41 %	136	70 %	47	47 %	616	41 %
6–9 persons	425	46 %	136	45 %	55	28 %	49	49 %	666	44 %
10 and above persons	185	20 %	43	14 %	2	1 %	4	4 %	234	15 %
15 years and above	*2055*	*100 %*	*779*	*100 %*	*533*	*100%*	243	*100%*	3610	*100 %*
Employed	833	41 %	390	50 %	228	43 %	107	44 %	1558	43 %
Unemployed	540	26 %	207	27 %	75	14 %	91	37 %	913	25 %
Not in labor force	682	33 %	182	23 %	230	43 %	45	19 %	1139	32 %
Household	*868*	*100 %*	*296*	100 %	*226*	*100%*	*99*	*100%*	*1492*	*100 %*
With land	374	43 %	236	80 %	131	58 %	12	12 %	755	51 %
Without land	494	57 %	60	20 %	95	42 %	87	88 %	737	49 %

"participation/local organizational capacity". The "access to information" index has two individual indicators sets, the first one is "ways of accessing information" and the second is "quality of accessed information". The "capability" index has three individual variables and two indicators sets: "selling goods inside and outside the village", and "control over eight different kinds of decision". The "participation and local organizational capacity" has ten individual variables. These thematic indices were then combined into a single overall "composite index" of empowerment (Table 10.2). All indices are scaled between 0 and 1, where 0 represents a minimal state of empowerment. Missing values were replaced with the mean value of the variable.

The possibility of using weights in constructing the indices was examined. However, factor analysis suggested that most variables used in constructing the indices were highly independent. For example, for the "access to information" index, only two factors had eigenvalues higher than one and only six of the 14 variables had uniqueness values lower than 0.60 (both key criteria in defining the

Table 10.2 The sub-indicators of the empowerment index

Individual variable	Indicator set	Thematic indicator	Composite index
Do you have access, whenever you want to:			
TV, Radio, Newspapers/Magazines, Books, Internet, Phone and Mobile, Educational Centers, Health Centers, Family members, Neighbors/Friends, Local Community, Neighboring Villages, Main Cities, Governmental Inst.	Ways of accessing information	Access to Information	*Empowerment Index*
The quality of information is: good, appropriate, sufficient, at the right time for each of the following:	Quality of accessed information	Capability	
General, Health, Legal, Commercial, Science, Agriculture, Social, Entertainment			
Level of control over your own money			
Selling a good inside the village	Accessing markets		
Selling a good outside the village			
In situations when problems arise, do you have access to an appropriate problem-solver?	Solving problems		
In situations when you have problems to resolve, do you feel that you are treated fair?			
Your control is high, medium, low, no control, over each of the following:	Controlling over decisions		
Household expenditures, Education, Marriage, Health, Work, Move within the village, Move outside the village, Buying			
If you needed money, would members of your local community provide you with it?		Participation and Local Organizational Capacity	
Usually, do you participate in local elections?			
How interested are you in these elections?			
Do you think that things will improve in your village?			
Do you expect to play a role in improving your village?			
Do you expect to play an effective role in improving your village, if you want?			
Do you have local organizations in your villages?			
How you are satisfied with the local organizations' activities?			
In your opinion, how much do these organizations positively contribute in the development process?			
In your opinion, is it easy to play an effective role in these organizations?			

factors). In other words, eight variables show a high level of independence. Applying weights would have resulted in indices that are biased against poorly correlated variables (Mishra 2007). Since theoretical and empirical grounds for weighting the factors were weak, the study applied equal weights in constructing most indices. However, in some cases, such as the "way of access to information" sub-index, the variables were weighted according to the number of users (justification is given below).

To construct the empowerment index and the sub-indices, the study uses the average linear summation of the weighted and normalized indicators as a method of additive aggregation. The importance of the indicators through inclusion/exclusion techniques was checked. For instance, the "participation and local organizational capacity" thematic index could be constructed using up to 14 indicators; yet, four indicators showed insignificant impact on the constructed thematic index value and irrelevancy to this sub-index and they were excluded in the final analysis.

10.5.1 Access to Information

Questions about access to information focus on the availability of various sources of information, the accessibility of these sources, the types of information obtained from each source, and the quality of the information obtained. We also include the improvement of roads from/to the villages as indicator of improved access.

Based on pre-survey discussions with local communities, the questionnaire includes 14 sources and eight types of information. The sources are: TV, radio, newspapers and magazines, books, internet, telephone and mobile, educational centers, health centers, family members, neighbors and friends, members from local community, neighboring villages, main cities, and governmental institutes. The types of information are: general, health, legal, commercial, scientific, agricultural, social, and entertainment. A descriptive analysis is conducted.

The results show that a high percentage of individuals have access to TV, family members, and neighbors/friends as sources of information. Newspapers/magazines, books, and internet are the less used sources of information (Table 10.3). Taking into consideration that the surveyed individuals could select more than one source of information, the table shows that:

- Around 85 % of the surveyed individuals consider TV as a source of general information, including political, economic, and sport news.
- Most (82 %) of the surveyed individuals see health centres as the main sources of health information.
- Governmental institutes are the primary source of legal information for 74 % of respondents, followed by local community with 36 %.
- Around 26 % of surveyed individuals believe that "main cities" are sources of commercial information, followed by TV with 19 %.

Table 10.3 Sources of information by accessibility degrees and types of info

Sources of Information	Individuals with access to the source of info (%)	Types of info obtained from these sources (%)[a]							
		General	Health	Legal	Commercial	Science	Agricultural	Social	Entertainment
TV	92	85	38	22	19	33	30	60	72
Radio	41	35	9	7	5	10	6	19	29
Newspapers/magazines	16	21	4	4	3	9	4	8	10
Books	18	15	6	4	3	18	3	7	6
Internet	5	6	1	1	1	6	1	4	5
Phone and mobile	66	37	10	5	8	6	8	44	17
Educational centers	54	16	5	4	2	50	3	10	2
Health centers	66	10	82	2	1	3	2	8	3
Family members	97	48	10	4	6	10	18	73	32
Neighbors/friends	93	46	8	4	8	8	18	66	39
Local community	77	39	9	36	7	14	10	38	7
Neighboring villages	72	34	11	6	19	9	24	41	14
Main cities	73	35	26	20	34	16	19	33	14
Governmental inst.	66	24	38	74	9	10	9	12	3

[a] Each participant could choose more than one type of info obtained from each source

Table 10.4 Quality of information

Quality of information	Individuals' opinions in the source of info by types (%)							
	General	Health	Legal	Commercial	Science	Agricultural	Social	Entertainment
Good	81	66	51	36	54	50	80	71
Appropriate	65	64	48	34	50	46	68	61
Sufficient	48	41	37	25	34	33	58	55
At the right time	41	31	30	19	27	25	50	47

- The educational centres are seen as sources of scientific information for 50 % of respondents, whereas, TV is a source of scientific information for 33 %.
- Interestingly, around 30 % of participants consider TV as one of the main sources agricultural information, followed by neighboring villages (24 %).
- Family members, neighbors/friends, and TV are the three main sources of social information.

Regarding the individuals' perceptions of the quality of different types of information available to them, a majority of the surveyed individuals believe that the information is 'good' and 'appropriate', but neither 'sufficient' nor 'at the right time' (Table 10.4). Commercial information has the lowest quality, particularly in terms of 'at the right time' indicator. This has important implications for the level of economic empowerment within villages. As for the level of road improvements from/to the surveyed villages, about 57 % believed that they were improving, while 12 % indicated that they were getting worse.

In order to come up with one index ranging between [0–1] and reflecting the degree of community empowerment with regard to access of information; two sets of questions were used. The "access to information" index is based on taking the average of these two partial indices.

1. Do you have access, whenever you want, to the (14) sources of information? Each answer takes (1) for "yes" and (0) for "no". The index was formed using the weighted sum of the answers, where weights are based on the relative frequency of the answer.[2]
2. The quality of the eight information types is: good, appropriate, sufficient, at the right time, resulting in 32 answers of "yes" or "no". Each answer takes (1) for "yes" and (0) for "no". The index was formed using the unweighted sum of the answers.[3]

[2] This reflects the assumption, for instance, that the person who has no TV when almost everybody else has one, is less empowered than the person who has no internet when almost everybody else also does not.

[3] It is worth nothing that we took the unweighted average of all answers in the second question because not all surveys identified which type of information was most important.

The average "access to information" index across all respondents equals 0.60. Table 10.6 shows that Idleb governorate has the highest index (0.614) whereas Rural Damascus governorate has the lowest (0.538).

10.5.2 Capability

The questions related to capability attempt to assess the individual's control over their own money, access to markets, ability to solve problems fairly, and control over personal decisions including: household expenditures, education, marriage, health care, work, movement within and outside the village, and purchasing decisions. The capability index also includes "accountability" indicators related solving problems fairly.

The results show that about 60 % of surveyed individuals had a high control over their own money, 29 % had some control, and the remaining 11 % has no control at all. As for selling goods in both internal and external markets, around 50 % of participants believed that selling goods in internal market is difficult and sometimes impossible, whereas about 40 % of participants said that selling goods in external markets is difficult and sometimes impossible. This could be due to the fact that external markets are bigger and have a higher demand than the internal market. In case of problems, around half the respondents prefer to solve problems through the head of their families, and around 20 % of them believe that problems or conflicts should be solved by police. Only 2.5 % of respondents thought that problems were resolved unfairly.

A second set of questions asked about the control of each individual over her/his own personal decisions, the majority of participants have complete or partial control over all decisions categories. Yet, 27 % of them have low or no control over decisions in terms of moving outside the village, and about 18 % of the surveyed individuals have low or no control over education and work decisions (Table 10.5).

In order to come up with one index ranging between [0–1] and reflecting the degree of empowerment in terms of capability; six such questions were covered:

1. The level of control over your own money takes (1) for "high control", (0.5) for "some control", and (0) for "no control".
2. Selling a good inside the village takes (1) for "easy", (0.5) for "somewhat easy", (0.25) for "difficult", and (0) for "impossible".
3. Selling a good outside the village takes (1) for "easy", (0.5) for "somewhat easy", (0.25) for "difficult", and (0) for "impossible".
4. In situations when problems arise, do you have access to an appropriate problem-solver takes (1) for "yes" and (0) for "no".
5. In situations when you have problems to resolve, you feel that you are treated: (1) for "completely fair", (0.5) for "somewhat fair", and (0) for "unfair".

Table 10.5 Individuals distribution by their control level over decisions (%)

Decisions	High control	Medium control	Low control	No control
Household expenditures	43	40	10	8
Education	47	36	10	7
Marriage	50	34	9	7
Health	64	28	6	2
Work	51	30	11	8
Move within the village	60	25	10	6
Move outside the village	47	25	13	14
Buying	54	34	7	5

Table 10.6 Empowerment indices by governorate

Empowerment elements	Governorates				Total
	Idleb	Aleppo	Lattakia	Rural Damascus	
Access to information	0.614	0.592	0.585	0.538	*0.601*
Capability	0.711	0.627	0.712	0.696	*0.693*
Org. capacity	0.746	0.798	0.740	0.693	*0.753*
Empowerment	*0.691*	*0.673*	*0.743*	*0.642*	*0.686*

6. According to each of the 8 "decision categories", your control over them takes (1) for "high", (0.5) for "medium", (0.25) for "low", (0) for "no control".

Given the large number of questions making up the index, missing values for each question were replaced with the sample mean. Missing values did not exceed 3 % of total observations for any question. The "capability" index was calculated by taking the un-weighted average of four groups of questions, the first one includes question (1), the second group is the average of questions (2) and (3), the third one is the average of question (4) and (5), and the last group is the average of eight decision categories. The questions were grouped based on subject similarity and correlation. The average "capability" index equals 0.693. The results show that Lattakia has the highest index (0.712) and Aleppo has the lowest (0.627).

10.5.3 Participation and Local Organizational Capacity

Questions related to participation and local organizational capacity include expected financial support from local community members in times of need, interest in local elections, participation, perceived/actual ability to change "current facts", satisfaction with the existing local organizations, and the possibility of playing an effective role within these organizations. As mentioned earlier, the number of questions related to local organizational capacity is relatively low.

Therefore, the study merged the two elements (participation/inclusion and local organizational capacity) into one index.

The results show that about 78 % of the surveyed individuals believe that the members of their local community would provide her/him with financial support in case she/he needed money. And more than 80 % of them depend on family members and friends in solving their financial problems, reflecting strong community cohesiveness in the solving personal and financial problems. Only 11 % of the villagers do not participate in local elections; however, about 20 % show no real interest in these elections. Around 90 % of the surveyed villagers see that there are many things that need to be improved in their villages, and only 6 % of them think that these things will never be improved. Around 30 % of respondents believed that they cannot play an effective role in the development process.

With regards to local organizational capacity, 16 % of surveyed individuals have no idea about the existence of local organizations. Among those who know about local organizations, about 54 % are satisfied with these organizations' activities, 4 % are unsatisfied, and the remaining 42 % have a "neutral" opinion. More than 60 % of individuals that are aware of the existence of local organizations believe that they can easily play an effective role in these organizations, whereas, about 40 % think that playing such a role is difficult if not impossible.

In order to come up with one index ranging between [0–1] and reflecting empowerment with regard to participation and organizational capacity; ten questions were identified:

1. If you needed money, would members of your local community provide you with it? (1) for "yes" and (0) for "no".
2. Usually, do you participate in local elections? (1) "yes" and (0) "no".
3. How interested are you in these elections? (1) for "very interested", (0.5) for "interested", (0.25) for "somewhat interested", and (0) for "not interested".
4. Do you think that things will improve in your village? (1) "yes" and (0) "no".
5. Do you expect to play a role in improving your village? (1) "yes" and (0) "no".
6. Do you expect to play an effective role in improving your village, if you want? (1) "yes" and (0) "no".
7. Do you have local organizations in your villages? (1) "yes" and (0) "no".
8. How satisfied are you with the local organizations' activities? (1) "satisfied", (0.5) "neutral", and (0) "not satisfied".
9. In your opinion, how much do these organisations positively contribute in the development process? (1) "high contribution", (0.5) "medium contribution", and (0) "low contribution".
10. In your opinion, is it easy to play an effective role in these organisations? (1) for "very easy", (0.75) "easy", (0.5) "difficult", and (0.25) "very difficult".

Because of the large number of questions, missing values were replaced in each variable with the sample mean. For the last three questions, missing values were replaced if the answer for question (7) was "yes"; otherwise, missing values were set equal to zero. For the first seven questions, missing values do not exceed 3 % of observations. The last three questions, however, contain a relatively high %age of missing values (30 %). Some people may think that these questions are related to politics, thus, they avoided answering. The "participation and local organizational capacity" index was calculated by taking the un-weighted average of the ten indicators above. The overall average "participation and local organizational capacity" index equals 0.753. The results show that Aleppo has the highest index (0.798) whereas Rural Damascus has the lowest one (0.693).

10.5.4 The Total Empowerment Index

The total empowerment index is taken as the average of the three indices covering "Access to Information", "Capability", and the "Participation and Local Organizational Capacity". The mean total empowerment index across all surveyed individuals equals 0.686. Lattakia Governorate has the highest empowerment index score (0.743), whereas Rural Damascus governorate (only one village from which are included and it does not reflect the situation in this governorate) has the lowest score at 0.642.

10.6 Determinants of Empowerment

OLS regression analysis is next used to identify the determinants/correlates of the various constructed empowerment indices. Robust regression analysis is carried out in this study, ensuring that standard errors are not biased downward. We discuss the results of each in turn.

With "access to information" index as the dependent variable, the analysis finds that women have significantly less access to information than men (Table 10.7). The access to information index increases with level of educational attainment and age. Respondents not in the labor tend to have a lower index score than those in the labor force. Those who own a land have a lower index score than those who do not.

With the "capability" index as the dependent variable, the regression analysis indicates that once again women have a significantly lower score than men (Table 10.7). As was the case with access to information, the capability index increases with level of educational attainment and age. However, it declines with household size. In other words, individuals living in larger households have not much control over resources and decisions. Employed respondents tend to have a higher index score than those who are unemployed. Finally, regression analysis shows that the association of the "capability" index with educational attainment is stronger for women than for men (results not shown).

Robust analysis, also finds that men have higher "community inclusion and local organizational capacity" than women (Table 10.7). Again, the index increases with

Table 10.7 Empowerment and its three elements in the 24 villages (Robust regression)

	Access to info	Capability	Org. Capacity	Empowerment
Female	−0.0646**	−0.0841**	−0.0690**	−0.0701**
	(0.00870)	(0.00777)	(0.00882)	(0.0059)
Education				
Omitted Illiterate				
Elementary and Below	0.0231**	0.0260**	0.0239**	0.0251**
	(0.00739)	(0.00673)	(0.00764)	(0.00501)
Preparatory	0.0550**	0.0603**	0.0181	0.0456**
	(0.0116)	(0.0102)	(0.0116)	(0.00792)
Secondary and Above	0.0741**	0.0676**	0.0505**	0.0652**
	(0.0112)	(0.00967)	(0.011)	(0.00757)
Age Group				
Omitted 15–24				
25–44	0.0233**	0.0682**	0.0484**	0.0467**
	(0.00764)	(0.00685)	(0.00778)	(0.00519)
45–64	0.0341**	0.0759**	0.0452**	0.0522**
	(0.00882)	(0.00789)	(0.00896)	(0.00598)
Household Size				
Omitted 1–5				
6–9	0.0101	−0.00664	0.0137	0.000762
	(0.0101)	(0.00900)	(0.0102)	(0.00683)
10 and Above	0.0128	−0.0597**	0.0369**	−0.00615
	(0.00933)	(0.00832)	(0.00946)	(0.00634)
Employment status				
Omitted employed				
Unemployed	0.0146	−0.0174*	−0.00906	−0.00824
	(0.00910)	(0.00826)	(0.00936)	(0.00617)
Not in labor force	−0.0146*	−0.00208	−0.0115	−0.0104*
	(0.00713)	(0.00637)	(0.00724)	(0.00484)
Owning land				
Omitted 'without land'				
With land	−0.0147*	0.0273**	0.0180**	0.0132**
	(0.00661)	(0.00596)	(0.00676)	(0.00448)
_cons	0.585**	0.695**	0.716**	0.668**
	(0.0121)	(0.0107)	(0.0122)	(0.00821)
N	3334	3734	3700	3311

Standards errors are in parentheses. * Significant at the 5 % level, ** Significant at the 1 % level

level of educational attainment and age. Like the access to information index, it also increases with household size. Finally, it increases with ownership of land.

As is expected, regression analysis on the overall empowerment index finds that men have a higher total empowerment index score than women (Table 10.7). The index increases strongly and consistently with level of educational attainment, among females in particular (Fig. 10.2). No other associations are significant except a positive association with land ownership. Those who are out of the labor force have a lower empowerment level than employed people, the result is statistically significant at the 1 % but not at the 5 % level. Among employed individuals, agricultural

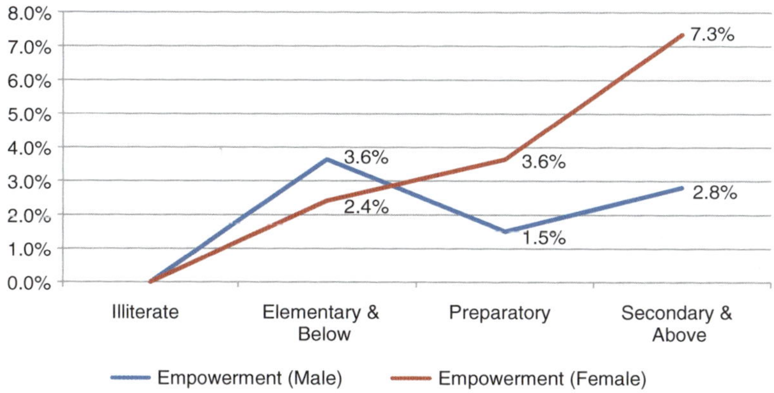

Fig. 10.2 Percentage changes in "Empowerment index score" by education and gender

workers are the least empowered and professionals (teachers, engineers, doctors, etc.…) and employees are more empowered. A possible reason behind this result is the fact that professionals and employees have a higher educational level than agricultural workers. Another possible reason is that agricultural jobs are seasonal whereas most professionals and employees have "regular" jobs.

10.7 Association between Empowerment and Poverty

As mentioned before, the concept of empowerment could reflect part of the non-monetary side of poverty (Maxwell 1999); and there is little theoretical or empirical evidence on the relationship between empowerment and money-metric poverty. The first thing to note, alluded to above in the context of economic development, is that monetary poverty is a social-economic concept whereas empowerment is a social-political concept. Thus, there is no strong theoretical reason as to why the two concepts *need be* empirically connected. A second issue to note while studying this relationship is that monetary poverty (as it is usually applied) is a household concept, whereas empowerment as we have defined it in this paper is an individual concept. The two concepts can still be compared; but, this distinction should be kept in mind.

The survey of the 24 villages collected data on income, spending, and assets. Most studies prefer to use household expenditures rather than income in constructing a poverty index for several reasons, including a tendency for respondents not to be as forthcoming about income and the fact that consumption is not as volatile as income— it tends to smooth out fluctuations and is considered a better representation of "permanent income". Indeed, the survey data collected from the villages shows that income has a higher standard deviation than spending. Thus, the study adopts spending per capita as the basis for identifying poverty status which equals total household spending divided by the number of household members.

A poverty index was developed based on a national poverty line from 2005 (UNDP 2005). The poverty line in the rural Syria that includes the surveyed villages was 1,459 SYP (roughly 30 USD) per capita per month. For simplicity, and taking into account an inflation rate of about 45 % between 2004 and 2008/09 (the time of conducting the surveys), we apply a poverty line of 2,115 SYP per capita. To facilitate the analysis, a value of (0) is given to individuals living in households under poverty line and (1) to those living in households above poverty line. About 48 % of households in the studied villages are poor. A logit regression was conducted to identify the determinants of poverty in the studied villages. The results (Table 10.8) show that a family tends to be above the poverty line if the head of this family has a high educational level (secondary and above) and/or owns a land. The probability for a family to be below the poverty line increases if the head of household is out of labor force and/or aged between 25 and 44 years. Finally, a large household size increases the possibility for a family to be poor.

The paper examines the relationship between empowerment and spending per capita (as a proxy for poverty) by using a regression analysis to control for possible confounding factors. The study includes fixed effects for villages. The total empowerment index is positively and significantly associated with the log spending per capita when no other variables are included and when we control for "gender", "education", and "age group". However, as soon as other variables (household size, employment status, and owning land) are controlled for, the association becomes weaker and loses its statistical significance (Table 10.9).

The above results could be affected by the fact that the questions asked of the surveyed individuals concentrated on "agency—the capacity to make meaningful choice" more than "opportunity structure—the presence and operation of formal and informal institutions or rules of the game". For instance, internet service is not always available in the villages, so poor and non-poor individuals both suffer from a lack of access to the internet. Also, since social norms support males to have high level of empowerment, poor, and non-poor females alike have lower empowerment levels. Knowing more about the overall environment of the surveyed individuals could explain why poverty is not associated significantly with empowerment.

Our empowerment indices must be examined in terms of validity and consistency. Future survey instruments should include a modified set of questions related to each element of empowerment. Each element has relatively low number of questions, thus, any ambiguous question or inaccurate answer may affect the overall results. Despite these shortcomings, the concluded results may reflect the real situation in the surveyed villages. Poor people are not necessarily empowered and vice versa, and policies aimed at reducing money-metric poverty might not lead to higher levels of empowerment.

For example, the agricultural sector in Syria suffers from substantial problems including poor irrigation methods, water shortages, and small land holding size. Between 2001 and 2009, agricultural contribution to GDP declined by nearly 25 %

Table 10.8 Poverty Determinants (Logit regression)

Poverty index		
Female	−0.167	(0.318)
Education		
Omitted Illiterate		
Elementary and below	0.214	(0.169)
Preparatory	0.260	(0.255)
Secondary and above	0.776**	(0.235)
Age group		
Omitted 15–24		
25–44	−0.660**	(0.226)
45–64	0.259	(0.224)
Household size		
Omitted 1–5		
6–9	−1.436**	(0.147)
10 and above	−2.607**	(0.210)
Employment status		
Omitted employed		
Unemployed	0.368	(0.465)
Not in labor force	−0.582**	(0.141)
Owning land		
Omitted 'without land'		
With land	0.633**	(0.136)
_cons	1.066**	(0.254)
N	1288	

Standards errors are in parentheses. * Significant at the 5 % level, ** Significant at the 1 % level

(World Bank 2010), while employment in the sector declined by around 35 % (CBS 2001–2010) and has been accompanied by high rates of rural–urban migration. The Syrian government has been attempting to reduce poverty and boost employment in rural areas by supporting the agricultural sector through subsidies and related policies. However, policies have been top-down and have not addressed the underlying issues facing the sector. As a result, government policies have not been able to slow the high rates of internal migration. Enhancing community empowerment through adopting a more participatory and inclusive approach to policy making and providing rural communities with better access to information and infrastructure could play a major role in heading toward a sustainable development process.

10.8 Conclusion

Syria is witnessing steady improvements in its socio-economic indicators, including a decline in poverty rates. However, evidence suggests that individuals in rural areas are not benefiting as much from economic growth compared to those

Table 10.9 Empowerment and spending per capita

Empowerment	(1)	(2)	(3)	(4)	(5)	(6)	(7)
Spending	0.00836**	0.00291**	0.00324**	0.00215**	0.00139	0.00141	0.00093
	(0.00059)	(0.00066)	(0.00066)	(0.00066)	(0.00074)	(0.00074)	(0.00083)
Female		−0.08464**	−0.07382**	−0.07528**	−0.06825**	−0.06890**	−0.06821**
		(0.00424)	(0.00447)	(0.00439)	(0.00487)	(0.00575)	(0.00575)
Education							
Omitted Illiterate							
Elementary and below			0.01841**	0.02440**	0.02496**	0.02498**	0.02540**
			(0.00477)	(0.00500)	(0.00499)	(0.00503)	(0.00504)
Preparatory			0.03155**	0.04312**	0.04360**	0.04376**	0.04440**
			(0.00691)	(0.00697)	(0.00696)	(0.00700)	(0.00701)
Secondary and above			0.05774**	0.06460**	0.06520**	0.06530**	0.06590**
			(0.00665)	(0.00675)	(0.00674)	(0.00688)	(0.00692)
Age Group							
Omitted 15–24							
24–44				0.04307**	0.04027**	0.03954**	0.04002**
				(0.00475)	(0.00486)	(0.00501)	(0.00502)
45–64				0.05235**	0.04936**	0.04880**	0.04846**
				(0.00543)	(0.00550)	(0.00558)	(0.00557)
Household size							
Omitted 1–5							
6–9					0.00856	0.00851	0.00750
					(0.00561)	(0.00561)	(0.00567)
10 and above					−0.01060	−0.01060	−0.01122
					(0.00604)	(0.00605)	(0.00606)
Employment status							

(continued)

Table 10.9 (continued)

Empowerment							
	(1)	(2)	(3)	(4)	(5)	(6)	(7)
Omitted employed							
Unemployed						0.00082	0.00112
						(0.00647)	(0.00648)
Not in labor force						−0.00297	−0.00270
						(0.00475)	(0.00474)
Owning land							
Omitted 'without land'							
With land							0.00693
							(0.00492)
_cons	0.66029**	0.72221**	0.69874**	0.66302**	0.67003**	0.67130**	0.67049**
	(0.01233)	(0.01274)	(0.01368)	(0.01350)	(0.01493)	(0.01552)	(0.01619)
N	3311	3311	3311	3311	3311	3311	3311

Standards errors are in parentheses. * Significant at the 5 % level, ** Significant at the 1 % level living in urban areas. Marginalized and disempowered groups, including rural communities, continue to face barriers to participation, and accessing opportunities. This paper operationalizes the concept of empowerment by constructing indices based on three empowerment elements: access to information, capability, participation, and local organizational capacity. A total empowerment index is constructed by aggregating the three indices. In order to move from theory to practice, the paper examines empowerment in 24 villages of rural Syria.

The empirical results suggest that men are more empowered than women, young people are less empowered than adults, and educational attainment is positively associated with empowerment. The association between empowerment and educational attainment is especially strong among women, helping to close the gender gap at higher levels of educational attainment. These core results are consistent across all elements of empowerment examined in the paper. The associations of the elements of empowerment differ with respect to household size. The analysis shows that individuals living in larger households have better participation and local organizational capacity, but not as much control over resources and decisions. Land ownership is correlated positively and significantly with empowerment. The strength of the association of empowerment and employment status differs across dimensions, but generally those who are not in labor force are less empowered than those who are.

Empowerment is positively correlated with household spending per capita. However, once other household factors are controlled for, the association between empowerment and poverty loses its statistical significance. Thus, policies aimed at reducing poverty might not lead to higher levels of empowerment (and vice versa). Care must be taken in applying these results. Our analysis presents a first attempt to integrate the concept of empowerment into a study of community development. The empowerment indices must be examined in terms of validity and consistency. Future surveys should include a modified set of questions related to each element of empowerment. Furthermore, other factors such as the "opportunity structure" of the villages should be triangulated using focus groups and interviews with key informants. Future research will aim to refine the analysis in order to verify and validate these results.

References

Alsop, R. & Heinsohn, N. (2005, February). *Measuring empowerment in practice: Structuring analysis and framing indicators*. World Bank Policy Research Working Paper 3510, Washington, DC.

Barout, J. (2010). *An introduction to mapping civil society organizations in Syria* (in Arabic). Background paper prepared for the NGO Platform Project in Syria, available through the Syrian Development Research Centre.

Central Bureau of Statistics. (2001–2010). Annual Labor Force Surveys, Damascus, Syria.

CIVICUS. (2010). Civil Society Index Country Reports Database, accessed May 2010.

Desai, R., Olofsgard, A., & Yousef, T. (2009). The logic of authoritarian bargains. *Economics and Politics, 21*, 93–125.

Hamyaran (Iran NGO Resource Center). (2006, April). *Community empowerment for sustainable development in Iran*. Paper presented at the Fifth MENA Development Forum, Beirut, Lebanon. Available in: www.info.worldbank.org/etools/mdfdb/docs/WP_TESEV1.pdf.

International NGOs Training and Research Center (INTRAC), (2001). *Evaluating empowerment: Reviewing the concept and practice*. In O Peter (Ed.), Oxford.

Khwaja, A. I. (2005). Measuring empowerment at the community level: An economist's perspective. In D. Narayan (ed.), *Measuring Empowerment: Cross-Disciplinary Perspectives*. Washington: World Bank.

Laverack, G. (2006). Using a 'Domains' approach to build community empowerment. *Community Development Journal, 41*(1), 4–12.

Maxwell, S. (1999). *The meaning and measurement of poverty*. London: Overseas Development Institute (ODI) Policy Briefing.

Mehchy, Z. & Kabbani, N. (2007, December). *Analysis of the characteristics of five villages near Ebla*. Syria Trust for Development Report. Available at: www.syriatrust.org/research.

Mishra, SK. (2007). *A comparative study of various inclusive indices and the index constructed by the principal components analysis*. Working papers series, Social Science Research Network (SSRN).

Rappaport, J. (1981). In praise of paradox: A social policy of empowerment over prevention. *American Journal of Community Psychology, 9*(1), 1–25.

Rauch, J. E., Kostyshak, S. (2009). The three arab worlds. *The Journal of Economic Perspectives, 23*(3), 165–188.

Sen, A. (1999). *Development as freedom*. New York: Knopf.

Sen, A. (2005). Human rights and capabilities. *Journal of Human Development, 6*(2), 151–166.

ul Haq, M. (2004). The birth of the human development index. In F. P. Sakiko & A. K. Shiva Kumar, (eds.), *Readings in human development: Concepts, measures and policies for a development paradigm*. UK: Oxford University Press.

UNDP. (2005). *Poverty in Syria: 1996–2004: Diagnosis and Pro-Poor Development Considerations*, by Heba El-Laithy and Khalid Abu-Ismail, June 2005.

UNDP. (2009). *Poverty and Distribution in Syria*, by Heba El-Laithy and Khalid Abu-Ismail, June 2009.

UNDP. (2009b). *Arab Human Development Report, 2009: Challenges to Human Security in the Arab Countries*. New York: UNDP.

UNDP. (2011). International Human Development Indicators: Human Development Index: Trends 1980—present, data accessed August 2011.

UNDP (United Nations Development Programme). (2002). *Arab Human Development Report, 2002: Creating Opportunities for Future Generations*. New York: UNDP.

Universes in Universe. (2011). Website accessed October 2011: http://universes-in-universe.org/eng/nafas/countries/middle_east/syr/(map).

World Bank. (2002). In N. Deepa (ed.), *Empowerment and poverty reduction: A sourcebook*, Washington: The World Bank.

World Bank. (2008). World Development Indicators 2008. Washington.

World Bank. (2010). World Development Indicators 2010. Washington.

World Bank. (2011). Worldwide Governance Indicators Database. Washington.

Chapter 11
Dynamics of Multidimensional Poverty and Trade Liberalization: Evidence from Panel Data for Egypt

Shireen AlAzzawi and Mona Said

Abstract This chapter investigates the dynamics of poverty in Egypt using monetary and non-monetary panel data, and the impact of trade liberalization on poor and low waged workers. We find a relatively low level of economic mobility in both income and non-income indicators, with the majority of those who were "poor" in 1998, whether in the monetary or non-monetary dimension, remaining so by 2006. Trade reform in the form of lower tariffs and increased export promotion exerted a small positive influence on the incomes of the poor; however, this came at the expense of greater informalization of workers and higher incidence of low quality jobs. We also find that private sector employment and gender have a much more important role in segregating the labor market.

S. AlAzzawi (✉)
Department of Economics, Santa Clara University,
500 El Camino Real, Santa Clara, CA95053, USA
e-mail: salazzawi@scu.edu

M. Said
Faculty affiliate of the London Middle East Institute, SOAS,
London University and Department of Economics, The American University in Cairo,
113 Kasr el Aini Street, 11511 Cairo, Egypt
e-mail: mona_said@aucegypt.edu

V. Bérenger and F. Bresson (eds.), *Poverty and Social Exclusion Around the Mediterranean Sea*, Economic Studies in Inequality, Social Exclusion and Well-Being 9,
DOI: 10.1007/978-1-4614-5263-8_11,

11.1 Introduction

Does trade liberalization help the poor? Is faster economic growth really good for the poor? International organizations and many prominent economists are increasingly supporting the idea that growth is the main channel through which poverty alleviation can be achieved on a wide scale.[1] At the same time, the idea that trade openness is one key channel through which growth can be achieved gained increasing support among a significant proportion of economists and many policymakers over the last two decades.[2] The links between poverty and trade liberalization are commonly discussed at the conceptual level, but empirical studies remain rare in Middle East and North Africa contexts. Furthermore, previous studies on the region and elsewhere often study these concepts solely from the monetary perspective, even though poverty is much more than lack of sufficient income. In fact, most of the Millennium Development Goals focus on the non-monetary aspects of achieving "development". In this chapter we will investigate the relationship between poverty and trade liberalization during a period of relatively high growth, distinguishing between monetary and non-monetary dimensions of poverty. Egypt presents an interesting and rarely studied case in this respect.

Egypt is the most highly populated country in the Middle East and North Africa, and with more than 40 % of the population living at under $2 a day,[3] it has one of the severest poverty problems, at least in terms of sheer numbers. Poverty and inequality in Egypt over the last 5 years have been on the rise, with the absolute (monetary) poverty rate increasing from 16.7 % in 2000 to 19.6 % in 2005, which constitutes a complete reversal of the decline in poverty incidence

---

[1] See for example various World Bank publications and Dollar and Kraay (2002), Kraay (2006), among others.

[2] Many influential papers such as Dollar (1992), Sachs and Warner (1995), Edwards (1998), Frankel and Romer (1999), claimed to find a strong correlation between trade openness and growth. Anne Krueger's words at her 1997 AEA Presidential address reflects the widespread belief in this link at that time: "It is now widely accepted that growth prospects for developing countries are greatly enhanced through an outer-oriented trade regime and fairly uniform incentives (primarily through the exchange rate) for production across exporting and import-competing goods (…). It is generally believed that import substitution at a minimum outlived its usefulness and that liberalization of trade and payments is crucial for both industrialization and economic development (…) the current consensus represents a distinct advance over the old one, in terms both of knowledge and of the prospects it offers for rapid economic growth" (Krueger, 1997, p.1). These views were not without critics, however. Rodriguez and Rodrik (1999)'s influential paper cast much doubt on the indicators for openness/lack thereof used in most of these studies. Rodriguez (2006) updated this study by focusing on more recent papers that dealt with openness-growth relationship, again arguing that their claims are not robust.

[3] World Bank (2007), page 6.

from 1996 to 2000.[4] "Near poverty,"[5] on the other hand, affects an additional 21 % of Egyptian households, bringing the total poor to over 40 % of the population, which constituted more than 28 million people in 2005. The few studies that have considered non-monetary poverty in Egypt have also found that it is quite high (Osman et al. 2006) and on the rise (Bérenger 2010).

In 2007, Egypt boasted a growth rate of almost 7 %[6] and many have attributed this to the increased pace of trade liberalization over the past few years. Egypt's dramatic, albeit slow, trade liberalization since 1991 is part of an ambitious economic reform and structural adjustment program that includes widespread privatization of public sector firms, and attempts at fiscal responsibility by reducing the deficits in the balance of payments and government budget. In the field of trade liberalization, average tariff rates were reduced by more than 50 % over a period of 15 years. The simple average tariff rate declined from 42.2 % in 1991 to 26.8 % in 1998, and further declined to 20 % in 2005 following the speeding up of the economic liberalization program after 2004 (WTO 1999, 2005). Trade liberalization may have stimulated economic growth, but may also have had adverse effects on the poor and the low skilled workers disproportionately concentrated in privatized firms, the informal sector, or industries that are experiencing increased competition from international markets.

The purpose of this chapter is, first, to investigate the extent of mobility in the Egyptian economy, and movement in and out of poverty over a period of high growth and trade liberalization, both from monetary and non-monetary perspectives. Second, to directly study the impact of trade liberalization on the wages and quality of jobs available to the poor and the low waged over this period. We use rich panel labor market survey data to trace the impact of these policies over time, using transition matrices to follow the various monetary and non-monetary welfare indicators for the same individuals between 1998 and 2006 and the dynamics of moving in and out of poverty. We study the impact of trade liberalization on the poor and low waged, using data on tariff reduction, export promotion and import orientation that has been linked to the individual data at the industry level.

The remainder of this chapter is organized as follows. Section 11.2 reviews the literature; Section 11.3 discusses the data sources and limitations. Section 11.4 examines the poverty dynamics from a monetary and non-monetary perspective. Section 11.5 analyzes the impact of trade liberalization on the poor and low waged. Section 11.6 concludes.

---

[4] World Bank (2007), page iv.

[5] In World Bank (2007), absolute poverty is defined as "spending less than needed to cover absolutely minimal food and non-food needs", and near poverty is defined as "spending barely enough to meet basic food [needs] and slightly more than essential non-food needs". (pages iii and iv)

[6] Economist Intelligence Unit, Country Profile (2007). Retrieved November 29, 2007 from www.economist.com.

11.2 Literature on Multidimensional Measures of Poverty and Trade Liberalization

The extension of the notion of poverty to incorporate elements of human wellbeing is based on the limitation that monetary indicators alone provide in identifying the poor. Human wellbeing rather than income and consumption methods offers a wider scope of consideration for monetary and non-monetary elements that affect the quality of life of the poor. Some attempts to explore the extent of multidimensional poverty in the case of Egypt include Bérenger (2010), Osman et al. (2006), and Bibi (2004). However, none of these studies use panel data that tracks the same individuals or households over two periods of time. Also, only the study by Bibi (2004) combines income and non-income dimensions of poverty, but again only for cross-sectional data.

The ability to track the income levels and poverty status of individuals over time can provide tremendous insight into the process by which poverty declines or increases over time, and the relationship of any such changes with growth and distributional changes. A considerable literature that investigates the dynamics of income and poverty mobility for advanced economies exists. In the USA and the UK, national longitudinal surveys have been collected for much longer time frames and have thus awarded researchers the opportunity to study patterns of mobility and moving in and out of poverty extensively.[7] However, these results may not be applicable to developing countries where economic, social and political conditions and constraints may make it either more or less difficult for mobility to take place, depending on the country under study and the time period.

Panel studies of poverty dynamics are quite rare for developing countries due to data limitations. Baulch and Haddinot (2000) provide a good overview of these studies and their results. As far as we know there is only one previous study that analyzes the dynamics of poverty in Egypt. Haddad and Ahmed (2003) study chronic and transitory poverty in Egypt between 1997 and 1999. Their analysis shows that 66.7 % of total poverty, 67.9 % of total rural poverty, and 63.9 % of total urban poverty is chronic, indicating a persistent poverty trend.

However, their study is limited to a very small sample size (347 households) that covers a 2-year period in the late 1990s, a period of decreasing absolute poverty overall in Egypt. The present study utilizes a large, nationally representative labor force survey of incomes across a longer time frame, within a period where the direction of poverty changes was mixed. This has also been a period of rapid economic growth and increased openness through trade liberalization and export promotion reform measures.

Trade liberalization is actively promoted by international organizations as a tool for economic growth and poverty reduction. Unfortunately, this policy may not be

[7] See for example Buchinsky and Hunt (1999), Steckel (1990) for the USA; Jarvis and Jenkins (1995, 1998) for the UK, and Zaidi and de Vos (2002), comparing the UK with Netherlands, to name a few.

the best strategy for achieving poverty reduction. Instead, it may actually increase unemployment, increase wage gaps between rich and poor, and reduce compliance of labor market standards. Additionally, trade liberalization may increase the prominence of part time work and temporary work as well as reduce the overall job quality of positions (Goldberg and Pavcnik 2004). There are several issues that arise from the study of trade liberalization which are pertinent to our study, namely, the effect on unskilled labor, compliance with labor market standards and industry wages. Trade policies can increase demand for skilled labor without reducing the demand for unskilled labor. As a result, wage inequality does not have to be linked to lower wages for poor or unskilled workers. Moreover, absolute wages do not necessarily determine poverty levels; instead, the focus should be on a country's income distribution and the position of unskilled workers within it. Beyond wages, several studies have also examined the impact of trade on informalization and quality of jobs held by the poor. For example, Bernard and Jensen (2007) and Bernard and Sjoholm (2003), argued that there is no direct relationship between trade liberalization and reduced minimum wage compliance or informal sector growth.

In this study, we incorporate both monetary and non-monetary dimensions of poverty to obtain a more comprehensive idea of the extent and dynamics of poverty in Egypt. We also study directly the effect of trade liberalization on the poor, from a monetary and non-monetary perspective using panel labor data linked to trade data at the industry level.

11.3 Data Description and Limitations

The data used in this study comes from the 1998 and 2006 Egypt Labor Market Panel Surveys (ELMS 1998 to ELMPS 2006). These nationally representative labor force sample surveys were conducted by the economic research forum (ERF) in cooperation with CAPMAS. The ELMPS 2006 tracks the labor market and demographic characteristics of the households and individuals interviewed in ELMS 1998, new households that have emerged as a result of splits in the original households, and a refresher sample of entirely new households.

The panel data consists of (1) 3,684 households from the original ELMS 1998 survey, (2) 2,167 new households that emerged as a result of splits in the original households. Of the 23,997 individuals interviewed in 1998, almost 22,987 were still alive or in the country in 2006 and 17,357 of those (75.5 %) were successfully re-interviewed in 2006.

We rely on data on real monthly earnings for wage workers as the main measure of monetary welfare for this study. Ideally, we would have liked to also use data on consumption as an additional monetary measure of welfare to conduct the analysis and hence make comparisons between results based on each measure. Unfortunately, actual consumption data are not available in the ELMS or ELMPS. In analyzing the dynamics of poverty and the extent of economic mobility, each type of welfare measure has its advantages and disadvantages. Incomes are in some cases more accurately reported than expenditures since they are easier to recall. Using incomes is also usually the only way we can identify the source of any mobility, for example, whether it is due to demographic or economic events (Woolard and Klasen 2005). Consumption, on the other hand, may be a better indicator of permanent income when households exercise consumption smoothing, which is common among the poor (Deaton 1997). Consumption is thus preferred if one is more interested in a long-term measure of mobility.

Although we do not have actual consumption data; fortunately, an attempt was made by Assaad and Roushdy (2008) to estimate a per capita consumption measure for the ELMS and ELMPS samples. They followed a dynamic cross survey micro-simulation methodology that is often used in the literature (see for example Grosse, Klasen and Spatz (2005) for a similar exercise) to overcome data availability problems, making use of common data variables from different surveys conducted at very close time intervals. They used the common characteristics of individuals from the labor market surveys of 1998–2006, and those from the Household Income, Expenditure and Consumption Surveys (HIECs) of 1999/2000 and 2004/2005 to estimate a consumption level for each individual. They estimated this estimated per capita consumption (EPCC) using a two-stage estimation technique where first an expenditure function is estimated using individual characteristics from the HIECs of 1999/2000 and 2004/2005. The estimated coefficients are then applied to the labor market data for 1998 and 2006 and the corresponding EPCC for each individual in the labor market surveys of 1998 and 2006 is predicted.[8]

This EPCC and the method used to estimate it suffer from a number of limitations of course, and we mention those here to allow the reader to put results based on EPCC in perspective. The method used to construct the EPCC in Assaad and Roushdy (2008) does not recover the original distribution of consumption in the data. It is designed to recover the mean consumption conditional on the X's used in the prediction equation. The estimated variance is therefore necessarily smaller than the original variance, because it does not include unexplained variation. This underestimation of the variance is likely to result in an underestimation of poverty, especially extreme poverty. Additionally, the fact that EPCC is predicted on the basis of observable characteristics of the household that are slow to change, such as ownership of durable assets, housing conditions, access to services, etc., means that the EPCC will be much less volatile than actual consumption expenditures. It may thus find more "chronic poverty" than actual consumption. Keeping these limitations in mind, we will use this additional monetary measure of welfare in our analysis, which will allow us to make a rough comparison between poverty

[8] Data from the HIECS are not made publically available except in published form by CAPMAS, and therefore we could not use it directly. Assaad and Roushdy kindly provided their constructed EPCC measure to us to help facilitate this research.

dynamics based on consumption and that based on income, and to point out any significant differences in conclusions and implications.

To study the non-monetary aspects of poverty dynamics we use an index of the quality of jobs. There are three main strands of the literature on job quality. The first, promoted by the ILO, focuses on the elaboration of a decent work index that focuses on several aspects of job quality. A second strand is based on studies of the relationship between job quality and job mobility, while the last one concentrates on quantifying particular non-wage aspects of employment using a compensating wage differential methodology. Assaad and Roushdy (2008) constitute the first attempt to address the challenge of measuring job quality using Egyptian data by formulating one such measure at the individual job level. They combine information on actual and estimated earnings with information on access to social insurance, regularity of employment, work hours, and nature of workplace into an index of job quality and investigate the individual and enterprise-level determinants of such an index. Their findings show that although on the whole job quality appears to have declined in Egypt over the 1998–2006 period, it has in fact increased among wage and salary workers in the private sector.

A review of the literature indicates that job quality is usually measured along four dimensions: (1) income security, (2) employment security, (3) voice representation, and (4) work security. Operationalizing the above measurement dimensions to the labor data we have, we construct an index of job quality that incorporates the following 11 criteria: (1) earnings, (2) nature of workplace, (3) adequate working hours, (4) commuting time to work, (5) has regular job, (6) has social insurance, (7) has health insurance, (8) has a contract, (9) has paid sick leave, (10) has paid annual leave, and (11) member of a trade union.

Based on the above criteria, three measures of job quality were constructed for use in this chapter: the institutional job quality measure that only incorporates criteria (6) to (11); a comprehensive measure entailing all criteria but is limited to only wage workers; and a less comprehensive one confined to only wage workers and disregards the earnings criteria, serving as a non-earnings measurement. We combine the different criteria for each job quality index into a single measure using factor analysis, and we let the correlation structure among variables produce the factor weights.[9] The advantage of these job quality measures is the comprehensiveness of the indices in comparison to previous studies that use a dummy for formality. Our job quality indices are informative about the quality of work as implied by the "decent work" literature.

[9] Future work is underway to use a compensating differential approach to arrive at a better way of weighting the components of job quality.

11.4 Poverty Dynamics Between 1998 and 2006

11.4.1 Transitions in and out of Monetary Poverty Between 1998 and 2006

In this section, we use transition matrices to examine the extent of movement in and out of monetary poverty between 1998 and 2006. Formally, a transition matrix can be defined as follows: Let $n \in N$ be the number of individuals in society, where N is the set of natural numbers. Let $y_i^{t_0}$ and $y_i^{t_1}$ be the income (or similar monetary or non-monetary variable) of individual i at two different points of time t_0 and t_1, with $y^{t_0} = (y_1^{t_0} \ldots y_n^{t_0}) \in R_+^n$ and $y^{t_1} = (y_1^{t_1} \ldots y_n^{t_1}) \in R_+^n$. Suppose that individuals have been divided into $m \leq n$ nonempty, exhaustive, and mutually exclusive classes in ascending order of income level. A transition matrix will be a square matrix $P = [p_{jk} \in R_+^{n \times m}]$, where p_{jk} denotes the proportion of individuals belonging to class j at t_0 that have shifted to class k at t_1 and where $\Sigma_{k=1}^{m} p_{jk} = 1$ for all $j \in \{1 \ldots m\}$ so that P is a stochastic matrix.

Transition matrices without set cutoff points provide information on the extent of movement up or down along the income distribution. However, they do not tell us anything about whether this movement was in fact sufficient to pull the poor out of their poverty, unless they are constructed in relation to specific poverty lines. We use poverty lines from World Bank (2007) for 2004/2005 updated to 2006 prices using the CPI to construct transition matrices relative to specific poverty lines. All income and consumption data are also expressed in 2006 LE using the CPI.[10]

These poverty lines are calculated based on the cost-of-basic-needs methodology, and account for differences in consumption patterns and prices across regions, and differences in household size and composition. The cost of the actual diet consumed by Egyptians of different ages and classes, not a hypothetical one based on caloric requirements, is used to calculate these poverty lines, and economies of scale in consumption or consumption sharing are taken into consideration. Following the World Bank (2007) study, we will also use three different poverty lines: the food poverty line (FPL), poverty line (PL), and upper poverty line (UPL). The FPL reflects the cost of the food bundle using the relative quantities observed in the diet of the poor (as proxied by the second quintile), and the prices they actually faced. Individuals and households whose consumption was below the FPL will be referred to as "extreme poor" (World Bank 2007). The PL was constructed by allowing for expenditure on essential non-food items in addition to the FPL. Specifically, the share of non-food expenditure was set to equal that of households whose *total* expenditure is at the food poverty line. This is designed to capture the extent of "non-food essentials" since households would

[10] We also tried using the Food Price Index (based on the subgroup of food and beverages) available from CAPMAS to update the poverty lines as well as the income and consumption data. The poverty lines and the constructed transition matrices were almost identical to those reported here.

Table 11.1 Distribution of Real Monthly Income and Estimated per Capita Consumption in EGP, 2006 prices

Years	Percentiles of real monthly income				
	10	25	50	75	90
2006	180	280	433.3	650	1000
1998	135.9	214.3	316.1	500.2	744
	Percentiles of estimated annual per capita consumption (EPCC)				
2006	1465.48	1852.15	2415.73	3206.37	4397.52
1998	1369.82	1823.15	2432.43	3330.58	4654.18

Source Authors' calculations based on ELMS 1998 and ELMPS 2006

have to lower their food expenditure to below the food basic needs to afford these non-food items. The UPL was calculated by setting the non-food share to equal that of households whose *food* expenditure is equal to the food poverty line.

To be able to use the income data to make comparisons about poverty status, comparable low earnings lines (LELs) were developed. These were calculated by first dividing each regional poverty line by 12 to account for the fact that income data are monthly while the expenditure-based poverty lines are annual. Second, since each worker supports several other household members, the monthly poverty line for each region was then scaled up by the regional average dependency ratio (calculated as the median ratio of household members to working age employed household members). This gives us three low earnings lines (LELs): the Food LEL (FLEL), based on the FPL; the LEL, based on the PL and the Upper LEL (ULEL), based on the UPL. These LELs and PLs are summarized in Table A.1 in the appendix by region.

Before describing the transition matrices based on poverty lines, it is useful to look at changes in the income[11] and consumption distributions that occurred between 1998 and 2006. These are reported in Table 11.1. Real monthly incomes increased at all levels of the distribution, and the increase was slightly higher at the middle of the distribution. On average, monthly incomes increased by about 30 % over the period. By contrast, estimated annual per capita consumption increased for only the lower 25 % of the distribution, while it decreased at higher levels. The increase for the bottom 10 % was only 7 % of EPCC, 1.6 % for the 25th percentile, and negative for all other groups. Note, of course, that monthly income is only earned by a subset of the population those of working age and able to find a job, while consumption data span the whole dataset, and therefore these two conclusions are not contradictory.

Transition matrices based on low earnings lines and poverty lines for both monthly incomes and the EPCC are reported in Tables 11.2 and 11.3. The sample is limited to those who were wage workers in both years and reported positive income. Table 11.2 shows the mobility pattern by real monthly income relative to the preset cutoff point z, where z is the Food Low Earnings Line (FLEL) in panel

[11] Data on income available from the ELMS and ELMPS are only available for wage workers, and refers to total wage earnings from all jobs. We therefore restrict our sample to wage workers.

Table 11.2 Transition Matrices based on Real Monthly Income, in relation to three Low Earnings Lines

Individuals with real monthly income in 1998	Individuals with real monthly income in 2006				
	Below z	Between z and 2z	Above 2z	Above 4z	Total
(a)					
Below z	24.85	51.32	18.76	5.07	100
Between z and 2z	7.429	40.94	42.91	8.73	100
Above 2z	2.86	14.8	55.61	26.73	100
Above 4z	1.02	15.31	31.63	52.04	100
(b)					
Below z	42.28	45.33	8.9	3.49	100
Between z and 2z	12.80	48.11	31.67	7.41	100
Above 2z	5.88	31.55	40.64	21.93	100
Above 4z	9.09	3.03	42.42	45.45	100
(c)					
Below z	58.01	34.20	5.17	2.61	100
Between z and 2z	17.53	52.6	20.78	9.09	100
Above 2z	17.00	32.00	34.00	17.00	100
Above 4z	0.00	0.00	60.00	40.00	100

Notes in panel (a) z is the FLEL; in panel (b) z is the LEL; and in panel (c) z is the ULEL
Source Authors' calculations based on ELMS 1998 and ELMPS 2006

(a); the Low Earnings Line (LEL) in panel (b); and the Upper Low Earnings Line (ULEL) in panel (c) provided in Table A.1 in the Appendix. Considering first panel (a) of Table 11.2, about 25 % of those whose income was below the FLEL in 1998 remained below this line in 2006. There was a considerable amount of movement along the income distribution, based on this FLEL, with 51 % moving above the FLEL but remaining below twice that number, 19 % moving to above twice the FLEL and only 5 % moving to above four times the FLEL. There is also considerable mobility downward for those who started out with higher incomes in 1998, with about 50 % of those who earned more than four times the FLEL in 1998, falling to below that level in 2006.

When the LEL is used instead (Table 11.2 panel (b)). a much larger number of people remain low earners of course: about 42 %, with most of the rest moving up to between LEL and 2*LEL in 2006. There is also slightly more mobility downward for those who started out at higher incomes in 1998. In Table 11.2 panel (c), the ULEL is used, and again slightly less movement out or into "poverty" or low earnings status, is detected. A whopping 58 % of those who were low earners in 1998 remained so in 2006 by this higher low earnings line, while the majority of those who moved up, only moved up to the next bracket (between ULEL and 2*ULEL). In general, these results point to a considerable amount of rigidity in status for low earners, and for wage earners in general, when compared with a fixed level of "minimum acceptable earnings line" that is based on a basic needs poverty line.

Table 11.3 Transition Matrices based on Estimated per Capita Consumption, in relation to three Poverty Lines

Individuals with estimated per Capita consumption in 1998	Individuals with estimated per Capita consumption in 2006				
	Below z	Between z and 2z	Above 2z	Above 4z	Total
(a)					
Below z	9.18	79.52	11.31	0	100
Between z and 2z	1.96	68.02	29.63	0.39	100
Above 2z	0	19.93	73.07	6.99	100
Above 4z	0	0.91	46.02	53.07	100
(b)					
Below z	41.48	55.89	2.62	0	100
Between z and 2z	8.8	91.2	0	0	100
Above 2z	0.37	38.39	55.43	5.81	100
Above 4z	0	7.66	44.01	48.33	100
(c)					
Below z	63.21	36.16	0.61	0.02	100
Between z and 2z	17.37	75.02	7.49	0.13	100
Above 2z	1.91	49.45	42.4	6.24	100
Above 4z	0	14.9	33.54	51.56	100

Notes in panel (a) z is the FPL; in panel (b) z is the PL; and in panel (c) z is the UPL
Source Authors' calculations based on ELMS 1998 and ELMPS 2006

To provide greater insight into the dynamics of poverty over this period and to allow for comparison with other studies based on consumption, we next report transition matrices based on the EPCC. Table 11.3 shows the mobility pattern by estimated per capita consumption relative to the preset cutoff point z, where z is the Food Poverty Line (FPL) in panel (a); the Poverty Line (PL) in panel (b); and the Upper Poverty Line (UPL) in panel (c) provided in Table A.1 in the Appendix. The picture is of course very different now. In Table 11.3 panel (a), we see that 9 % of those who were below the FPL in 1998 remained there in 2006. These are, people who are considered extreme poor, not being able to meet even the minimum caloric requirements for their healthy survival. Of those who managed to escape, almost all only moved between FPL and 2*FPL. There is also very little movement up for those who started out nonpoor in 1998, with 70 % or so staying in the same bracket for categories 2 and 3, and 50 % in the highest bracket.

Using the more reasonable poverty line that accounts for non-food essentials as well as food items, the percent of those who were chronically poor was much larger at 41 % of the population, and again those who escaped just barely managed to move up to the next bracket (between PL and 2*PL). There is very little movement up or down in the next bracket, while of those who were in brackets 3 and 4, almost half stayed in the same bracket while the other half moved down. The UPL paints an even bleaker picture: 63 % are chronically poor by this standard, and in the highest brackets if there is any movement at all, the majority is actually *downwards*.

Table 11.4 Transition Matrix based on Real Monthly Income in relation to multiple Low Earnings Lines (FLEL, LEL and ULEL)

Individuals with real monthly income in 1998	Individuals with real monthly income in 2006				
	Below FLEL	Between FLEL and LEL	Between LEL and ULEL	Above ULEL	Total
Below FLEL	21.68	20.84	13.26	44.22	100
Between FLEL and LEL	7.23	12.8	20.94	59.03	100
Between LEL and ULEL	5.62	7.93	11.35	75.1	100
Above ULEL	4.51	2.68	1.91	90.89	100
Total	6.03	4.68	3.83	85.46	100

Source Authors' calculations based on ELMS 1998 and ELMPS 2006

Another interesting question is to determine how many people passed each of the three low earning line or poverty line thresholds. These earning/consumption levels could be seen as representing milestones of welfare: being able to meet basic food needs, being able to meet food and essential non-food needs, and finally being able to meet food and more of the non-food needs actually consumed by families at similar levels of income/consumption. They are also easier to interpret, than the brackets used in Tables 11.2 and 11.3, especially that the second bracket in these tables (2*LEL and 2*PL) are likely to be very wide.[12]

To overcome this problem, we construct transition matrices using each of the three LELs (Table 11.4) and PLs (Table 11.5) as cutoff points. Approximately 21 % who were "extreme poor" in 1998 based on real monthly income remained so in 2006, another 21 % managed to escape extreme poverty, but still remained poor by the LEL standard, while a total of 44 % of those who were extreme poor escaped poverty completely by earning income above the ULEL. Individuals who were poor by the other two LELs in 1998 were more likely to move upwards. These movements are likely a reflection of labor market conditions and the availability of formal versus informal jobs. Hence, we see those at the lowest income levels again being more likely to remain there as these have less stable income streams and may only be seasonally/partly employed.

Table 11.5 shows the transition matrix based on multiple poverty lines using the EPCC data. About 10 % of those who were extreme poor in 1998, remained so in 2006. Of those who escaped extreme poverty between 1998 and 2006, almost 50 % were still poor by the poverty line standard, while another 30 % were still poor by the upper poverty line standard. Only 13 % managed to escape poverty all together in this group. Of those who were poor in 1998 but above the FPL, about 4 % *fell into extreme poverty* by 2006, 40 % remained poor but only by the upper poverty line standard, while 21 % escaped poverty altogether. A much larger number of individuals who were poor by the UPL standard in 1998, escaped poverty altogether by

[12] Consider the PLs in Table A.1 for example: 2*FPL for all Egypt is 2, 059 EGP, which is 125 EGP *above the UPL*. Similarly, 2* FLEL for All Egypt is 528 EGP, 32 EGP *above the ULEL*. It is thus no surprise that we see the majority of individuals in both Tables 11.2 and 11.3 only moving up by one bracket.

Table 11.5 Transition Matrix based on Estimated per Capita Consumption in relation to multiple Poverty Lines (FPL, PL and UPL)

Individuals with estimated per Capita consumption in 1998	Individuals with estimated per Capita consumption in 2006				
	Below FPL	Between FPL and PL	Between PL and UPL	Above UPL	Total
Below FPL	9.9	48.2	28.6	13.3	100
Between FPL and PL	3.69	35.72	39.17	21.42	100
Between PL and UPL	1.59	16.37	35.35	46.69	100
Above UPL	0.07	2.29	12.24	85.4	100
Total	0.93	9.64	20.02	69.41	100

Source Authors' calculations based on ELMS 1998 and ELMPS 2006

2006 (47 %), while 18 % fell far deeper into poverty, as their consumption plunged below the PL and FPL. Finally, 85 % of the nonpoor by the UPL remained so in 2006, while we still saw 12 % falling below the UPL and the PL out of that group.

The difference in interpretation when monthly incomes and corresponding low earnings lines are used, versus estimated consumption and corresponding poverty lines are used, points to the importance of using these two measures together to produce a complete picture of movement along the welfare distribution over this period. Even though income data may be more accurate here since it is the data actually collected in this panel survey, it will only tell us about the welfare as related to wages, which are much less stable especially for those in informal jobs that work only seasonally. The EPCC complements the picture by allowing us to explore what happened to an intrinsically smoother measure of welfare. In either case, the conclusions based on these two measures of welfare are only different when the food poverty or food low earnings lines are used. Conclusions based on the other two poverty lines are very similar. The higher extreme poverty levels by income (based on the FLEL) are likely a reflection of the fact that these income earners may be in and out of work, and therefore do not have a stable stream of income, while they are still able to exercise consumption smoothing, either through savings from previous months, or through in-kind donations, which are quite common in Egypt.

11.4.2 Transitions In and Out of Non-Monetary Poverty Between 1998 and 2006

Transitions over time in the quality of jobs, using the institutional job quality index, are shown in Table 11.6 panels (a), (b) and (c).[13] A job is classified as having "poor" job quality if the constructed institutional job quality index is more

[13] Transition matrices based on the other two job quality measures described at the end of Sect. 11.3 were almost identical, and are not shown to save on space.

Table 11.6 **a**. Transition matrix for the job quality index, all wage workers. **b** Job quality index, male wage workers. **c** Job quality index, female wage workers

Job quality in 1998	Job quality in 2006		
	Poor	Fair	Good
(a)			
Poor	0.74	0.15	0.11
Fair	0.33	0.44	0.23
Good	0.04	0.04	0.92
(b)			
Poor	0.73	0.16	0.11
Fair	0.34	0.45	0.22
Good	0.05	0.05	0.89
(c)			
Poor	0.87	0.07	0.06
Fair	0.21	0.29	0.5
Good	0.01	0.02	0.97

Source Authors' calculations based on ELMS 1998 and ELMPS 2006

than one standard deviation below the mean, "good" if it is more than one standard deviation above the mean, and "fair" if it is in between.[14] The transition matrix for all wage workers in panel (a) shows that only 25 % of those who started out with poor quality jobs in 1998 were able to escape to better quality jobs by 2006. Of those who were in "fair" jobs, however, a larger percentage moved down to "poor" quality jobs, than those who moved up to "good" jobs, while close to 90 % of those with "good" jobs were able to maintain that quality in 2006. Thus, there is a relatively low level of mobility where it comes to quality of jobs, especially mobility that allows for better quality jobs.

Splitting the sample into males and females, the transition matrices in Table 11.6 panels (b) and (c) respectively show that females tend to be stuck more often in the poor quality jobs; however, if they were in fair jobs in 1998 they were more likely to move to good jobs by 2006 than their male counterparts. Females with good quality jobs were almost assured of staying with a similar quality job between 1998 and 2006, while about 10 % of males in that category moved down to fair or poor jobs by 2006. These results again point to the very low chance a person who starts out with a low quality job has of "escaping" to a better job,

[14] We define the classes for job quality with respect to deviations from each year's mean. As the descriptive statistics show, the mean job quality index declined dramatically between 1998 and 2006 (from 0.085 to 0.055). If these classes had been defined with respect to the mean for both years, it would have appeared that a much larger number of people fell to "poor" quality jobs in 2006. We decided that a "relative' notion of job quality, with respect to what the average worker was facing in each respective year, would be more impartial and provide a better insight into the changes in job quality that were actually going on in the labor market during that period.

especially if they are female. However, females with good jobs are almost assured of maintaining that quality. These results can be interpreted in the context of the fact that most of the "good" jobs are public sector jobs ("good" in the sense of the existence of a contract, paid sick/maternity leaves, social security, etc.). These tend to be highly sought after by females given their dual responsibilities inside and outside the home, due to short working hours and the ability to take extended leaves for child bearing and rearing, despite the relatively low wages. It is also almost impossible to get "fired" from these jobs. Males, on the other hand, are more likely to try to find better paying jobs to support their families financially even at the expense of other "low quality" job aspects.

11.5 Monetary and Non-monetary Effects of Trade Liberalization and Privatization on the Poor and Low Waged Manufacturing Workers

In this section, we investigate the effect of trade reform on incomes of low waged and poor workers in the Egyptian manufacturing sector, over a period of significant trade liberalization and privatization between 1998 and 2006.[15] Trade liberalization might impact poverty through two channels: the first one is direct through the wage changes and the second one indirectly through consumption price changes. Due to the nature of our data, we concentrate here on the former channel.

Although Egypt has taken a gradual approach to trade reform, the once highly restrictive trade regime has been reversed with the initiation of reforms in 1986 which built up to the signing of the WTO agreements in 1994/1995 and other multi and bilateral trade agreements in the mid-1990s. To further open the Egyptian economy, the government, starting in 2004, announced a reduction of average unweighted tariff rate from 27 to 20 % and rationalization of the tariff structure. The number of products subject to nontariff barriers was also substantially reduced. After the 2004 reform, it became evident that both nominal and effective protections have declined for almost all manufacturing sectors with most of trade liberalization efforts concentrated in the area of intermediate and capital goods.

[15] Although by limiting this study to the manufacturing sector, we might be missing the bulk of the non waged poor concentrated in rural and agricultural setting, our analysis is still informative about the potential for trade reforms to change the income and poverty status of the groups of workers in Egypt that are most likely to be directly affected by those measures, namely skilled and unskilled wage workers working in manufacturing industry.

11.5.1 Empirical Methodology and Variable Definitions

We use panel regression estimation methods to measure the impact of such reforms on income and non-income dimensions of welfare of the poor. This allows us to examine the interaction between various industry characteristics (including trade openness and whether the firm is privately or publicly owned) and individual characteristics (such as education and labor market experience) in influencing wages, as well as non-monetary outcomes for the poor and low waged as measured by a composite measure of quality of jobs.

As mentioned above, job quality is a multifaceted concept, and a wide range of indicators have been proposed by several studies. Although we used three different measures of job quality we report results mainly based on the institutional job quality index.[16] The advantage of the use of the institutional job quality measure as dependent variable in the empirical analysis lies in the comprehensiveness of the index in comparison to previous studies that use a dummy for formality.

Our estimation equations look as follows. Let i = 1, 2, 3...I_j index workers in industry j. We regress the log of worker i's wages at time t ($\ln(w_{ijt})$) and the index of their job quality (JQI_{ijt}) on a vector of worker characteristics H_{ijt} (age, age squared, gender, education indicators, sector indicators, and region of residence) and a set of variables (P_{jt}) reflecting institutional sector (private or public enterprise), the degree of protection and international trade at the industry level.

Thus we estimate the following equations:

$$\ln(w_{ijt}) = H_{ijt}\beta_{Ht} + P_{jt}\beta_{Pt} + \varepsilon_{ijt} \tag{11.1}$$

$$\mathrm{JQI}_{ijt} = H_{ijt}\alpha_{Ht} + P_{jt}\,\alpha_{Pt} + u_{ijt} \tag{11.2}$$

for $i = 1, \ldots I_j$; $j = 1, \ldots, J$ and $t = 1998$ or 2006.

We estimate the above equations for all waged workers in the manufacturing sector as well as for workers classified as low waged and poor. In each case, we use panel data methods and undertake Hausman and Taylor specification tests to decide whether a fixed effects (FE) or a random effects (RE) model should be applied. It should be noted that the sample includes those that were employed in 1998 and 2006. This means that it is not possible to take into account creation and destruction of jobs due to trade liberalization.

The low waged are identified by their real monthly income falling below the LEL as defined in Table A.1 in the appendix and described in detail in Sect. 11.4.1

[16] Recall from Sect. 11.3 that this incorporates the following six criteria: whether the worker (1) has social insurance, (2) has health insurance (3) has a contract, (4) has paid sick leave, (5) has paid annual leave, and (6) is a member of a trade union. Results based on the other two job quality indices were almost identical and are not shown to save on space.

above. The poor are identified by their EPCC falling below the PL. These LELs and PLs are region specific and take family size and composition into consideration as mentioned earlier.

The working sample is restricted to manufacturing sector workers, between the ages of 15 and 65, who report positive monthly earnings. Hourly real wages are calculated as the sum of wages earned in the reference month from primary jobs, adjusted for average number of work days per month and average hours per day. For comparability purposes, wages of 1998 are inflated to 2006 Egyptian pounds using the consumer price index (inflation factor is 1.43 from 1998 to 2006). We are primarily interested in the effect of trade policy changes on wages, and job quality, where trade policy is measured by tariff levels and changes in tariffs. We also include measures of the degree of export orientation (exports/output) and import penetration (imports/(output +imports–exports)) as controls, and to facilitate the comparison of our results with other studies. [17]

Table 11.7 presents summary descriptive statistics of variables used in regressions. As noted in the introduction, the sample of estimation is limited to wage workers currently employed in the manufacturing sector. The panel comprised of two waves 1998 and 2006 has a total of 2,556 observations, of which 972 workers are classified as low waged and 275 are classified as poor.

All average figures are properly weighted to reflect the population distribution. A comparison of 1998 and 2006 figures reveals that on average, real hourly wages increased, whereas job quality recorded a decline over this period. Although log hourly wages are higher for the poor sample than low waged workers, the low waged compare favorably to them in terms of their job quality index. This calls attention to the importance of separating monetary and non-monetary aspects of welfare, when analyzing poverty in the manufacturing sector.

As for trade and industry related variables, the average figures indicate a 5 % points decline in average tariffs, a 6 % points increase in export orientation and a 2 % points increase in import penetration between 1998 and 2006. The share of public enterprises in the manufacturing sector in Egypt has declined from 29 to 19 %, reflecting the extent of privatization effort in that sector over the same period. It is worth noting that for the poor and low wage subsamples; there is a higher proportion of workers concentrated in private enterprises than is the case for the whole manufacturing sector. This might make them more vulnerable to private sector practices than the average manufacturing worker.

Figure 11.1 shows the relationship between change in wage and job quality index and change in the three trade policy variables: export orientation, import penetration, and tariffs in a scatter diagram across industries with a fitted linear

---

[17] Export orientation and import penetration measures are arguably endogenous since they depend on factor costs, (see for example, Dutta 2007), yet it has been acknowledged that it is also not possible to satisfactorily address the issue due to absence of appropriate instruments in datasets similar to ours. To counter this, we avoid the use of contemporaneous trade flows and instead regress on lagged trade variables. In interpretation we also rely mostly on tariffs as the main measure of trade policy changes.

Table 11.7 Summary statistics of variables used in regressions

Variable	1998						2006					
	All manufacturing		Poor		Low wage		All manufacturing		Poor		Low wage	
	Mean	Std. Dev.	Mean	Std. Dev.	Mean	Std. Dev.	Mean	Std. Dev.	Mean	Std. Dev.	Mean	Std. Dev.
Log real hourly wage	0.249	0.665	−0.180	0.648	−0.229	0.472	0.727	0.776	0.389	0.562	0.139	0.487
Job quality index	0.085	0.901	−0.306	0.810	−0.867	0.773	−0.055	0.861	−0.406	0.690	−0.334	0.722
Tariff rate	21.425	9.037	21.951	8.114	21.474	8.232	16.000	9.650	16.386	8.763	16.134	9.747
Import penetration	32.914	25.748	37.623	29.005	32.832	25.405	35.287	26.979	32.029	27.919	33.576	27.682
Export orientation	15.020	13.337	15.726	13.303	14.418	13.152	21.090	13.968	19.683	11.875	19.344	12.504
Public sector	0.292	0.455	0.206	0.406	0.270	0.445	0.193	0.395	0.094	0.292	0.125	0.331
Private sector	0.708	0.455	0.794	0.406	0.730	0.445	0.807	0.395	0.906	0.292	0.875	0.331
Experience	18.142	13.017	17.368	12.915	13.647	12.350	16.808	12.157	17.074	11.656	12.273	10.919
Experience squared/100	4.984	5.852	4.668	5.204	3.384	5.231	4.302	5.427	4.266	5.160	2.696	4.505
Male	0.887	0.317	0.881	0.326	0.844	0.364	0.860	0.347	0.848	0.360	0.765	0.424
Female	0.113	0.317	0.119	0.326	0.156	0.364	0.140	0.347	0.152	0.360	0.235	0.424
Illiterate	0.196	0.397	0.431	0.498	0.270	0.444	0.176	0.381	0.385	0.488	0.206	0.405
Read and write	0.146	0.354	0.204	0.405	0.161	0.368	0.085	0.279	0.130	0.338	0.081	0.273
Primary	0.167	0.373	0.190	0.395	0.164	0.371	0.138	0.345	0.198	0.400	0.152	0.359
Preparatory	0.081	0.274	0.038	0.193	0.075	0.263	0.072	0.259	0.076	0.266	0.060	0.237
General secondary	0.014	0.117	0.003	0.053	0.010	0.101	0.010	0.099	0.000	0.000	0.006	0.077
Vocational education	0.266	0.442	0.134	0.342	0.259	0.439	0.354	0.479	0.194	0.396	0.412	0.493
Post secondary institute	0.048	0.213	0.000	0.000	0.047	0.211	0.048	0.215	0.015	0.122	0.026	0.158
University and above	0.081	0.273	0.000	0.000	0.014	0.118	0.116	0.321	0.002	0.044	0.057	0.231
Greater Cairo	0.281	0.450	0.064	0.246	0.174	0.380	0.259	0.438	0.090	0.287	0.154	0.361
Alexandria	0.116	0.320	0.048	0.215	0.093	0.291	0.126	0.332	0.039	0.194	0.094	0.292
Rural Upper Egypt	0.152	0.359	0.532	0.501	0.184	0.388	0.156	0.363	0.462	0.500	0.173	0.379
Urban Upper Egypt	0.050	0.218	0.124	0.331	0.045	0.206	0.049	0.217	0.174	0.380	0.056	0.230
Rural Lower Egypt	0.265	0.442	0.204	0.405	0.384	0.487	0.283	0.450	0.183	0.388	0.402	0.491
Urban Lower Egypt	0.135	0.342	0.029	0.168	0.121	0.326	0.127	0.333	0.052	0.223	0.121	0.326

Source Authors' calculation based on ELMS 1998 and ELMPS 2006

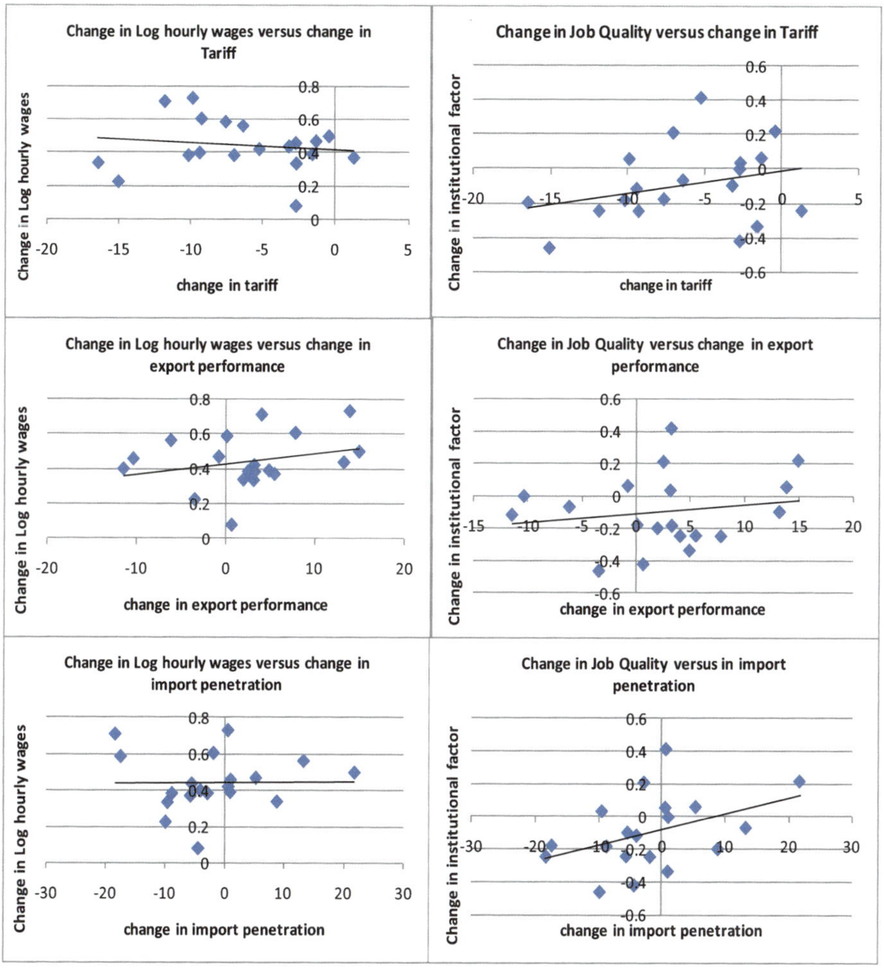

Fig. 11.1 Scatter diagrams for change in wage and job quality index versus trade variables in Egyptian manufacturing sector, 1998–2006. Notes: (*i*) Solid lines are fitted linear trends. (*ii*) *Source* Authors' calculation based on ELMS 1998 and ELMPS 2006

trend. These reveal that there is only a clear positive relationship between the change in export orientation and wages, the other two trade variables association appears to be more ambiguous. On the job quality side, the most interesting result reveals that the change in import penetration is positively correlated to the change in job quality. Therefore, in order to compensate the wage decline due to more import penetration, many industries increase the quality of their jobs.

Sample statistics for the other explanatory variables show that the manufacturing industries also become more feminized on average, but disproportionately in lower paying jobs; hence, a more substantial rise of the share of women in the

Table 11.8 Panel regression results for all wage workers in manufacturing

	(1)	(2)	(3)	(4)
	Log real hourly wage		Job quality index	
Variables	Fixed effects	Random effects	Fixed effects	Random effects
Tariff rate	−0.019***	−0.020***	−0.003	−0.007***
	(0.004)	(0.002)	(0.003)	(0.002)
Import penetration	−0.001	−0.001	−0.003***	−0.004***
	(0.002)	(0.001)	(0.001)	(0.001)
Export orientation	0.011***	0.012***	−0.002	−0.004***
	(0.002)	(0.001)	(0.002)	(0.001)
Private sector (indicator var.)	−0.008	0.008	−0.948***	−1.017***
	(0.083)	(0.035)	(0.064)	(0.031)
Female	0.055	−0.065	−0.111	−0.132***
	(0.111)	(0.046)	(0.073)	(0.037)
Experience	0.049***	0.050***	0.022***	0.025***
	(0.008)	(0.004)	(0.005)	(0.003)
Experience squared divided by 100	−0.053***	−0.065***	−0.000	−0.000***
	(0.017)	(0.009)	(0.000)	(0.000)
Read and write	−0.047	0.026	0.089	0.152***
	(0.109)	(0.058)	(0.070)	(0.045)
Primary	0.057	0.131**	0.171**	0.179***
	(0.102)	(0.052)	(0.069)	(0.042)
Preparatory	0.098	0.190***	0.205**	0.218***
	(0.132)	(0.064)	(0.092)	(0.052)
General secondary	0.118	0.387***	0.441*	0.349***
	(0.316)	(0.140)	(0.225)	(0.110)
Vocational education	0.182*	0.336***	0.433***	0.453***
	(0.101)	(0.048)	(0.071)	(0.039)

(continued)

Table 11.8 (continued)

	(1)	(2)	(3)	(4)
	Log real hourly wage		Job quality index	
Variables	Fixed effects	Random effects	Fixed effects	Random effects
Post secondary institute	0.283	0.410***	0.595***	0.451***
	(0.172)	(0.071)	(0.128)	(0.063)
University and above	0.558***	0.853***	0.715***	0.699***
	(0.135)	(0.060)	(0.099)	(0.051)
Alexandria	−0.038	−0.025	0.501***	0.156***
	(0.131)	(0.044)	(0.099)	(0.040)
Rural Upper Egypt	−0.235	−0.136**	−0.236**	−0.261***
	(0.145)	(0.053)	(0.092)	(0.044)
Urban Upper Egypt	−0.178	−0.136**	0.017	−0.166***
	(0.174)	(0.055)	(0.126)	(0.046)
Rural Lower Egypt	−0.161*	−0.135***	−0.102	−0.166***
	(0.094)	(0.043)	(0.066)	(0.036)
Urban Lower Egypt	−0.150	−0.126***	−0.181**	−0.146***
	(0.115)	(0.045)	(0.082)	(0.039)
Observations	2024	2024	2556	2556
R-squared within	0.275	0.26	0.450	0.433
Between	0.307	0.35	0.529	0.571
Overall	0.302	0.32	0.537	0.566
Hausman test (Chi 2)	16.34		33.83	
Hausman test (*P* value)	0.6345		0.0192	

Standard errors in parentheses; ***$p < 0.01$, ** $p < 0.05$, * $p < 0.1$. *Source*: Authors' calculation based on ELMS 1998 and ELMPS 2006

Table 11.9 Panel regression results for low wage workers in manufacturing sector

	(1) Log real hourly wage	(2)	(3) Job Quality Index	(4)
Variables	Fixed effects	Random Effects	Fixed effects	Random Effects
Tariff rate	−0.013**	−0.016***	−0.004	−0.002
	(0.005)	(0.002)	(0.006)	(0.003)
Import penetration	0.003	−0.001	−0.002	−0.003***
	(0.002)	(0.001)	(0.002)	(0.001)
Export orientation	0.006*	0.006***	0.001	−0.005***
	(0.003)	(0.001)	(0.004)	(0.002)
Private sector (indicator var.)	−0.027	0.009	−0.923***	−1.090***
	(0.127)	(0.042)	(0.157)	(0.051)
Female	0.023	−0.005	0.054	−0.128***
	(0.111)	(0.045)	(0.125)	(0.049)
Experience	0.023**	0.021***	0.020*	0.019***
	(0.010)	(0.005)	(0.012)	(0.005)
experience squared divided by 100	−0.035	−0.028**	−0.000	−0.000
	(0.023)	(0.011)	(0.000)	(0.000)
Read and write	0.004	−0.075	−0.007	0.222***
	(0.142)	(0.063)	(0.159)	(0.072)
Primary	0.084	0.044	0.095	0.145**
	(0.117)	(0.056)	(0.134)	(0.064)
Preparatory	−0.022	0.048	0.107	0.224***
	(0.159)	(0.072)	(0.185)	(0.084)
General secondary	0.708	0.184	0.100	0.221
	(0.484)	(0.182)	(0.603)	(0.212)
Vocational education	0.238**	0.150***	0.366***	0.353***
	(0.118)	(0.050)	(0.136)	(0.058)

(continued)

Table 11.9 (continued)

	(1) Log real hourly wage	(2)	(3) Job Quality Index	(4)
Variables	Fixed effects	Random Effects	Fixed effects	Random Effects
Post secondary institute	0.115	0.035	0.066	0.313***
	(0.266)	(0.087)	(0.309)	(0.106)
University and above	0.108	0.158*	0.478*	0.378***
	(0.204)	(0.093)	(0.251)	(0.109)
Alexandria	0.082	0.023	0.285	0.082
	(0.173)	(0.057)	(0.213)	(0.070)
Rural Upper Egypt	0.003	0.029	–0.211	–0.227***
	(0.177)	(0.064)	(0.185)	(0.072)
Urban Upper Egypt	–0.149	0.006	–0.170	–0.281***
	(0.260)	(0.064)	(0.273)	(0.074)
Rural Lower Egypt	0.083	0.096**	–0.113	–0.135**
	(0.118)	(0.049)	(0.144)	(0.059)
Urban Lower Egypt	–0.149	0.025	–0.184	–0.128*
	(0.147)	(0.055)	(0.176)	(0.067)
Observations	849	849	972	972
R-squared within	0.216	0.175	0.309	0.278
Between	0.081	0.135	0.47	0.53
Overall	0.109	0.152	0.457	0.501
Hausman test (Chi 2)	11.68		12.73	
Hausman test (*P* value)	0.8987		0.851	

Standard errors in parentheses; ***$p < 0.01$, ** $p < 0.05$, * $p < 0$
Source Authors' calculation based on ELMS 1998 and ELMPS 2006

Table 11.10 Panel regression results for poor wage workers in manufacturing sector

	(1) Log real hourly wage	(2)	(3) Job Quality Index	(4)
Variables	Fixed effects	Random effects	Fixed effects	Random effects
Tariff rate	−0.034**	−0.023***	−0.003	−0.004
	(0.014)	(0.005)	(0.010)	(0.004)
Import penetration	0.011	−0.002	−0.004	−0.003***
	(0.007)	(0.002)	(0.004)	(0.001)
Export orientation	0.014	0.010***	−0.007	−0.005*
	(0.011)	(0.003)	(0.007)	(0.003)
Private sector (indicator var.)	−0.453*	0.022	−1.322***	−1.343***
	(0.245)	(0.103)	(0.197)	(0.089)
Female	0.158	−0.169	−0.116	−0.118
	(0.325)	(0.130)	(0.224)	(0.086)
Experience	0.012	0.040***	0.022	0.016*
	(0.028)	(0.011)	(0.019)	(0.008)
Experience squared divided by 100	0.024	−0.056**	−0.000	−0.000
	(0.049)	(0.024)	(0.000)	(0.000)
Read and write	0.028	−0.116	−0.001	0.121
	(0.269)	(0.114)	(0.206)	(0.090)
Primary	0.032	0.097	0.095	0.132
	(0.345)	(0.107)	(0.221)	(0.084)
Preparatory	1.416***	0.365**	0.167	0.132
	(0.338)	(0.152)	(0.242)	(0.117)
General secondary	0.000	0.145	0.000	0.119
	(0.000)	(0.547)	(0.000)	(0.493)
Vocational education	0.171	0.103	0.630**	0.372***
	(0.393)	(0.108)	(0.274)	(0.090)

(continued)

Table 11.10 (continued)

	(1) Log real hourly wage	(2)	(3) Job Quality Index	(4)
Variables	Fixed effects	Random effects	Fixed effects	Random effects
Post secondary institute	0.000	0.753*	0.000	0.879***
	(0.000)	(0.386)	(0.000)	(0.286)
University and above	0.254	0.678	2.101***	1.891***
	(0.635)	(0.544)	(0.503)	(0.430)
Alexandria	0.000	−0.163	0.000	−0.035
	(0.000)	(0.199)	(0.000)	(0.167)
Rural Upper Egypt	−0.673**	−0.089	−0.141	0.033
	(0.297)	(0.147)	(0.230)	(0.119)
Urban Upper Egypt	−0.246	−0.123	−0.220	0.034
	(0.338)	(0.148)	(0.270)	(0.121)
Rural Lower Egypt	−0.490	−0.213	0.111	0.075
	(0.365)	(0.164)	(0.273)	(0.132)
Urban Lower Egypt	−0.531	−0.157	−0.040	0.008
	(0.499)	(0.189)	(0.374)	(0.160)
Constant	0.637	0.174	0.803*	0.698***
	(0.682)	(0.250)	(0.416)	(0.187)
Observations	221	221	275	275
R-squared within‘	0.645	0.42	0.639	0.608
Between	0.023	0.314	0.56	0.641
Overall	0.099	0.33	0.57	0.635
Hausman test (Chi 2)	30.18		5.47	
Hausman test (*P* value)	0.0171		0.992	

Standard errors in parentheses; *** $p < 0.01$, ** $p < 0.05$, * $p < 0.1$ *Source* Authors' calculation based on ELMS 1998 and ELMPS 2006 low wage subsample. Trade openness and privatization over this period also coincided with a slight decline in average years of experience from 18 to 16.8, a more notable reduction in the share of workers with a university degree from 8 to 12 %, and a substantial increase in the proportion of workers with vocational secondary degrees, from 26.6 to 35 %. The low wage group has much lower years of experience and a much higher representation of vocational school graduates than the rest, and that proportion has been on the rise. As expected, more than half of the poor have only primary schooling or less; and there are virtually no university graduates amongst them.

Regionally, figures in both years reflect a high concentration of manufacturing activity in greater Cairo and rural Lower Egypt, together accounting for over 50 % of employment in the sector. There is a high concentration of low wage workers in rural Lower Egypt and of poor workers in Lower and Upper Egypt than is the case for average manufacturing.

11.5.2 Panel Estimation Results

Tables 11.8–11.10 present panel fixed effects and random effects regression results for the two dependent variables (log hourly wages and job quality index) for all workers in the manufacturing sector, for low wage workers and poor workers, respectively. Our discussion below is mostly based on fixed effects regression estimates, as Hausman specification test results indicate that it is justifiable to use the random effects model (i.e., that it gives both more efficient and consistent results) in only four cases: (1) log hourly wage model for all manufacturing workers, (2) log hourly wage model for low wage workers, (3) the job quality model for low wage workers, and (4) the job quality model for poor worker. Only for these latter four cases, therefore, do we rely on results of the random effects model.

Based on goodness of fit statistics, the human capital model appears to explain well the wage setting, and the job quality outcome in the manufacturing sector in Egypt. As expected, there is a large and significant negative differential in all three dependent variables, associated with being female and for working for a private sector firm. The experience wage profiles follow the expected inverted U-shape implied by human capital theory. There are also increasing returns to education, and they particularly jump at the university or above levels. Finally, there are mostly negative and significant differentials due to residence outside of greater Cairo.

Now turning to our central concern with the impact of trade and private sector variables on monetary and non-monetary welfare indicators for workers, the results indicate that there is a negative significant relationship between tariff and log hourly wages for the manufacturing sector as a whole, and the effect is stronger for poor workers. The results imply, for example, that a 34 % reduction in tariffs (which is the mean change for all industries over this period) would lead to approximately 0.7 % increase in wages for the average manufacturing worker and

1 % increase for the poor worker. There is also a positive relationship between export orientation and wages, and it is the same whether we are dealing with all of the sector or just the poor.

When the variable of interest is job quality there is a significant negative relationship between job quality and all trade variables for the manufacturing sector. This exists also for low wage workers with regard to import penetration and export orientation, and for poor workers with respect to import penetration. Thus, especially for lower wage and poor workers, there appears to be a tradeoff resulting from trade openness in form of higher wages, but at the expense of lower quality jobs. However these effects are relatively small. For example, a 70 % increase in export intensity (which is the mean change for all industries over this period) would lead to 0.35 % reduction in the job quality index for a low wage worker.

Third, and in fact the most significant finding, is that for manufacturing sector as a whole, gender and sector remain the main axes of segregation of the labor market, all resulting in a negative differential in wages, and job quality. For low wage manufacturing workers, the negative effect for females only exists with regard to job quality and disappears once we look at the poor workers (it is negative but not significant).

11.6 Conclusion and Policy Recommendations

In this chapter we have examined the dynamics of poverty from a monetary and non-monetary perspective during a period of trade liberalization in Egypt using panel data that follows the same individuals over time. Furthermore, we have directly examined the effect of trade liberalization on the poor and low waged. A number of interesting conclusions emerge from this analysis, and it is useful to summarize them here.

First, a large percentage of monetary poverty is chronic, with many of those initially poor unable to escape poverty by a variety of poverty lines and by both income and consumption measures. There is, however, more mobility by income than by consumption. Second, exploring the question of poverty also reveals some interesting distinctions based on the measure of wellbeing and also on the poverty line used. Using the food poverty line which reveals extreme poverty, there are about two and half times as many poor by income than by consumption. On the other hand, using the poverty and upper poverty lines, the numbers of poor based on these two more commonly used yardsticks are almost the same whether we use income or consumption data. The higher extreme poverty levels by income are likely a reflection of the fact that these income earners may be in and out of work, and therefore do not have a stable stream of income, while they are still able to exercise consumption smoothing, either through savings from previous months, or through in-kind donations, which are quite common in Egypt.

Third, non-monetary poverty as measured by job quality is also largely chronic, with a very low probability of "escaping" to a better job once one starts out with a low quality job, especially for females. However, females with good jobs are almost assured of maintaining that quality.

This chapter also investigates the impact of trade liberalization on welfare measures in a direct way, using data on tariff reduction, export orientation, and import penetration that has been linked to the individual data at the industry level. The results show that trade reform based on lower tariff protection and export promotion appears to exert a positive influence on incomes of poor workers in manufacturing, but this appears to come at the expense of greater casualization of workers and higher incidence of low quality jobs, although the effects on job quality are quite weak for the poor and low waged. The fact that trade liberalization variables have sometimes opposite effects on wage and job quality has important implications for global effect of trade. Our results highlight that greater insights into the trade off between monetary and non-monetary dimensions of poverty can sometimes be derived from separate analysis of such measures as opposed to aggregate multi-dimensional measures which are commonly used in the current literature. It is also evident that all trade-related variables play a rather minor role in determining monetary and non-monetary outcomes for low waged and poor workers in comparison to the rather significant role played by institutional sector and gender in segregating the labor market at that level.

Results of the mobility and poverty analysis suggest that it is relatively difficult to escape poverty as measured by income or expenditure or to find a more "decent job" if you are already in a "poor quality" one. These findings point to several policy recommendations. First, better targeting of the poor should be a top priority of poverty alleviation programs. Egypt already devoted more than LE35 billion to direct and indirect subsidies in addition to more than LE15 billion in social assistance grants in 2006 for example. Yet most of these fail to directly reach the poor (Osman et al. 2006). Better targeting of vital programs such as food and fuel subsidies to reach those who are truly poor should be on the top of reforms to the current subsidy system. This will provide a strong push to attempts to reach the poor and vulnerable in a more effective way. The analysis with respect to the impact of trade liberalization also highlights that, from a poverty alleviation point of view, public policy needs to be directed toward improvement of labor standards, especially in the private sector. Trade promotion policies that improve the export competitiveness of Egyptian manufacturing such as production subsidies, tax exemptions, and special credit lines, could also exert a positive influence on the incomes of the poor and low waged.

Acknowledgments We thank Ali Rashed and Randa Hamza for able research assistance and are grateful to Ragui Assaad, Valerie Bérenger, Florent Bresson and two anonymous referees for valuable comments and suggestions. The usual disclaimer applies.

Appendix

Table A.1 Low earnings lines and poverty lines in 2006 prices by region, in EGP

Region	FLEL	LEL	ULEL	FPL	PL	UPL
Metropolitan	265.83	377.12	498.50	1060.49	1504.45	1988.69
Lower Egypt Urban	252.86	364.04	479.76	1008.73	1452.28	1913.95
Lower Egypt Rural	284.85	411.89	525.64	1023.12	1479.40	1887.97
Upper Egypt Urban	255.22	367.49	496.66	1018.15	1466.05	1981.34
Upper Egypt Rural	258.20	365.41	470.29	1030.05	1457.77	1876.17
All Egypt	263.86	377.87	495.41	1029.31	1473.74	1933.35

References

Assaad, R., & Roushdy, R. (2008). *Measuring and explaining job quality for household enterprise workers*. ECES Working Paper no. 127.

Baulch, B., & Hoddinott, J. (2000). Economic mobility and poverty dynamics in developing countries. *Journal of Development Studies, 36*(6), 1–24.

Bérenger, V. (2010). Multidimensional fuzzy poverty and pro-poor growth measures in non-monetary dimensions in Egypt between 1995 and 2005. *Middle East Development Journal, 2*(1), 15–38.

Bernard, A., & Jensen, B. (2007). Firm structure, multinationals, and manufacturing plant deaths. *Review of Economics and Statistics, 89*(2), 193–204.

Bernard, A., & Sjoholm, F. (2003). *Foreign owners and plant survival*. National Bureau for Economic Research, NBER Working Paper No. 10039

Bibi, S. (2004). *Comparing multidimensional poverty between Egypt and Tunisia*. Cahiers de Recherche du CIRPEE, No. 0416.

Buchinsky, M., & Hunt, J. (1999). Wage mobility in the United States. *The Review of Economics and Statistics, 81*(3), 351–368.

Deaton, A. (1997). *The analysis of household surveys*. Maryland: The John Hopkins University Press for the World Bank.

Dollar, D. (1992). Outward-oriented developing economies really do grow more rapidly: Evidence from 95 LDCs, 1976–85. *Economic Development and Cultural Change, 40*(3), 523–544.

Dollar, D., & Kraay, A. (2002). Trade, growth, and poverty. *Economic Journal, 114*(493), F22–F49.

Dutta, P. (2007). Trade protection and industry wages in India. *Industrial and Labor Relations Review, 60*(2), 268–286.

Edwards, S. (1998). Openness, productivity and growth: What do we really know? *Economic Journal, 108*(447), 383–398.

Frankel, J., & Romer, D. (1999). Does trade cause growth? *American Economic Review, 89*(3), 379–399.

Grosse, M., Klasen, S., & Spatz, J. (2005). *Creating national poverty profiles and growth incidence curves with incomplete income or consumption expenditure data*. Background paper for the study: Operationalizing pro-poor growth—Country Case study Bolivia. IAI Discussion Papers No. 129, Ibero America Institute for Economic Research, University of Gottingen, Gottingen.

Goldberg, P. K., & Pavcnik, N. (2004). *Trade, inequality and poverty: What do we know? Evidence from recent trade liberalization episodes in developing countries*. Paper prepared for the 2004 Brookings Trade Forum on "Globalization, Poverty and Inequality: What do we know? Where are we going?" Washington DC, May 13–14, 2004.

Haddad, L., & Akhter, A. (2003). Chronic and transitory poverty: Evidence from Egypt, 1997–99. *World Development, 31*(1), 71–85.

Jarvis, S., & Jenkins, S. (1995). *Do the poor stay poor? New evidence about income dynamics from the British Household panel survey*. Occasional Paper 95-2, ESRC Research Centre on Micro-Social Change.

Jarvis, S., & Jenkins, S. P. (1998). How much income mobility is there in Britain? *Economic Journal, 108*, 428–443.

Kraay, A. (2006). When is growth pro-poor? Evidence from a panel of countries. *Journal of Development Economics, 80*, 198–227.

Krueger, A. (1997). Trade policy and development: How we learn. *American Economic Review, 87*(1), 1–22.

Osman, M., Zakareya, E., & Mahrous, W. (2006). *Targeting the poor in Egypt: A ROC approach*. Presented at the 13th ERF Annual Conference "Oil: Its Impact on the Global Economy", Kuwait, December 16–18.

Rodríguez, F. (2006). *Openness and growth: What have we learned? Wesleyan economics working papers 2006–011*. Department of Economics: Wesleyan University.

Rodríguez, F., & Rodrik, D. (1999). Trade policy and economic growth: A skeptic's guide to the cross-national evidence. NBER Working Paper #7081.

Sachs, J. D., & Warner, A. (1995). Economic reform and the process of global integration. *Brookings Papers on Economic Activity, 1*, 1–118.

Steckel, R. H. (1990). Poverty and prosperity: A longitudinal study of wealth accumulation, 1850–1860. *The Review of Economics and Statistics, 72*(2), 275–285.

Woolard, I., & Klasen, S. (2005). Determinants of income mobility and household poverty dynamics in South Africa. *The Journal of Development Studies, 41*(5), 865–897.

World Bank. (2007). *Arab Republic of Egypt: A poverty assessment update*. Report No. 39885-EGT.

Zaidi, M. A., & De Vos, K. (2002). *Income mobility of the elderly in Great Britain and The Netherlands: A comparative investigation*. Economics Series Working Papers 107, University of Oxford, Department of Economics.

Printed by Printforce, the Netherlands